IS-95 CDMA and cdma2000

Cellular/PCS Systems Implementation

ISBN 0-13-087112-5

90000

9 780130 871121

Prentice Hall Communications Engineering and Emerging Technologies Series

Theodore S. Rappaport, Series Editor

IS-95 CDMA and cdma2000

Cellular/PCS Systems Implementation

Vijay K. Garg, PhD, PE

Department of Electrical Engineering and
Computer Science
University of Illinois at Chicago
Chicago, Illinois U.S.A.

PH
PTR

Prentice Hall PTR
Upper Saddle River, NJ 07458
www.phptr.com

Library of Congress Cataloging-in-Publication Data

Garg, Vijay Kumar, 1938–
    IS-95 CDMA and cdma2000 : cellular/PCS systems
implementation / Vijay K. Garg.
        p. cm. — (Prentice Hall communications engineering and
emerging technologies series)
    Includes bibliographical references and index.
    ISBN 0-13-087112-5 (case)
        1. Code division multiple access. 2. Cellular telephone
systems. 3. Personal communication service systems. I. Title.
II Series.

TK5103.452 G37 1999
621.3845—dc21   99-049860
                                                                    CIP

Editorial/production supervision: *BooksCraft, Inc., Indianapolis, IN*
Acquisitions editor: *Bernard Goodwin*
Editorial assistant: *Diane Spina*
Marketing manager: *Lisa Konzelmann*
Manufacturing manager: *Alexis Heydt*
Cover design director: *Jerry Votta*
Cover designer: *Talar Agasyan*
Project coordinator: *Anne Trowbridge*

© 2000 by Prentice Hall PTR
Prentice-Hall, Inc.
Upper Saddle River, NJ 07458

Prentice Hall books are widely used by corporations and government agencies for
training, marketing, and resale.

The publisher offers discounts on this book when ordered in bulk quantities.
For more information, contact:
Corporate Sales Department
Prentice Hall PTR
One Lake Street
Upper Saddle River, NJ 07458
Phone: 800-382-3419   Fax: 201-236-7141
E-mail: corpsales@prenhall.com.

Printed in the United States of America

10   9   8   7   6   5   4   3   2   1

ISBN: 0-13-087112-5

Prentice-Hall International (UK) Limited, *London*
Prentice-Hall of Australia Pty. Limited, *Sydney*
Prentice-Hall Canada Inc., *Toronto*
Prentice-Hall Hispanoamericana, S.A., *Mexico*
Prentice-Hall of India Private Limited, *New Delhi*
Prentice-Hall of Japan, Inc., *Tokyo*
Pearson Education Asia Pte. Ltd., Singapore
Editora Prentice-Hall do Brasil, Ltda., *Rio de Janeiro*

I dedicate this book to my lovely grandchildren:
Adam Dorr, Renu Dorr, Monica Taneja, and Nevin Taneja.

Contents

3   Speech and Channel Coding   41

4   Diversity, Combining, and Antennas   57

10 Soft Handoff and Power Control in IS-95 CDMA 181

11 Security and Identification in IS-95 CDMA

12 RF Engineering and Network Planning

13 Reverse and Forward Link Capacity of IS-95 CDMA System   283

Preface

The global mobile communications market is booming. There are almost 250 million users worldwide and should be nearly 1 billion by early next century. Code Division Multiple Access (CDMA) is the fastest-growing digital wireless technology, tripling its worldwide subscriber base between 1997 and 1998. There are already 30 million CDMA customers and, at the current growth rate, there will be 50 million by the millennium. The major markets for CDMA are North America, Latin America, and Asia (particularly Japan and Korea). In total, CDMA has been adopted by almost 50 countries around the world.

It is not hard to see the reasons for the success of CDMA. CDMA is an advanced digital technology that can offer about 7 to 10 times the capacity of analog technologies and up to 6 times the capacity of digital technologies such as Time Division Multiple Access (TDMA). The speech quality provided by CDMA systems is far superior to any other digital cellular technology, particularly in difficult radio environments such as dense urban areas and mountainous regions. In both initial deployment and long-term operation, CDMA provides the most cost-effective solution for cellular operators. After an 18-month of market rollout, Personal Communications Services (PCS) providers have adequately demonstrated the power of CDMA technology to support a marketing strategy based on low prices and superior performance in key areas such as voice quality, system reliability, and handset battery life.

CDMA service providers have a strong advantage when pursuing the market to the minutes-of-use model, given the longevity of CDMA handset battery life and the higher quality of the voice signal. A recent analysis of wireless platform performance by the Telecommunications Research and Action Center (TRAC) found that CDMA outperformed other digital and analog technologies on every front, including signal quality, security, power consumption, and reliability. Although analog technology came out ahead in availability, all three digital services (GSM, IS-136 TDMA, and IS-95 CDMA) were rated equally over analog with respect to availability of enhanced service features. The TRAC study found CDMA to be superior in signal security and voice quality over the other digital air interface standards. According to TRAC, CDMA has several advantages for consumers. Lower power consumption enables CDMA handsets to support up to 4 hours of talk time or 48 hours of standby time on a single battery charge. It has also been found that the soft-handoff characteristics of CDMA lead to fewer dropped calls than with GSM and IS-136 TDMA. One possible drawback for some CDMA customers is that there are some limitations on roaming capabilities. Some PCS operators with cellular affiliates are supporting dual-mode handsets to allow roaming between CDMA and analog platforms.

CDMA technology is constantly evolving to offer customers new, advanced services. The mobile data speeds offered through CDMA phones are increasing, and new voice codecs provide speech quality close to wireline. Internet access is now available through CDMA terminals. The time will soon be at hand when CDMA service providers can further exploit the enhanced service potential of their platforms. There has been much talk of so-called third-generation (3G) data capabilities, where PCS providers will be able to compete with wireline service providers at high access speeds. PCS providers are looking ahead toward providing a range of service categories such as Internet and intranet access, multimedia applications, high-speed business transactions, and telemetry. The CDMA network offers operators a smooth evolutionary path to 3G mobile systems.

The IS-95B standard is quite flexible, enabling service providers to allocate data in increments of 8 kilobits per second (kbps) within the 1.25-megahertz (MHz) CDMA channel bandwidth based on how service providers configure software download to already-installed network controllers. This means operators can implement return data speeds at rates much lower than 64 kbps, ensuring much lower power consumption in handsets than would be the case at a full 64-kbps return rate. While operators in GSM and IS-136 TDMA sectors are making efforts to ensure they won't be left behind as data becomes a factor, CDMA appears to have a clear edge in its ability to go to relatively high speeds over the existing infrastructure.

The opportunity to use the CDMA platform to add a fixed wireless service feature represents an added advantage for operators. Because CDMA has ample spectrum to provide a fixed service on top of mobile, several operators are exploring using terminals that would be able to shift the handset between fixed and mobile service, depending on where the user is. The universal handset would serve as a cordless phone in the home and as a mobile handset outside the home. The evolution to 3G will open the wireless local loop (WLL) with Public-Switched Telephone Network (PSTN) and Public Data Network (PDN) access, while providing more convenient control of applications and network resources. It will also open the door to convenient global roaming, service portability, zone-based ID and billing, and global directory access. The 3G technology is even expected to support seamless satellite interworking.

With the cornucopia of benefits surrounding CDMA, it is evident that operators using this platform will have every opportunity to grow the business once the community-based strategy begins to unfold. The question is, when will they get serious about bringing these new capabilities to market?

Recently an enhanced hybrid technology combining the CDMA air interface with the GSM network has been built, tested, and evaluated. GSM operators can save over 60% in cumulative capital costs using a GSM-CDMA overlay for network expansion of the GSM network using IS-95 CDMA radio access in addition to, or as a substitute for, TDMA radio access. This combines the spectral efficiency of CDMA with all GSM features, including seamless roaming and network services. The GSM-CDMA technology provides operators with a way to serve multiple market segments economically and to offer various services on one network platform. In addition to being a cost-effective network expansion solution, GSM-CDMA also paves an evolutionary path to 3G services including high-speed data, multimedia, and mobile/fixed convergence services.

CDMA is the selected approach for the 3G system, as evidenced by the proposals submitted by the European Telecommunications Standards Institute (ETSI), the Association Radio Industry Business (ARIB), and the Telecommunications Industry Association (TIA). The 3G cdma2000 uses a CDMA air interface based on the existing IS-95B standard to provide wireline-quality voice service and high-speed data services, ranging from 144 kbps for mobile users to 2 megabits per second (Mbps) for stationary users. It is important to note that cdma2000 is a core proposal of the TIA for International Mobile Telecommunications-2000 (IMT-2000). Moreover, support for cdma2000 is not limited to North America; Korean carriers have a great opportunity to provide 3G-like service with today's existing CDMA technology. Mobile data rates of up to 114 kbps and fixed peak rates beyond 1.5 Mbps are within reach before the end of the decade with today's CDMA technology. These capabilities will be provided without degrading the systems' voice transmission capabilities or requiring additional spectrum. This will have tremendous implications for the majority of operators that are spectrum constrained. A doubling of capacity and a 1.5-Mbps data rate capability within a 1.25-MHz channel structure look very appealing.

This book is an extension of the book *Applications of CDMA in Wireless Communications* (Garg, Smolik, and Wilkes, Prentice Hall, 1997). In that book, the primary focus was on the CDMA systems standardized by TIA and American Telecommunications Industries Standards (ATIS) as standards IS-95 and IS-665. Since the publication of that book, CDMA technology has undergone major changes and has become a viable technology for 3G systems. In this book, I discuss those aspects of CDMA that are essential to understanding system capacity. I also provide guidelines for system parameters of a CDMA network. The book outlines a migration path for CDMA to a 3G cdma2000 system.

In writing this book, I addressed the needs of practicing engineers and engineering managers by explaining CDMA concepts, system capacity, radio frequency (RF) engineering, and other important aspects of the CDMA network. Students studying courses in telecommunications will also find this book useful as they prepare for careers in the wireless industry. I included a sufficient amount of mathematics so that you can understand the operation of the CDMA network, but I tried not to overwhelm you with very complex mathematical derivations.

This book can be used by practicing telecommunications engineers involved in the design and operation of CDMA-based cellular/PCS networks as well as by senior or graduate students in electrical engineering, telecommunications engineering, and computer engineering curricula. I assume that you have some basic background in mobile communications and CDMA technology. If you don't, the book mentioned above by Garg, Smolik, and Wilkes can provide that understanding. By selectively reading pertinent chapters of that book, telecommunications managers who are engaged in managing CDMA networks but who have little or no technical background can gain enough of an understanding of CDMA systems to read this book.

This book can be divided into four segments: Chapters 1 through 4 provide a foundation for understanding the material in subsequent chapters. Chapters 5 through 11 deal with IS-95 CDMA standards, and chapters 12 and 13 provide design aspects of a CDMA system. Chapters 14 and 15 focus on the data applications in CDMA and the evolution of IS-95 (2G system) to cdma2000 (3G system) in order to satisfy ITU IMT-2000 specifications. The following is a synopsis of the subjects covered in each chapter.

- **Chapter 1.** Major attributes of CDMA and the access technologies used for cellular/PCS systems.
- **Chapter 2.** The different types of Spread Spectrum (SS) systems that are used. The main focus is on the Direct Sequence Spread Spectrum (DSSS) techniques that are employed in CDMA. I provide a relationship to calculate the performance of a CDMA system.
- **Chapter 3.** Speech and channel coding applications in the IS-95 CDMA system.
- **Chapter 4.** The concepts of diversity reception used to improve signal-to-noise ratio (SNR) of the system; various combining schemes used to combine the signals; some practical antennas used in the cellular telephone industry.
- **Chapter 5.** Functional entities of the wireless network and the TIA-standardized interfaces between the entities. I examine the activities of the International Telecommunication Union (ITU) to add Intelligent Network (IN) to wireless systems.
- **Chapter 6.** A high-level description of the IS-95 CDMA air interface, including important aspects of the forward link (base station to mobile) and reverse link (mobile to base station) and modulation parameters for the channels.
- **Chapter 7.** Modulation schemes, bit repetition, block interleaving, and channel coding; these are used in processing logical channels on the IS-95 CDMA forward and reverse links. Details about information processing, message types, and message framing are presented for the pilot, sync, paging, and traffic channels on the forward link. Similar details are provided for the access and traffic channels on the reverse link
- **Chapter 8.** IS-95 CDMA call processing states that a mobile station (MS) goes through in getting to a traffic channel; idle handoff, slotted paging operation, CDMA registration, and authentication procedures; call flows for CDMA call origination, call termination, call release, and authentication.
- **Chapter 9.** The layering concept used to develop the protocols for IS-95 CDMA; the standardized interfaces between the functional entities, mainly the *A-Interface* and TIA IS-634-defined MSC-BS messages, message sequencing, and mandatory timers at the

BS and the MSC. The chapter also provides call flow diagrams for typical supplementary services, handoff scenarios, and Over-The-Air Service Provisioning (OTASP).

- **Chapter 10.** Handoff strategy used in IS-95 CDMA; power control schemes for the reverse and forward links.
- **Chapter 11.** Various parameters used to identify an MS including International Mobile Station Identity (IMSI), Mobile Station Number (MDN), Electronic Serial Number (ESN), and station class mark. I focus on authentication procedures, including the authentication of MS registration, MS originations, MS terminations, MS data bursts, and Temporary Mobile Station Identity (TMSI) assignment. Also discussed are unique challenge response procedures.
- **Chapter 12.** Basic guidelines for engineering a CDMA system, including a discussion of propagation models, link budgets, the transition from analog operation to CDMA operation, radio link capacity, facility engineering, border cells on a boundary between two service providers, and interfrequency handoff.
- **Chapter 13.** Procedures for calculating the capacity of the reverse and forward link of a CDMA system; a procedure to develop a link safety margin parameter for each of the forward link channels.
- **Chapter 14.** Standards for data services supported by CDMA cellular/PCS systems; highlights of the TIA IS-99, TIA IS-637, and TIA IS-657 standards. I describe the architecture for each of the four data services (e.g., packet data, asynchronous data, facsimile, and short message services) and the protocol stacks supported by these services.
- **Chapter 15.** The cdma2000, 3G evolution of IS-95. The cdma2000 Radio Transmission Technology (RTT) is a wideband, SS radio interface that uses CDMA technology to satisfy the needs of 3G wireless communication systems.

Appendix A presents traffic tables for a variety of blocking probabilities and channel numbers. **Appendix B** comprises a list of abbreviations I introduce in the text and that are common to the industry. The references cited in **Appendix C** are papers and texts that I have found useful and, when considered in addition to those cited in the text, provide a rich background for readers interested in looking into digital wireless technology in greater depth.

I suggest chapters 1–11 for those who are interested in IS-95 standards but who do not have much background in digital communications. Those who have adequate background in digital communications may skip chapters 1–4.

I recommend chapters 1, 2, 4–10, 12, and 13 for those who are involved with the design of a CDMA system. The engineering managers should use chapters 1 and 5–12 to achieve adequate knowledge of IS-95 CDMA.

I suggest chapters 1–8, 10–12, 13, and 15 for a one-semester graduate course in IS-95 CDMA and its evolution to cdma2000.

I would like to thank the many people who helped me prepare the material in this book. Bernard Goodwin provided his encouragement in motivating me to write the book. Professor Ted Rappaport of Virginia Tech took me under the banner of his new series. I acknowledge the many helpful suggestions I received from my many friends.

Finally, I acknowledge the assistance of my wife, Pushpa Garg, and the staff of BooksCraft, Inc. during the production of this book.

Introduction to Access Technologies

1.1 Introduction

After 18 months of market rollouts, digital cellular and Personal Communications Services (PCS) providers have adequately demonstrated the power of Code Division Multiple Access (CDMA) technology to support a marketing strategy based on low prices and superior performance in key areas such as voice quality, system reliability, and handset battery life. According to the CDMA Development Group (CDG), the market base for IS-95 platform providers has grown from only 200,000 customers in 1997 to about 3.0 million in 1999.

CDMA service providers have a strong advantage in pursuing the market to the minutes-of-use model, given the longevity of CDMA handset battery life and the higher quality of the voice signal. A recent analysis of wireless platform performance by Telecommunications Research and Action Center (TRAC) found that CDMA beat other digital and analog technologies on every front, including signal quality, security, power consumption, and reliability. Although analog was more readily available, all three digital systems (GSM, IS-136 TDMA, and IS-95 CDMA) were rated equally over analog with respect to enhanced service features. The TRAC study found CDMA to be superior in signal security and voice quality over the other digital air interface standards. According to TRAC, CDMA has several advantages for consumers. Lower power consumption enables CDMA handsets to support up to 4 hours of talk time or 48 hours of standby time on a single battery charge. It has also been found that the soft handoff characteristics of CDMA lead to fewer dropped calls than GSM and IS-136 TDMA. One possible drawback for some CDMA customers is that there are some limitations on roaming capabilities. Some PCS operators with cellular affiliates are supporting dual-mode handsets to allow roaming between the CDMA and analog platforms.

The IS-95 standard is quite flexible, enabling service providers to allocate data in increments of 8 kilobits per second (kbps) within the 1.25-megahertz (MHz) CDMA channel bandwidth based on how service providers configure software download to already-installed network controllers. This means operators can implement return data speeds at rates much lower than 64 kbps, ensuring much lower power consumption in handsets than would be the case at a full 64-kbps return rate. While operators in GSM and IS-136 TDMA sectors are making efforts to ensure they won't be left behind as data becomes a factor, CDMA appears to have a clear edge in its ability to go to relatively high speeds over the existing infrastructure.

The opportunity to use the CDMA platform to add a fixed wireless service feature represents an added advantage for operators. With ample spectrum to provide a fixed service on top of mobile, several operators are exploring using terminals that would be able to shift the handset between fixed and mobile service, depending on where the user is. The universal handset would serve as a cordless phone in the home and as a mobile handset outside the home.

CDMA offers about eight times the capacity of analog, which implies that there's enough capacity in 30 MHz of spectrum alone to serve the entire mobile market now and into the future for a country like Canada. With the cornucopia of benefits surrounding CDMA, it is evident that operators using this platform have every opportunity to grow the business once the community-based strategy begins to unfold. The question is, when will they get serious about bringing these new capabilities to market?

Recently, an enhanced hybrid technology combining the CDMA air interface with the GSM network was built, tested, and evaluated. GSM operators can save over 60% in cumulative capital costs using a GSM-CDMA overlay for network expansion of the GSM network using IS-95 CDMA radio access in addition to, or as a substitute for, TDMA radio access. This combines the spectral efficiency of CDMA with all GSM features, including seamless roaming and network services. The GSM-CDMA technology provides operators the flexibility to serve multiple market segments economically and to offer various services on one network platform. In addition to being a cost-effective network expansion solution, GSM-CDMA also paves an evolution path to third-generation services including high-speed data, multimedia, and mobile/fixed convergence services.

In this chapter, we discuss major attributes of CDMA and describe the access technologies used for cellular/PCS systems, outlining their advantages and disadvantages. We also discuss the use of CDMA for future third-generation systems and outline modes of operation for cellular/PCS systems.

1.2 Major Attributes of CDMA Systems

The major attributes of IS-95A/J-STD-008 CDMA systems [5] are as follows:

- **System Capacity.** The projected capacity of CDMA systems is much higher than that of existing analog/digital systems. The increased system capacity is due to an improved coding gain/modulation scheme, voice activity, three-part sectorization, and reuse of the same spectrum in every cell and all sectors.
- **Quality of Service.** CDMA improves the quality of service by providing robust operation in fading environments and transparent (soft) handoffs. CDMA takes advantage of

multipath fading to enhance communications and voice quality. By using RAKE receivers and other improved signal-processing techniques, each mobile station selects the three strongest multipath signals and coherently combines them to produce an enhanced signal. Thus, the multipath fading of the radio channel is used to an advantage in CDMA whereas, in narrowband systems, fading causes a substantial degradation of signal quality. By using soft handoff, CDMA eliminates the ping-pong effect that occurs when the mobile is near a border between cells and the call is rapidly switched between the two cells. The ping-pong effect results in handoff noise, increases the load on switching equipment, and increases the chance of a dropped call. In CDMA's soft handoff, a connection is made to the target cell while maintaining the connection with the serving cell, both operating on the same carrier frequency. This procedure ensures a smooth transition between cells, one that is undetectable to the subscriber. In comparison, many analog and other digital (TDMA) systems use a break-before-make connection and require a change in mobile frequency that increases handoff noise and the chance of a dropped call.

- **Economies.** CDMA is a cost-effective technology that requires fewer cell sites and no costly frequency reuse pattern. The average power transmitted by CDMA mobile stations averages 6 to 7 megawatts (mW), which is significantly lower than the average power typically required by FM and TDMA phones. Transmitting less power means that average battery life will be longer.

1.3 Third-Generation Systems

The time will soon be at hand when CDMA service providers can further exploit the enhanced service potential of their platforms. There has been much talk of so-called third-generation (3G) data capabilities, where PCS providers will be able to compete with wireline service providers at high access speeds. PCS providers are looking ahead to providing a range of service categories, such as Internet and intranet access, multimedia applications, high-speed business transactions, and telemetry.

Taking into account the limitations imposed by the finite amount of radio spectrum available, the focus of third-generation mobile systems is on economy of network and radio transmission design to provide seamless service from the customer's perspective. Third-generation mobile systems have to provide their users with seamless access to the fixed data network. They are perceived as the wireless extension of future fixed networks, as well as an integrated part of the fixed network infrastructure.

In Europe, three related network platforms are currently the subject of intensive research—Future Land Public Mobile Telephone Systems (FLPMTS, now known as IMT-2000), Mobile Broadband Systems (MBS), and Wireless Local Area Network (WLAN). One major distinction of IMT-2000 relative to second-generation systems is its hierarchical cell structure, which is designed to support a wide range of multimedia broadband services by using advanced transmission and protocol technologies. Second-generation systems mainly use a one-layer cell structure and employ frequency reuse within adjacent cells in such a way that each single cell manages its own radio zone and radio circuit control within the mobile network, including traffic management and handoff procedures. The amount of traffic supported in each cell is fixed because of frequency limitations and also because of limited flexibility of radio transmission which is optimized mainly for voice and low-data-rate transmissions. Increasing traffic leads to costly cellular reconfigurations such as cell splitting and cell sectorization.

The multilayer cell structure in IMT-2000 aims to overcome these problems by overlaying—discontinuously—picocells and microcells on the macrocell structure with wide area coverage. Global/satellite cells can be used in the same sense to provide area coverage where macrocell constellations are not economical to deploy and/or to support long distance traffic.

With low mobility and small delay spread profiles in picocells, high bit rates and high traffic densities can be supported with low complexity as opposed to low bit rates and low traffic load in macrocells that support high mobility. Users expect service to be selected in a uniform manner with consistent procedures, irrespective of whether access to these services is fixed or mobile. Freedom of location and means of access will be facilitated by *smart cards*, which allow customers to register on different terminals with varying capabilities (speech, multimedia, data, short messaging).

The choice of a radio interface parameter set corresponding to a multiple access scheme is a critical issue in terms of spectral efficiency, taking into account the ever increasing market demand for mobile communications and the fact that radio spectrum is a very expensive and scarce resource. A comparative assessment of several different schemes has been carried in the framework of the Research in Advanced Communications Equipment (RACE) program. One possible solution is to use a hybrid CDMA/TDMA/FDMA technique, integrating the advantages of each and meeting the varying requirements on channel capacity, traffic load, and transmission quality in different cellular/PCS layouts. Disadvantages of such hybrid access schemes are the high complexity and difficulties in achieving simplified low-power, low-cost transceiver design as well as efficient flexibility management in the several cell layers.

CDMA is the selected approach for third-generation systems, as evidenced by the proposals in ETSI, ARIB (Japan), and the TIA. In Europe and Japan, Wideband CDMA (W-CDMA) has been proposed to avoid IS-95 intellectual property rights (IPR). In North America, cdma2000 uses a CDMA air-interface based on the existing IS-95 standard to provide wireline-quality voice service and high-speed data services ranging from 144 kbps for mobile users to 2 megabits per second (Mbps) for stationary ones. The 64-kbps data capability of IS-95 will provide high-speed Internet access in a mobile environment, a capability that cannot be matched by other narrowband digital technologies.

Mobile data rates of up to 114 kbps and fixed peak rates beyond 1.5 Mbps are within reach before the end of the decade using wideband CDMA technologies. These services will be provided without degrading the systems' voice transmission capabilities or requiring additional spectrum. This will have tremendous implications for the majority of operators that are spectrum constrained.

1.4 Multiple Access Technologies

There are two types of access technologies [6]—narrowband and wideband. The narrowband access technologies are *Frequency Division Multiple Access* (FDMA) and *Time Division Multiple Access* (TDMA).

In FDMA, signals from various users are assigned different frequencies (see Fig. 1-1). Guard bands are used between adjacent signal spectra to minimize crosstalk between channels. The advantages of FDMA are that

- Capacity increases can be obtained by reducing the information bit rate and using efficient digital codes.
- Technological advances required for implementation are simple. A system can be configured so that improvements in terms of speech coder bit-rate reduction could be readily incorporated.

The main disadvantages of FDMA are that

- It does not differ significantly from analog systems; capacity improvement depends on reducing signal-to-interference ratio, or signal-to-noise ratio (SNR).
- FDMA involves narrowband filters, and, because these are not realized in very-large-scale integrated (VLSI) digital circuits, this may set a high cost floor for terminals even under volume production conditions.

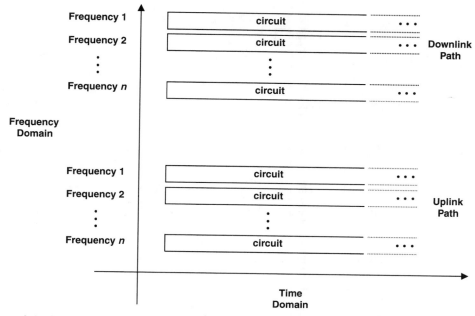

Figure 1-1   Frequency Division Multiple Access (FDMA) with Frequency Division Duplex (FDD).

- The maximum bit rate per channel is fixed and small, inhibiting the flexibility in bit-rate capability that may be a requirement for computer file transfer in some applications.

In a TDMA system, information from each user is conveyed in time intervals called time slots (see Fig. 1-2). Several time slots make up a time frame. Each time slot consists of a preamble plus information bits. The functions of the preamble are to provide identification and incidental information and to allow synchronization of the time slot at the intended receiver. Guard times are provided between each user's transmission to minimize crosstalk between channels. Most TDMA systems time divide a frame into multiple slots used by different transmitters. This approach is called Time Division Multiplex (TDM). The information is transmitted via a radio carrier from a base station (BS) to several active mobile stations (MSs) in the downlink. In the uplink, transmissions from mobile stations to a BS are time-sequence synchronized on a common frequency for TDMA. The advantages of TDMA are that it

- Allows a flexible bit rate, not only for multiples of a basic single channel but also sub-multiples for low-bit-rate broadcast-type traffic.
- Potentially integrates in VLSI without narrowband filters, giving a low cost floor in volume production.
- Offers the opportunity for frame-by-frame monitoring of signal strength/bit error rates to enable either mobiles or base stations to initiate and execute handoffs.

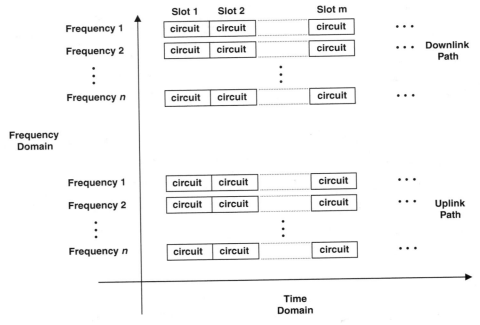

Figure 1-2 Time Division Multiple Access (TDMA) with FDD

- Uses bandwidth more efficiently because no frequency guard band is needed between channels.
- Transmits each signal with sufficient guard time between time slots and accommodates time inaccuracies caused by clock instability, delay spread, transmission time delay because of propagation distance, and the "tails" of signal pulses in TDMA because of transient responses.

The main disadvantage of TDMA are:

- Particularly for mobile hand sets, on uplink TDMA demands high peak power in transmit mode. This reduces battery life.
- TDMA requires a substantial amount of signal processing for matched filtering and correlation detection for synchronizing with a time slot.

In wideband technologies, the entire bandwidth is made available to each mobile user; this bandwidth is many times larger than the bandwidth required to transmit information. Such systems are called *spread spectrum* (SS) systems. The primary advantage of wideband technology is its ability to tolerate a fair amount of signal interference compared to FDMA and TDMA, which typically cannot tolerate any such interference. Because of the interference tolerance of wideband systems, the problems of frequency band assignment and adjacent cell interference are greatly simplified. Flexibility in system design and deployment are improved since interference to others is not a problem. FDMA and TDMA radios must be carefully assigned a frequency or time slot to assure that there is no interference with other similar radios. In wideband systems, adjacent cells share the same frequencies; with TDMA/FDMA it is not feasible for adjacent cells to share same frequencies because of interference. There is no need for frequency planning in wideband systems.

1.5 Modes of Operation in Wireless Communications

There are two operational modes in wireless communications: Frequency Division Duplex (FDD) and Time Division Duplex (TDD). In FDD, the transmitter and receiver operate simultaneously on different frequencies. Separation is provided between the downlink and uplink channels to avoid interference between the transmitter and the receiver (see Fig. 1-2). In TDD, a bidirectional flow of information is achieved using a simplex-type scheme by automatically changing in time the direction of transmission on a single frequency. The basic advantages of TDD are that

- There is no need for a dedicated duplex stage (duplexer). The only requirements are that a fast switching synthesizer, RF filter paths, and fast antenna switch be available.
- It increases battery life or reduces battery weight.
- It provides phones with better quality and at a lower cost.
- It requires a single frequency band instead of the two in FDD.

1.6 Summary

In this chapter, we focused on the growth of CDMA technology in wireless communications. We discussed the suitability of CDMA for future 3G systems to provide higher data rate access to Internet/intranet services because it is the most suitable digital technology to handle future wireless communications demand.

We briefly described narrowband channelized and wideband nonchannelized technologies and listed their advantages and disadvantages. Finally, we described modes of operations in wireless communications.

1.7 References

1. Balston, D. M., "The Pan-European Cellular Technology," *IEE Conference Publication*, 1988.

2. Balston, D. M., and Macario, R. C. V., *Cellular Radio Systems*, Artech House, Boston, 1993.

3. Dasilva, J. S., Ikonomou, D., and Erben, H., "European R&D Programs on Third-Generation Mobile Communications Systems," *IEEE Personal Communications* 4(1), February 1997, pp. 46–52.

4. "The European Path Towards UMTS," *IEEE Personal Communications*, special issue, February 1995.

5. Garg, V. K., Smolik, K. F., and Wilkes, J. E., *Applications of CDMA in Wireless/Personal Communications*, Prentice Hall, Upper Saddle River, NJ, 1997.

6. Garg, V. K., and Wilkes, J. E., *Wireless and Personal Communications Systems*, Prentice Hall, Upper Saddle River, NJ, 1996.

7. Marley, N., "GSM and PCN Systems and Equipment," *JRC Conference*, Harrogate, 1991.

8. Rapeli, J., "UMTS: Targets, System Concepts, and Standardization in a Global Framework," *IEEE Personal Communications*, special issue, February 1995.

9. Salmasi, A., "An Overview of Code-Division Multiple Access (CDMA) Applied to the Design of Personal Communications Networks," in *Third Generation Wireless Information Network*, Nanda, S., and Goodman, D. J., ed., Kluwer Academic Publishers, Boston, 1992, pp. 277–98.

Direct Sequence Spread Spectrum and Spreading Codes

2.1 Introduction

In this chapter we describe the different types of Spread Spectrum (SS) systems that are used and then focus on the Direct Sequence Spread Spectrum (DSSS) technique that is employed in CDMA IS-95 systems. We develop the necessary relationships to evaluate the performance of a DSSS system with Binary Phase-Shift Keying (BPSK) and Quadrature Phase-Shift Keying (QPSK) modulation schemes and provide a relationship to calculate the performance of a CDMA system.

2.2 Types of Techniques Used for Spread Spectrum

Since the late 1940s, SS techniques have been used for military applications in which clandestine operation is a major objective. SS techniques provide excellent immunity to interference—possibly the result of intentional jamming—and allow transmission to be hidden within background noise. Recently, SS systems have been adopted for civilian applications in wireless telephony systems.

There are three general approaches to implementing SS systems.

1. **Direct Sequence Spread Spectrum (DSSS)** where a carrier is modulated by a digital code in which the code bit rate is much larger than the information signal bit rate (see Fig. 2-1). These systems are also called pseudonoise (PN) systems.
2. **Frequency Hopping Spread Spectrum (FHSS)** where the carrier frequency is shifted in discrete increments in a pattern generated by a code sequence (see Fig. 2-2). Sometimes, the codes are chosen to avoid interference to or from other non-SS systems. In an FHSS system, the signal frequency remains constant for a specified time duration, referred to as a time chip, T_c. The FHSS system can be either a fast-hop or a slow-hop system. In a fast-hop system, the frequency hopping occurs at a rate that is faster than

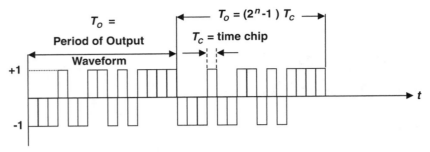

Figure 2-1   Direct Sequence Spread Spectrum Approach

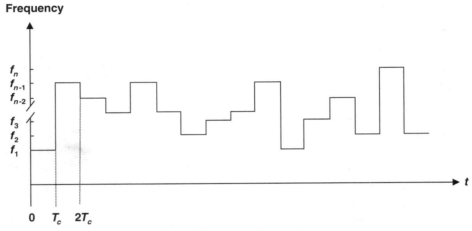

Figure 2-2   Frequency Hopping Spread Spectrum Approach

the message bit rate. In a slow-hop system, the hop rate is slower than the message bit rate. There is, of course, an intermediate situation in which the hop rate and message bit rate are the same.

FHSS radio systems experience occasional strong bursty errors, while DSSS radio systems experience continuous but lower-level random errors. With DSSS radio systems, single errors are dispersed randomly over time, whereas, with FHSS radio systems, errors are distributed in clusters. Bursty errors are attributable to fading or single frequency interference, which is time and frequency dependent. DSSS spreads the information in both the time and frequency domains, thus providing time and frequency diversity and minimizing the effects of fading and interference.

3. **Time-Hopped (TH)** system, where the transmission time is divided into intervals called frames (see Fig. 2-3). Each frame is divided into time slots. During each frame,

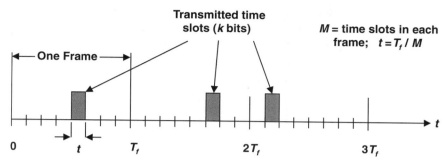

Figure 2-3 Time Hopping Spread Spectrum Approach

one and only one time slot is modulated with a message. All of the message bits accumulated in previous frames are transmitted.

2.3 The Concept of Spread Spectrum System

The theoretical capacity of any communications channel is defined by C. E. Shannon's channel capacity formula [7], Eq. (2.1):

$$C = B_w \log_2\left[1 + \frac{S}{N}\right]$$ (2.1)

where B_w = bandwidth in Hertz,
  C = channel capacity in bits per second,
  S = signal power, and
  N = noise power.

Eq. (2.1) gives the relationship between the theoretical ability of a channel to transmit information without errors for a given SNR and a given bandwidth of a channel. Channel capacity is increased by increasing the channel bandwidth, the transmitted power, or a combination of both.

Shannon modeled the channel at baseband. However, Eq. (2.1) is applicable to a Radio frequency (RF) channel by assuming that the Intermediate Frequency (IF) filter has an ideal (flat) band-pass response with a bandwidth that is at least $2 \times B_w$. This bound assumes that channel noise is Additive White Gaussian Noise (AWGN). AWGN is often adopted in the modeling of an RF channel. This assumption is justified since the total noise is generated by random electron fluctuations. The central limit theorem provides us with the assumption that the output of an IF filter has a Gaussian distribution and is frequency independent. For most communications systems that are limited by thermal noise, this assumption is true. For interference-limited systems, this assumption is not true and the results may be different. The Shannon equation does not provide a method to achieve the bound. Approaching the bound requires complex channel coding and modulation techniques. In many cases, achieving an implementation that provides performance near this bound is impractical due to the resulting complexity.

An analog cellular system is typically engineered to have an SNR of 17 decibels (dB)[*] or more. CDMA systems can be engineered to operate at much lower SNRs since the channel bandwidth can be traded for the SNR to achieve good performance at a very low SNR.

Next we rewrite Eq. (2.1) as Eq. (2.2).

$$\frac{C}{B_w} = 1.44 \log_e\left[1 + \frac{S}{N}\right] \tag{2.2}$$

Since

$$\log_e\left(1 + \frac{S}{N}\right) = \frac{S}{N} - \frac{1}{2}\left(\frac{S}{N}\right)^2 + \frac{1}{3}\left(\frac{S}{N}\right)^3 - \frac{1}{4}\left(\frac{S}{N}\right)^4 + \cdots$$

we use the logarithmic expansion and assume that the SNR is small, e.g., SNR ≤ 0.1, so we can neglect the higher-order terms to rewrite Eq. (2.2) as

$$B_w \approx \frac{C}{1.44} \times \frac{N}{S} \tag{2.3}$$

For any given SNR we can have a low information error rate by increasing the bandwidth used to transmit the information. As an example, if we want a system to operate on a link in which the information rate is 10 kbps and SNR is 0.01, we must use a bandwidth of

$$B_w = \frac{10 \times 10^3}{1.44 \times 0.01} = 0.69 \times 10^6 \text{ Hz or 690 kHz}$$

Information can be modulated into the spread spectrum signal by several methods. The most common method is to add the information to the spectrum-spreading code before it is used for modulating the carrier frequency (Fig. 2-4). This technique applies to any SS system that uses a code sequence to determine RF bandwidth. If the signal being sent is analog (voice, for example), the signal must be digitized before being added to the spreading code.

2.3.1 System Processing Gain

One of the major advantages with an SS system is its robustness to interference. The system processing gain (G_p) quantifies the degree of interference rejection. The system processing gain is the ratio of RF bandwidth to the information rate and is given as

$$G_p = \frac{B_w}{R} \tag{2.4}$$

---

[*] This assumes a fading radio environment, which is typical for analog cellular systems that use Frequency Modulation. In the absence of fading, good FM performance is achievable at lower SNR.

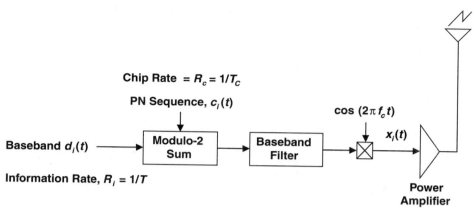

Figure 2-4   Basic DSSS System Transmitter

Typical processing gains for SS systems lie between 20 and 60 dB. With an SS system, the noise level is determined both by the thermal noise and by interference. For a given user, the interference is processed as noise. The input and output SNRs are related as

$$\left(\frac{S}{N}\right)_o = G_p \left(\frac{S}{N}\right)_i \tag{2.5}$$

It is instructive to relate the SNR to the $E_b/N_0{}^*$ ratio where E_b is the energy per bit and N_0 is the noise power spectral density.

$$\left(\frac{S}{N}\right)_i = \frac{E_b \times R}{N_0 \times B} = \frac{E_b}{N_0} \times \frac{1}{G_p} \tag{2.6}$$

From Eq. (2.6) we can express E_b/N_0 as

$$\frac{E_b}{N_0} = G_p \times \left(\frac{S}{N}\right)_i = \left(\frac{S}{N}\right)_o \tag{2.7}$$

EXAMPLE   2.1

Calculate the processing gain for a DSSS system that has a 10 Megachips per second (Mcps) code clock rate and 4.8-kbps information rate. How much improvement in the processing gain will be achieved if the code generation rate is changed to 50 Mcps? Is there an advantage in going to a higher code generation rate with a 4.8-kbps information rate?

---

* The noise power spectral density actually consists of both the thermal noise and interference. Unless stated explicitly, N_0 represents the thermal noise. However, common usage of this ratio assumes that N_0 includes both the thermal noise and interference. With SS systems, interference is transformed into noise.

We assume that the DSSS waveform has a voltage distribution of $(\sin x)/x$. The power distribution has a form of $[(\sin x)/x]^2$. The bandwidth of the main lobe is equal to the spreading code clock rate.

$$G_p = \frac{1.0 \times 10^7}{4.8 \times 10^3} = 2.1 \times 10^3 = 33.1 \text{ dB}^*$$

With 50 Mcps

$$G_p = \frac{5 \times 10^7}{4.8 \times 10^3} = 1.04 \times 10^4 = 40.2 \text{ dB}$$

By increasing the code generation rate from 10 to 50 Mcps, we get only 7dB improvement in the processing gain. The effort required to get five times the operating speed of a circuit may be much more demanding compared to an improvement of 7 dB in the processing gain.

2.4 The Performance of DSSS

2.4.1 The DSSS System

The DSSS system is a wideband system in which the entire bandwidth of the system is available to each user. A system is defined as a DSSS system if it satisfies the following requirements:

1. The spreading signal has a bandwidth much larger than the minimum bandwidth required to transmit the desired information, which for a digital system is the baseband data.
2. The spreading of the data is performed by means of a *spreading signal*, often called a *code signal*. The code signal is independent of the data and of a much higher rate than the data signal.
3. At the receiver, despreading is accomplished by the cross-correlation of the received spread signal with a synchronized replica of the same signal used to spread the data.

2.4.2 Coherent Binary Phase-Shift Keying

The simplest form of a DSSS communications system employs coherent BPSK for both the data modulation and the spreading modulation. But the most common form uses BPSK for data modulation and Quadrature Phase-Shift Keying (QPSK) for spreading modulation. We first consider the simplest case.

The encoded DSSS BPSK signal is given by

$$x(t) = c(t)s(t) = c(t)d(t)\sqrt{2S}\cos\omega_c t \qquad (2.8)$$

where $s(t) = d(t)\sqrt{2S}\cos\omega_c t$,

$d(t)$ = the baseband signal at the transmitter input and receiver output,

---

* dB = $10 \log G$

$c(t)$ = the spreading signal,
S = the signal power, and
ω_c = the carrier frequency.

In Eq. (2.8), we represent the modulo-2 addition of $c(t)$ and $d(t)$ as a multiplication because the binary signals 0 and 1 represent values of 1 and −1 into the modulator.

The signal $s(t)$ has a $[(\sin x)/x]^2$ spectrum of bandwidth roughly $1/T$ (where T is the periodicity at baseband), while the SS signal $x(t)$ has a similar spectrum but with a bandwidth of approximately $1/T_c$ (where T_c is the periodicity of the spreading signal). From Eq. (2.4), the processing gain of the system is $G_p = (B_w/R) = T/T_c$. If the interfering signal is represented by $I(t)$, then in the absence of noise (assuming that the interferer limits the system performance—in other words, that the interferer's power level exceeds the thermal noise power), the signal at the receiver is given as

$$[r(t)]^* = x(t) + I(t) \qquad (2.9)$$

The receiver multiplies this by the PN waveform to obtain the signal

$$r(t) = c(t)[x(t) + I(t)] = c(t)[c(t)s(t)] + c(t)I(t) = s(t) + c(t)I(t) \qquad (2.10)$$

since $c(t)^2 = 1$. $c(t)I(t)$ is the effective noise waveform due to interference.

The conventional BPSK detector output is given as

$$r = d\sqrt{E_b} + n \qquad (2.11)$$

where d = the data bit for the T second interval,
E_b = the bit energy, and
n = the equivalent noise component.

The spreading-despreading operation does not affect the signal, nor does it affect the spectral and probability density function of the noise. For this reason the bit error probability P_b associated with the coherent BPSK SS signal is the same as with the BPSK [8] signal and is given as

$$P_b = \frac{1}{2}\mathrm{erfc}\left(\sqrt{\frac{E_b}{N_0}}\right) \qquad (2.12)$$

2.4.3 Quadrature Phase-Shift Keying

For QPSK modulation, we denote the in-phase and quadrature data waveforms as $d_c(t)$ and $d_s(t)$, respectively, and the corresponding PN binary waveform as $c_c(t)$ and $c_s(t)$. We can represent a QPSK signal as (see Fig. 2-5)

$$x(t) = c_c(t)d_c(t)\sqrt{S}\cos\omega_c t + c_s(t)d_s(t)\sqrt{S}\sin\omega_c t \qquad (2.13)$$

where each QPSK pulse is of duration $T_s = 2T$.

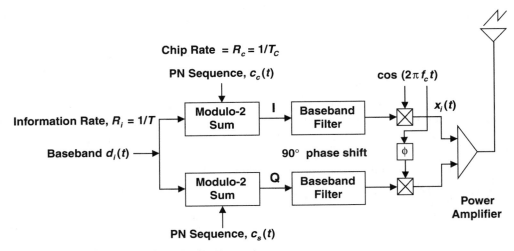

Figure 2-5   DSSS System with QPSK Transmitter

The in-phase output component is

$$r_c = d_c\sqrt{E_b} + n_c \tag{2.14}$$

where $n_c = \sqrt{\dfrac{2}{T_s}}\displaystyle\int_0^{T_c} c_c(t)I(t)\cos\omega_c t\, dt$,

and the quadrature component is

$$r_s = d_s\sqrt{E_b} + n_s \tag{2.15}$$

where $n_s = \sqrt{\dfrac{2}{T_s}}\displaystyle\int_0^{T_c} c_s(t)I(t)\sin\omega_c t\, dt$.

QPSK modulation can be viewed as two independent BPSK modulations. Thus the net data rate is doubled. We consider a special case of QPSK modulation where $c_c(t)$ and $c_s(t)$ are equal and have a value of c. The QPSK symbol energy is also the bit energy (one bit per QPSK signal).

For this case Eqs. (2.14) and (2.15) have the form

$$r_c = d\sqrt{\frac{E_b}{2}} + n_c \tag{2.16}$$

and

$$r_s = d\sqrt{\frac{E_b}{2}} + n_s \tag{2.17}$$

where n_c and n_s are 0 mean independent with conditional variances

$$\text{Var}\langle n_c|\theta\rangle = (IT_c)(\cos\theta)^2 \tag{2.18}$$

and

$$\text{Var}\langle n_s|\theta\rangle = (IT_c)(\sin\theta)^2 \tag{2.19}$$

Next we use

$$r = \frac{(r_c + r_s)}{2} = d\sqrt{\frac{E_b}{2}} + \frac{(n_c + n_s)}{2} \tag{2.20}$$

as the statistic for decision rule; then

$$\text{Var}\langle (n_c + n_s)/2|\theta\rangle = \frac{1}{4}[IT_c(\cos\theta)^2 + IT_c(\sin\theta)^2] = \frac{IT_c}{4} \tag{2.21}$$

The final expression for narrowband interference, $G_j(f)$, at the demodulator baseband output is given as

$$G_j(f) = \frac{IT_c}{4} = \frac{I}{4R_c} \tag{2.22}$$

For baseband systems, we define the baseband interference, $I(f)$, as

$$I(f) = 2G_j(f); \qquad 0 \le f \le R_c \tag{2.23}$$

The bit error probability for AWGN is given [8] as

$$P_b = \frac{1}{2}\text{erfc}\left(\sqrt{\frac{E_b}{N_0}}\right) = Q\left(\sqrt{\frac{2E_b}{N_0}}\right) \tag{2.24}$$

where $Q(u) \approx \dfrac{e^{-u^2/2}}{\sqrt{2\pi}u} \quad u \gg 1$

We assume the demodulated baseband interference,[*] I, is represented by AWGN. For coherent PSK demodulation we have

---

[*] This is not strictly true since the noise is known. It is sufficient for the purposes of this discussion.

$$P_b = \frac{1}{2}\text{erfc}\left(\sqrt{\frac{E_b}{N_0}}\right) = \frac{1}{2}\text{erfc}\left(\sqrt{\frac{E_b}{(2I)/(4R_c)}}\right) = \frac{1}{2}\text{erfc}\left[\sqrt{2\left(\frac{S}{I}\right)\left(\frac{R_c}{R_b}\right)}\right] \quad (2.25)$$

$I_{\text{eff}} = \dfrac{I}{2(R_c/R_b)}$ is referred to as the effective interference power.

The effective interference power, in comparison with the signal power, determines the bit error rate probability P_b of the SS system. Note that the effective interference power is reduced by the ratio of the bandwidth expansion between the baseband signal and the transmitted signal, (R_c/R_b).

2.5 Bit Scrambling

Referring to Table 2-1, we consider the following activities at a given transmitter location (see Fig. 2-6).

1. An arbitrary data sequence $s_i(t)$ is generated by a digital source.
2. An arbitrary code sequence $c_i(t)$ is generated by a direct spread (DS) generator.
3. Two sequences are modulo-2 added and transmitted to a distant receiver assuming there is no propagation delay.
4. At the distant location, the resulting sequence (assuming no propagation delay) is picked up by the receiver (see Fig. 2-7).
5. The code $c_i(t)$ used at the transmitter is also available at the receiver.
6. The original data sequence is recovered by modulo-2 adding the received sequence with the locally available code $c_i(t)$.

Table 2-1   Operations with Modulo-2 Addition

Transmitter	1	$s_i(t)$	1	1	0	1	0	0	1	1	1	1
	2	$c_i(t)$	1	0	0	1	1	1	0	1	0	0
	3	$s_i(t)\,c_i(t)$	0	1	0	0	1	1	1	0	1	1
Receiver	4	$s_i(t)\,c_i(t)$	0	1	0	0	1	1	1	0	1	1
	5	$c_i(t)$	1	0	0	1	1	1	0	1	0	0
	6	$s_i(t)\,c_i(t)\,c_i(t) = s_i(t)$	1	1	0	1	0	0	1	1	1	1

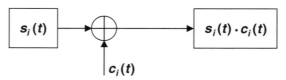

Figure 2-6   Mobile Receiver

Table 2-2 Operations without Modulo-2 Addition

Transmitter	1	$s_i(t)$	−1	−1	1	−1	1	1	−1	−1	−1	−1
	2	$c_i(t)$	−1	1	1	−1	−1	−1	1	−1	1	1
	3	$s_i(t)\,c_i(t)$	1	−1	1	1	−1	−1	−1	1	−1	−1
Receiver	4	$s_i(t)\,c_i(t)$	1	−1	1	1	−1	−1	−1	1	−1	−1
	5	$c_i(t)$	−1	1	1	−1	−1	−1	1	−1	1	1
	6	$s_i(t)\,c_i(t)\,c_i(t) = s_i(t)$	−1	−1	1	−1	1	1	−1	−1	−1	−1

Next, referring to Table 2-2, we consider the following set of activities at the given transmitter location.

1. An arbitrary data sequence $s_i(t)$ is generated by a digital source. In this case, we use +1s and −1s to represent 0s and 1s (Fig. 2-6).
2. An arbitrary code sequence $c_i(t)$ is generated by a DS generator.
3. We multiply $s_i(t)$ and $c_i(t)$. The output of the multiplier is transmitted to a distant receiver.
4. At the distant location, the resulting sequence (again assuming no propagation delay) is picked up by the receiver (Fig. 2-7).
5. The code $c_i(t)$ used at the transmitting location is assumed to be available at the receiver.
6. The original data sequence is recovered by multiplying the received sequence by the locally available code $c_i(t)$.

From Tables 2-1 and 2-2 we conclude that the modulo-2 addition using 1s and 0s binary data is equivalent to multiplication using −1 and 1 binary data as long as we remain consistent in mapping 0s to +1s and 1s to −1s as shown in Table 2-2. (For circuit implementation, the modulo-2 addition is preferred since exclusive OR gates are cheaper than multiplication circuits. However, for modeling purposes, the multiplication method is usually easier to formulate and understand than the modulo-2 approach).

We notice that, for the output of the receiver to be identical to the original data, the following relationship must be satisfied:

$$s_i(t) \cdot c_i(t) \cdot c_i(t) = s_i(t) \tag{2.26}$$

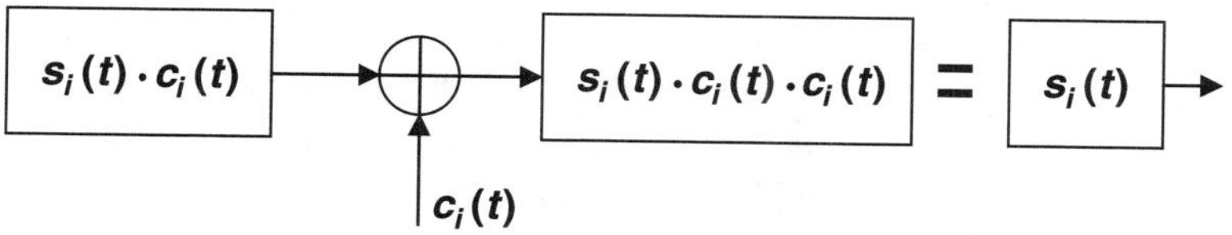

Figure 2-7 Mobile Receiver

In other words, $c_i(t) \cdot c_i(t)$ must be equal to unity. Note that $c_i(t)$ is a binary sequence made up of 1s and –1s; therefore

$$\text{if } c_i(t) \text{ is } 1, c_i(t) \cdot c_i(t) = 1 \tag{2.27}$$

$$\text{if } c_i(t) = -1, c_i(t) \cdot c_i(t) = 1 \tag{2.28}$$

In our previous discussion we assume that no propagation delay and no other processing delay occurs between the transmitter and receiver input. Thus the code copy used at the receiver is perfectly lined up with the initial code used at the transmitter. The two codes are said to be in phase or in synchronization (synch). In practice, however, a propagation delay and other processing delay occur between the transmitter and the receiver input. Therefore, the receiver may be time shifted relative to the initial code at the transmitter, and the two codes are no longer in synch. As a result, the output of the receiver will no longer be identical to the original data, $s_i(t)$.

In order to recover the original data $s_i(t)$, we must *tune* the receiver code sequence to that of the incoming code from the transmitter. In other words, we must time shift the receiver code in order to line it up with the incoming code. It should be noted that, by synchronizing or tuning the receiver code to the phase of the incoming code, the original data (shifted by propagation delay) can now be recovered at the output of the receiver. In these examples, the data sequence and code sequence have the same length (one code bit for each data bit) and are used for encrypting the data bits. This is referred to as *bit scrambling* and does not result in spectrum spreading.

2.6 The Performance of a CDMA System

A traditional narrowband system based on FDMA or TDMA is a dimension-limited system. The number of dimensions is determined by the number of nonoverlapping frequencies for FDMA or by the number of time slots for TDMA. In a TDMA system, once all time slots are assigned, no additional users can be added. Thus, it is not possible to increase the number of users beyond the dimension limit without causing an intolerable amount of interference to a mobile station's reception at the cell-site receiver.

Spread spectrum systems can tolerate some interference, so the introduction of each additional active radio increases the overall level of interference to the cell-site receivers receiving CDMA signals from mobile station transmitters. Each mobile station introduces a unique level of interference that depends on its received power level at the cell site, its timing synchronization relative to other signals at the cell site, and its specific cross-correlation with other CDMA signals.

The number of CDMA channels in the network depends on the level of total interference that can be tolerated in the system. Thus, the CDMA system is limited by interference, and the quality of system design plays an important role in its overall capacity. A well-designed system will have a required bit error probability with a higher level of interference than a poorly designed system. Forward Error Correction (FEC) coding techniques improve tolerance for interference and increase overall CDMA system capacity.

We assume that, at the cell site, the received signal level of each mobile user is the same and that the interference seen by each receiver is modeled as Gaussian noise. Each modulation method has a relationship that defines the bit error rate as a function of the E_b/N_0 ratio. If we know the performance of the coding methods used on the signals and the tolerance of the digitized voice and the data-to-errors ratio, we can define the minimum E_b/N_0 ratio for proper system operation. If we maintain operation at this minimum E_b/N_0, we can obtain the best performance of the system. The relationship between the number of mobile users, M, the processing gain, G_p, and the E_b/N_0 ratio is therefore given as

$$M \approx \frac{G_p}{(E_b/N_0)}$$

(2.29)

For a given bit error probability, the actual E_b/N_0 ratio depends on the radio system design and error correction code. It may approach but never equal the theoretical calculations.

The best performance that can be obtained is defined by the Shannon limit[*] in AWGN. In Eq. (2.2), if we note that

$$\log_e\left(1 + \frac{S}{N}\right) = \frac{S}{N} - \frac{1}{2}\left(\frac{S}{N}\right)^2 + \frac{1}{3}\left(\frac{S}{N}\right)^3 - \frac{1}{4}\left(\frac{S}{N}\right)^4 + \ldots < \frac{S}{N}$$

then, from Eq. (2.2)

$$\frac{C}{B_w} < \frac{1}{\log_e 2}\left(\frac{S}{N}\right)$$

and

$$\frac{C}{B_w} < \frac{1}{\log_e 2}\left(\frac{E_b}{N_0}\right)\left(\frac{C}{B_w}\right)$$

Thus

$$\frac{E_b}{N_0} \geq \log_e 2 = 0.69 = -1.59 \text{ dB}$$

(2.30)

provides error-free communications.

For the Shannon limit, the number of users we can have is

$$M = \frac{G_p}{0.69} = 1.45 G_p$$

(2.31)

This theoretical Shannon limit shows that CDMA systems can have more users per cell than traditional narrowband systems that are limited by number of dimensions. This limit is theoretical; in practice a wireless system is typically engineered such that $E_b/N_0 \approx 6$ dB. However, due to practical limitations on CDMA radio design, it is difficult to accommodate as many users in a

---

[*] This limit is a lower bound [7]. It is assumed that the channel coding has an infinite length to achieve this bound.

single cell as given by Eq. (2.29). The CDMA cell capacity depends upon many factors. As seen by Eq. (2.29), the upper bound theoretical capacity of an ideal noise-free CDMA channel is limited by the processing gain G_p. In an actual system, the CDMA cell capacity is much lower than the theoretical upper bound value. The CDMA cell capacity is affected by the receiver modulation performance, power control accuracy, interference from other non-CDMA systems sharing the same frequency band, and other effects.

CDMA transmissions in neighboring cells use the same carrier frequency and therefore cause interference that we account for by introducing a factor β. This reduces the number of users in a cell since the interference from users in other cells must be added to the interference generated by the other mobiles in the user's cell. The practical range for β is 0.4 to 0.55. The power control accuracy is represented by a factor α. The practical range for α is 0.5 to 0.9. We designate the reduction in the interference due to voice activity by a factor υ. The practical range for υ is 0.45 to 1. If directional antennas are used rather than omnidirectional antennas at the base station, the cell is sectorized with A sectors. Each of the antennas used at the cell radiates into a sector of $360/A$ degrees and we have an interference improvement factor of λ. For a three-sector cell, the practical value of the improvement factor λ is 2.55. The average values for β, α, υ, and λ are 0.5, 0.85, 0.6, and 2.55, respectively.

Introducing β, α, υ, and λ into Eq. (2.29) we get

$$M \approx \frac{G_p}{E_b/N_0} \times \frac{1}{1+\beta} \times \alpha \times \frac{1}{\upsilon} \times \lambda \qquad (2.32)$$

EXAMPLE 2.2

Estimate the number of mobile users that can be supported by a CDMA system using an RF bandwidth of 1.25 MHz to transmit data at 9.6 kbps. Assume: $E_b/N_0 = 6$ dB; the interference from neighboring cells $\beta = 60\%$; the voice activity factor $\upsilon = 50\%$; the power control accuracy factor $\alpha = 0.8$.

$$G_p = \frac{1.25 \times 10^6}{9.6 \times 10^3} = 130$$

$$\frac{E_b}{N_0} = 6 \text{ dB} = 3.98$$

$$M = \frac{130}{3.98} \times \frac{1}{1+0.6} \times \frac{1}{0.5} \times 0.8 = 32.64 \approx 33 \text{ mobile users per sector}$$

The results of this example can be compared with the capacity of an analog FM system with the same frequency allocation, i.e., 41 FM channels. Typically, an analog system is engineered with a frequency reuse pattern equal to 7. With a three-sector configuration and a reuse factor of 7, the number of channels per sector equals $41/(7 \times 3) \approx 2$. This comparison suggests that a DSSS system offers a greater than tenfold improvement in the channel capacity. It is interesting to note that the processing gain of a DSSS system is directly proportional to spectrum expansion while the processing gain of an FM system is proportional to the square of the frequency expansion.[*] This would seem to imply that the FM system should perform better than the CDMA system; yet it doesn't. There are several reasons for this CDMA performance:

- DSSS techniques take advantage of the voice activity.
- DSSS techniques use the concept of orthogonality to multiple users on a common frequency channel. This concept is applicable across different base stations and sectors.
- DSSS techniques synchronize transmission for all base stations so that soft handoffs (see chapter 10) can be implemented. This approach reduces the level of interference.

EXAMPLE 2.3

For the CDMA system (TIA IS-95), a chip rate[†] of 1.2288 Mcps is specified for the data rate of 9.6 kbps. E_b/N_0 is taken as 6.8 dB. Estimate the average number of subscribers that can be supported by a sector of the 3-sector cell. Assume: interference from neighboring cells $\beta = 50\%$; the voice activity factor $\upsilon = 60\%$; the power control accuracy factor $\alpha = 0.85$; and the improvement from sectorization $\lambda = 2.55$.

$$M \approx \frac{G_p}{E_b/N_0} \times \frac{1}{1+\beta} \times \alpha \times \frac{1}{\upsilon} \times \lambda$$

$$G_p = \frac{(1/9.6)}{(1/1.2288 \times 10^3)} = 128, \quad E_b/N_0 = 6.8 \text{ dB} = 4.7863$$

$$M = \frac{128}{4.7863} \times \frac{1}{1.5} \times \frac{1}{0.6} \times 0.85 \times 2.55 = 64.4$$

$$\text{Subscriber/Sector} = \frac{64.4}{3} = 21.46 \approx 21$$

EXAMPLE 2.4

A total of 40 equal-power mobile stations are to share a frequency band through a CDMA system. Each mobile station transmits information at 9.6 kbps with a DSSS BPSK modulated signal. Calculate the minimum chip rate of the psudorandom noise (PN) code in order to maintain a bit error probability of 10^{-3}. Assume: the interference factor β from the other base stations = 60%; voice activity $\upsilon = 50\%$; and power control accuracy factor $\alpha = 0.8$. What will chip rate be if the probability of error is 10^{-4}?

$$P_b = Q\left(\sqrt{\frac{2E_b}{N_0}}\right) \approx \frac{e^{-E_b/N_0}}{2\sqrt{\pi(E_b/N_0)}}$$

---

[*] For an FM system, the frequency expansion is specified by the deviation ratio.
[†] The chip rate is the frequency of the code clock.

$$\frac{e^{-E_b/N_0}}{2\sqrt{\pi(E_b/N_0)}} = 10^{-3}$$

$$\frac{E_b}{N_0} \approx 4.8 = 6.8 \text{ dB}$$

$$M = \frac{G_p}{(E_b/N_0)} \times \frac{1}{1+\beta} \times \frac{1}{\upsilon} \times \alpha$$

$$\frac{G_p}{4.8} \times \frac{1}{1.6} \times \frac{1}{0.5} \times 0.8 = 40$$

$$\therefore G_p = 192$$

$$\therefore R_c = 192 \times 9.6 \times 10^3 = 1.843 \text{ Mcps}$$

$$\text{For } P_b = 10^{-4}, \frac{E_b}{N_0} = 8.43 \text{ dB} = 6.9663$$

$$\therefore G_p = 278.6 \text{ and } R_c = 2.675 \text{ Mcps}$$

2.7 Pseudorandom Noise Sequences

In CDMA systems, PN sequences are used to

- Spread the bandwidth of the modulated signal to the larger transmission bandwidth.
- Distinguish between the different user signals by utilizing the same transmission bandwidth in the multiple access scheme.

PN sequences are not random; they are deterministic, periodic sequences. The following are the three key properties of an ideal PN sequence [2]:

1. The relative frequencies of 0 and 1 are each 1/2.
2. The run lengths (of 0s or 1s) are: 1/2 of all run lengths are of length 1; 1/4 are of length 2; 1/8 are of length 3; and so on.
3. If a PN sequence is shifted by any nonzero number of elements, the resulting sequence will have an equal number of agreements and disagreements with respect to the original sequence.

PN sequences are generated by combining the outputs of feedback shift registers. A feedback shift register consists of consecutive two-stage memory or storage stages and feedback logic. Binary sequences are shifted through the shift register in response to clock pulses. The contents of the stages are logically combined to produce the input to the first stage. The initial contents of the stages and feedback logic determine the successive contents of the stages. A feedback shift register and its output are called linear when the feedback logic consists entirely of modulo-2 adders.

To demonstrate the properties of a PN a binary sequence, we consider a linear feedback shift register (see Fig. 2-8) that has a four-stage register for storage and shifting, a modulo-2 adder, and a feedback path from adder to the input of the register. The operation of the shift register is controlled by a sequence of clock pulses. At each clock pulse the contents of each stage in the register is shifted by one stage to the right. Also, at each clock pulse the contents of stages X_3 and X_4 are modulo-2 added, and the result is fed back to stage X_1. The shift register sequence is defined to be the output of stage X_4. We assume that stage X_1 is initially filled with a 0 and the other remaining stages are filled with 0, 0, and 1; i.e., the initial state of the register is 0 0 0 1. Next, we perform the shifting, adding, and feeding operations, where we obtain the results after each cycle that is shown in Table 2-3.

We notice that the contents of the registers repeat after $2^4 - 1 = 15$ cycles. The output sequence is given as 0 0 0 1 0 0 1 1 0 1 0 1 1 1 1 (see Fig. 2-9) where the left-most bit is the earliest bit. In the output sequence, the total number of 0s is 7 and total number of 1s is 8; the numbers differ by 1.

If a linear feedback shift register reached the 0 state at some time, it would always remain in the 0 state and the output sequence would subsequently be all 0s. Since there are exactly $2^n - 1$ nonzero states, the period of a linear n-stage shift register output sequence cannot exceed $2^n - 1$.

The output sequences are classified as either *maximal length* or *nonmaximal length*. Maximal-length sequences are the longest sequences that can be generated by a given shift register of a given length. In the binary shift register sequence generators, the maximal length sequence is $2^n - 1$ chips, where n is the number of stages in the shift registers. Maximal-length sequences have this property for an n-stage linear feedback shift register: the sequence repetition period in

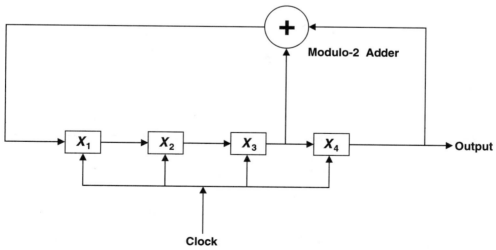

Figure 2-8 Four-Stage Linear Feedback Shift Register

Table 2-3 Results of Shifting after Each Cycle

Shift	Stage X_1	Stage X_2	Stage X_3	Stage X_4	Output Sequence
0	0	0	0	1	1
1	1	0	0	0	0
2	0	1	0	0	0
3	0	0	1	0	0
4	1	0	0	1	1
5	1	1	0	0	0
6	0	1	1	0	0
7	1	0	1	1	1
8	0	1	0	1	1
9	1	0	1	0	0
10	1	1	0	1	1
11	1	1	1	0	0
12	1	1	1	1	1
13	0	1	1	1	1
14	0	0	1	1	1
15	0	0	0	1	1
16	1	0	0	0	0

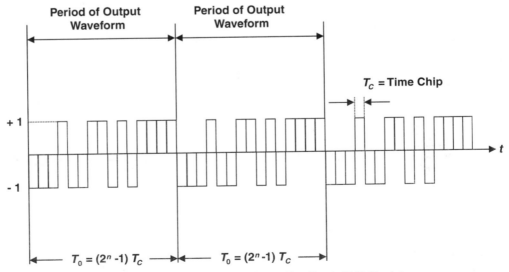

Figure 2-9 Output Waveform for Four-Stage Linear Feedback Shift Register

clock pulses is $T_0 = 2^n - 1$. If a linear feedback shift register generates a maximal sequence, then all of its nonzero output sequences are maximal, regardless of the initial stage. A maximal sequence contains $(2^{n-1} - 1)$ 0s and (2^{n-1}) 1s per period.

2.7.1 Properties of a Maximal-Length Pseudorandom Sequence

When an n-stage shift register (see Fig. 2-10) is configured to generate a maximal-length sequence, the sequence has the following properties:

- The number of binary 0s differs from number of 1s by one chip at most. The number of binary 1s is 2^{n-1} and the number of 0s is $2^{n-1} - 1$, where n is the number of stages in the code generator, and the code length is $2^n - 1$ chips.
- A run is defined as a sequence of a single type of binary digits. The appearance of the alternate digit in a sequence starts a new run. The length of the run is the number of digits in the run. The statistical distribution of 1s and 0s is well defined and always the same. Relative positions of the runs vary from code sequence to code sequence, but the number of each run length does not.
- A modulo-2 addition of a maximal linear code with a phase-shifted replica of itself results in another replica with a phase shift different from either of the originals.
- If a period of the sequence is compared term by term with any cyclic shift itself, it is best if the number of agreements differs from the number of disagreements by not more than one count.
- If we transform the binary $(0,1)$ sequence of the shift register output to a binary $(+1, -1)$ sequence by replacing each 0 by $+1$, and each 1 by -1, then the periodic correlation function of the sequence is given by

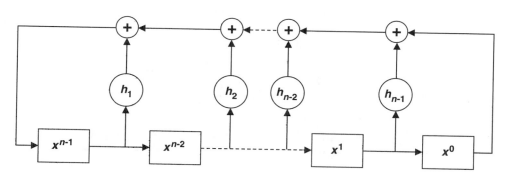

Note: $h_i = 1$ represents a closed circuit

$h_i = 0$ represents an open circuit

Figure 2-10 n-Stage Linear Feedback Shift Register

$$\theta(\tau) = \begin{bmatrix} 2^n - 1, & \tau = 0 \\ -1, & \tau \neq 0 \end{bmatrix}$$

where τ = the shift in increments of one chip (see Figure 2-11 as an example of $n = 3$), and

n = the number of stages in the shift register.

In the region between $\tau = 0$ and $\tau = \pm 1$, the correlation function decreases linearly from $2^n - 1$ to -1 so that the autocorrelation function for a maximal-length pseudorandom sequence is triangular with a maximum value at $\tau = 0$ (see Example 2.5). With this property, two or more communicators can operate independently if their codes are phase shifted more than one chip. For other code sequences, the autocorrelation properties may be markedly different than the properties of the maximal-length sequences.

- Every possible state of a given n-stage generator exists at some time during the generation of a complete code cycle. Each state exists for one and only one clock interval. The exception is that the all-0s state does not normally occur and is not allowed to occur.

It has been shown [11] that there are exactly $2^{n-(p+2)}$ runs of length p for both 1s and 0s in every maximal sequence (except that there is only one run containing n 1s and one containing $(n - 1)$ 0s; there are no runs of 0s of length n or 1s of length $(n - 1)$. The distribution of runs for $(2^4 - 1)$ chip sequence is given in Table 2-4.

Whether an n-stage linear feedback shift register generates only one sequence with period $2^n - 1$ depends upon its connection vector (see Fig. 2-10). Let $h(x)$ be the nth-order polynomial given by

$$h(x) = h_0 + h_1 x + h_2 x^2 + \ldots + h_n x^n \qquad (2.33)$$

We refer to $h(x)$ as the associated polynomial of the shift register with feedback coefficient (h_0, h_1, h_2, ... h_n). Here $h_0 = h_n = 1$ and other feedback coefficients take values 0 and 1. Thus, the polynomial for the four-stage linear feedback shift register as shown in Fig. 2-8 is given by

$$h(x) = 1 + x^3 + x^4 \qquad (2.34)$$

Table 2-4   Distribution of Runs for a $2^4 - 1$ Chip Sequence

Run Length	1s	0s	Number of Chips Included
1	2	2	$1 \times 2 + 1 \times 2 = 4$
2	1	1	$1 \times 2 + 1 \times 2 = 4$
3	0	1	$0 \times 3 + 1 \times 3 = 3$
4	1	0	$1 \times 4 + 0 \times 4 = 4$
Total No. of Chips			15

When $h(x)$ is an irreducible (not factorable) primitive polynomial of degree n, then all sequences generated by $h(x)$ have a maximum period of $2^n - 1$. For an n-stage register, there are $N_p(n)$ maximal sequences that can be generated [11]. $N_p(n)$ is the number of primitive polynomial of degree n.

$$N_p(n) = \left[\frac{2^n - 1}{n} \right] \prod_{i=1}^{k} \frac{P_i - 1}{P_i} \tag{2.35}$$

where P_i is the prime decomposition of $2^n - 1$.

Table 2-5 gives the number of maximal sequences available from register lengths 2 through 10 and provides an example of a primitive polynomial of degree n.

2.7.2 Autocorrelation

The autocorrelation function for a signal $x(t)$ is defined as

$$R_x(\tau) = \int_{-\infty}^{\infty} x(t)x(t + \tau)dt \tag{2.36}$$

Table 2-5 Number of Maximal Sequences Available from Register Lengths 2 Through 10

No. of Stage n	$2^n - 1$	Prime Decomposition of $2^n - 1$	No. of n-sequence $N_p(n)$	Example of Primitive Polynomial of degree n $h(x)$
2	3	3	$\frac{3}{2} \cdot \frac{2}{3} = 1$	$1 + x + x^2$
3	7	7	$\frac{7}{3} \cdot \frac{6}{7} = 2$	$1 + x + x^3$
4	15	3×5	$\frac{15}{4} \cdot \frac{2}{3} \cdot \frac{4}{5} = 2$	$1 + x + x^4$
5	31	31	$\frac{31}{5} \cdot \frac{30}{31} = 6$	$1 + x^2 + x^5$
6	63	$3 \times 3 \times 7$	$\frac{63}{6} \cdot \frac{2}{3} \cdot \frac{6}{7} = 6$	$1 + x + x^6$
7	127	127	$\frac{127}{7} \cdot \frac{126}{127} = 18$	$1 + x^3 + x^7$
8	255	$3 \times 5 \times 17$	$\frac{255}{8} \cdot \frac{2}{3} \cdot \frac{4}{5} \cdot \frac{16}{17} = 16$	$1 + x^2 + x^3 + x^4 + x^8$
9	511	7×73	$\frac{511}{9} \cdot \frac{6}{7} \cdot \frac{72}{73} = 48$	$1 + x^4 + x^9$
10	1023	$3 \times 11 \times 31$	$\frac{1023}{10} \cdot \frac{2}{3} \cdot \frac{10}{11} \cdot \frac{30}{31} = 60$	$1 + x^3 + x^{10}$

Autocorrelation refers to the degree of correspondence between a sequence and a phase-shifted replica of itself. An autocorrelation plot shows the number of agreements minus disagreements for the overall length of the two sequences being compared, as the sequences assume every shift in the field of interest. If $x(t)$ is a periodic pulse waveform representing a PN sequence, we refer to each fundamental pulse as a PN sequence symbol or a chip. For such a PN waveform of unit chip duration and period $T_0 = 2^n - 1$ chips, the normalized autocorrelation function is expressed as

$$R_x(\tau) = \frac{1}{T_0} \begin{bmatrix} \text{number of agreements} - \text{number of disagreements in a comparison of one full} \\ \text{period of sequence with a } \tau \text{ position cyclic shift of the sequence} \end{bmatrix}$$

The normalized autocorrelation function $R_x(\tau)$ of a periodic waveform $x(t)$ with period T_0 is given as

$$R_x(\tau) = \frac{1}{R_x(0)} \frac{1}{T_0} \int_{-T_0/2}^{T_0/2} x(t) x(t + \tau) dt \qquad \text{for } -\infty < \tau < \infty \qquad (2.37)$$

where $R_x(0) = \dfrac{1}{T_0} \displaystyle\int_{-T_0/2}^{T_0/2} x^2(t) dt$

2.7.3 Cross-Correlation

The cross-correlation function between two signals, $x(t)$ and $y(t)$, is defined as the correlation between two different signals $x(t)$ and $y(t)$ and is given as

$$R_c(\tau) = \int_{-T_0/2}^{T_0/2} x(t) y(t + \tau) dt \qquad \text{for } -\infty < \tau < \infty \qquad (2.38)$$

EXAMPLE   2.5

Consider a three-stage shift register generator that is generating a seven-chip maximal linear code. The reference sequence is 1 1 1 0 0 1 0. Sketch the autocorrelation function if the chip rate is 10 Mcps.

Table 2-6 provides the sequence after each shift and shows the corresponding agreements (A) and disagreements (D) with the reference sequence.

It can be noted that the net correlation A − D is −1 for all shifts except for the 0-shift or synchronous condition. This is typical of all n-sequences. In the region between 0 and plus or minus one chip shift ($\tau = \pm 1/10^6$ seconds), the correlation increases linearly so that the autocorrelation function for an n-sequence is triangular as shown in Fig. 2-11. This characteristic of autocorrelation is used to great advantage in communication systems. A channel can simultaneously support multiple users if the corresponding codes are phase shifted more than one chip.

Table 2-6 Agreements and Disagreements with Reference Sequence

Shift	Sequence	Agreement (A)	Disagreement (D)	A – D
1	0 1 1 1 0 0 1	3	4	–1
2	1 0 1 1 1 0 0	3	4	–1
3	0 1 0 1 1 1 0	3	4	–1
4	0 0 1 0 1 1 1	3	4	–1
5	1 0 0 1 0 1 1	3	4	–1
6	1 1 0 0 1 0 1	3	4	–1
0	1 1 1 0 0 1 0	7	0	7

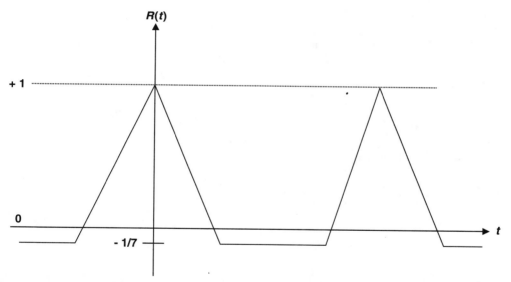

Figure 2-11 Autocorrelation of a Three-Stage Linear Feedback Shift Register

2.7.4 Orthogonal Functions

Orthogonal functions are employed to improve the bandwidth efficiency of an SS system. Each mobile user uses one member of a set of orthogonal functions representing the set of symbols used for transmission. While there are many different sequences that can be used to generate an orthogonal set of functions, the Walsh and Hadamard sequences make useful sets for CDMA.

Two different methods can be used to modulate the orthogonal functions into the information stream of the CDMA signal. The orthogonal set of functions can be used as the spreading code or to form modulation symbols that are orthogonal.

With orthogonal symbol modulation, the information bit stream can be divided into blocks, with each block representing a nonbinary information symbol that is associated with a particular transmitted code sequence. If there are b bits per block, one of the set of $K = 2^b$ functions is transmitted in each symbol interval. The signal at the receiver is correlated with the set of K matched filters, each matched to the code function of one symbol. The outputs from corelators are compared and the symbol with the largest output is taken as the transmitted symbol.

If we assume a simple one-path channel with perfect power control and negligible additive noise, and if we include the interference due to multipath, multiple users, and the decision process of the correlators, the E_b/N_0 ratio can be given as [8]

$$\frac{E_b}{N_0} \approx \frac{G_p}{(M-1)+(K-1)} \tag{2.39}$$

where M = number of mobile users,

G_p = processing gain of the system, and

$K-1$ = noise from the outputs of correlators other than one corresponding to the correct symbol.

We rewrite Eq. (2.39) as

$$M = \frac{G_p}{E_b/N_0} - K + 2 \tag{2.40}$$

Next we introduce factors β, α, υ, and λ (see Section 2.6) in Eq. (2.40) to get

$$M \approx \frac{G_p}{E_b/N_0} \times \frac{1}{1+\beta} \times \alpha \times \frac{1}{\upsilon} \times \lambda - K + 2 \tag{2.41}$$

$$\eta = \frac{MR}{B_w} = \frac{M \cdot \log_2 K R_s}{B_w} = \frac{M(\log_2 K)}{G_p} \tag{2.42}$$

where η = bandwidth efficiency, and

R_s = symbol transmission rate.

EXAMPLE   2.6

Calculate the bandwidth efficiency of the system using the data in Example 2.2 and assuming an orthogonal code with $K = 2$ symbols. If an orthogonal code with $K = 16$ symbols is used for the system, how many simultaneous mobile users can be supported and what is the bandwidth efficiency of the system?

• $K = 2$ symbols

$$G_p = \frac{1.25 \times 10^6}{9.6 \times 10^3} = 130, \frac{E_b}{N_0} = 6 \text{ dB} = 3.98$$

$$M \approx \frac{130}{3.98} \times \frac{1}{1+0.6} \times \frac{1}{0.5} \times 0.8 - 2 + 2 = 32.64 \approx 33 \text{ users}$$

$$\eta = \frac{M(\log_2 K)}{G_p} = \frac{33(\log_2 2)}{130} = 25.4\%$$

• $K = 16$ symbols

$$M \approx \frac{130}{3.98} \times \frac{1}{1+0.6} \times \frac{1}{0.5} \times 0.8 - 16 + 2 = 18.64 \approx 19 \text{ users}$$

$$\eta = \frac{M(\log_2 K)}{G_p} = \frac{19(\log_2 16)}{130} = 58.5\%$$

The bandwidth efficiency of the system is improved by 33.1%. The disadvantage of the orthogonal signaling scheme is the complexity of the receiver design. In this example, we need 16 receiver correlators per user channel instead of only one required in the simplest design.

The TIA IS-95 CDMA system uses orthogonal functions for the spreading code on the forward channel and orthogonal functions for the modulation on the reverse channel.[*] One of 64 possible modulation symbols is transmitted for each group of 6 code symbols. The modulation symbol is one member of the set of 64 mutually orthogonal functions. The orthogonal functions have the following characteristic:

$$\sum_{k=0}^{M-1} \phi_i(k\tau)\phi_j(k\tau) = 0 \qquad i \neq j \tag{2.43}$$

where $\phi_i(k\tau)$ and $\phi_j(k\tau)$ are the ith and jth orthogonal members of an orthogonal set,
$\quad M$ is the length of the set, and
$\quad \tau$ is the symbol duration.

Walsh functions are generated by code-word rows of special square matrices called *Hadamard matrices*. These matrices contain one row of all 0s, with the remaining rows each having an equal number of 1s and 0s. Walsh functions can be constructed for block length $N = 2^j$, where j is an integer.

The TIA IS-95 CDMA system uses a set of 64 orthogonal functions generated by using Walsh functions. The modulated symbols are numbered from 0 through 63.

The 64×64 matrix can be generated by using the following recursive procedure:

$$H_1 = \begin{bmatrix} 0 \end{bmatrix} \qquad H_2 = \begin{bmatrix} 0 & 0 \\ 0 & 1 \end{bmatrix} \tag{2.44}$$

---

[*] The IS-665 wideband CDMA system uses the orthogonal codes for spreading in both directions; see chapter 8.

$$H_4 = \begin{bmatrix} 0\ 0\ 0\ 0 \\ 0\ 1\ 0\ 1 \\ 0\ 0\ 1\ 1 \\ 0\ 1\ 1\ 0 \end{bmatrix} \qquad H_{2N} = \begin{bmatrix} H_N\ H_N \\ H_N\ \overline{H_N} \end{bmatrix} \qquad (2.45)$$

where N is a power of 2 and $\overline{H_N}$ is the negative of H_N.

The period of time needed to transmit a single modulation symbol is called a Walsh symbol interval and is equal to 1/4800 second (208.33μs). The period of time associated with 1/64 of the modulation symbol is referred to as a Walsh chip and is equal to 1/307,200 second (3.255μs). Within a Walsh symbol, Walsh chips are transmitted in the order 0, 1, 2, …, 63.

For the forward channel, Walsh functions (Figs. 2-12 and 2-13) are used to eliminate multiple access interference among users in the same cell. On downlink, all Walsh functions are synchronized in the same cell and have zero correlation between each other. The following steps are used:

- The input user data (e.g., digital speech) is multiplied by an orthogonal Walsh function (TIA IS-95 standard uses the first 64 orthogonal Walsh functions).
- The user data is then spread by the BS pilot PN code and transmitted on the carrier.
- At the receiver, after removing the coherent carrier, the mobile receiver multiplies the signal by the synchronized PN code (associated with the base station).

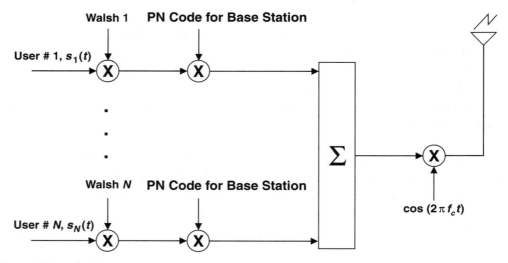

Figure 2-12 Applications of Walsh Functions and Offset Code at the Base Station

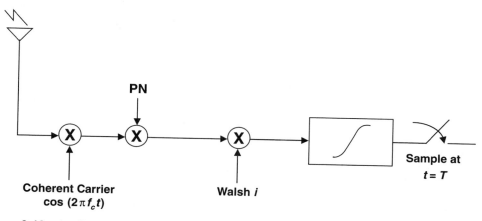

Figure 2-13 Applications of Walsh Functions and Offset Code in the Mobile Station

- The multiplication by the synchronized Walsh function for the *i*th user eliminates the interference due to transmission from the BS to other users.

The Walsh functions form an ordered set of rectangular waveforms taking only two amplitudes: +1 and −1. They are defined over a limited time interval T_L, known as the time base. If ϕ_i represents the *i*th Walsh function and T_L is the time base, then

$$\frac{1}{T_L} \int_0^{T_L} \phi_i(t)\phi_j(t)dt = 0 \qquad \text{for } i \neq j \tag{2.46}$$

and

$$\frac{1}{T_L} \int_0^{T_L} \phi_i^2(t)dt = 1 \qquad \text{for all } is \tag{2.47}$$

To correlate the Walsh codes at the receiver requires that the receiver be synchronized with the transmitter. In the forward direction the base station can transmit a pilot signal to enable the receiver to recover synchronization. Walsh symbol modulation is used from the mobile station to the base station.

EXAMPLE 2.7

We consider a case where 8 chips are used per bit to generate the Walsh functions. Specify these functions, sketch them, and show that they are orthogonal to each other.

$$H_8 = \begin{bmatrix} H_4 & H_4 \\ H_4 & \overline{H_4} \end{bmatrix} = \begin{bmatrix} 0 & 0 & 0 & 0 & 0 & 0 & 0 & 0 \\ 0 & 1 & 0 & 1 & 0 & 1 & 0 & 1 \\ 0 & 0 & 1 & 1 & 0 & 0 & 1 & 1 \\ 0 & 1 & 1 & 0 & 0 & 1 & 1 & 0 \\ 0 & 0 & 0 & 0 & 1 & 1 & 1 & 1 \\ 0 & 1 & 0 & 1 & 1 & 0 & 1 & 0 \\ 0 & 0 & 1 & 1 & 1 & 1 & 0 & 0 \\ 0 & 1 & 1 & 0 & 1 & 0 & 0 & 1 \end{bmatrix} = \begin{bmatrix} \phi_1 \\ \phi_2 \\ \phi_3 \\ \phi_4 \\ \phi_5 \\ \phi_6 \\ \phi_7 \\ \phi_8 \end{bmatrix}$$

Figure 2-14 shows the sketches of the eight Walsh functions. We consider ϕ_2 and ϕ_4 to show orthogonality.

$$\frac{1}{T_L}\int \phi_2(t)\phi_4(t)dt = \frac{1}{T_L}[-1 \times -1 + 1 \times 1 + 1 \times -1 + 1 \times (-1) + (-1) \times (-1) + 1 \times 1 + 1 \times -1 + 1 \times -1] = 0$$

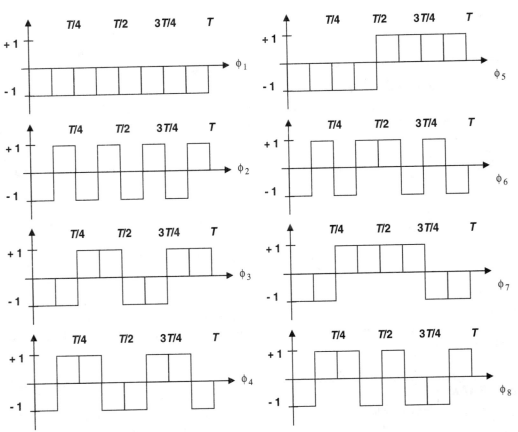

Figure 2-14 Plots of Walsh Functions

and

$$\frac{1}{T_L}\int\phi_1^2(t)dt = \frac{1}{T_L}[T_L] = 1$$

Similarly, we can show that all eight Walsh functions are orthogonal to each other.

EXAMPLE 2.8

We consider a case where 8 chips are used per bit to generate the Walsh functions. Stations A, B, C, and D are assigned the chip sequence 0 1 0 1 0 1 0 1, 0 0 1 1 0 0 1 1, 0 1 1 0 0 1 1 0, and 0 0 0 0 1 1 1 1, respectively. The stations use the chip sequence to send a 1 bit and negative chip sequences to send a 0 bit (e.g., station A uses 1 0 1 0 1 0 1 0 to send the 0 bit and so on). All chip sequences are pairwise orthogonal. This implies that the normalized correlation of any two distinct chip sequences is 0 and the normalized correlation of any chip sequence with itself is 1. We assume that all stations are synchronized in time, so all chip sequences begin at the same instant. When two or more stations transmit simultaneously, their bipolar signals add linearly. For example, if in one chip period three stations output +1 and one station outputs –1, the net result is +2. We consider five different cases when one or more stations transmit. We want to show that the receiver recovers the bit stream of station C by computing the normalized inner products of the received sequences with the chip sequence of station C.

Chip Sequence	Binary Values of Chip Sequence
A: 0 1 0 1 0 1 0 1	A: (–1 +1 –1 +1 –1 +1 –1 +1)
B: 0 0 1 1 0 0 1 1	B: (–1 –1 +1 +1 –1 –1 +1 +1)
C: 0 1 1 0 0 1 1 0	C: (–1 +1 +1 –1 –1 +1 +1 –1)
D: 0 0 0 0 1 1 1 1	D: (–1 –1 –1 –1 +1 +1 +1 +1)

Normalized inner products are

$$(S_1 \cdot C)/8 = (1 + 1 + 1 + 1 + 1 + 1 + 1 + 1)/8 = 1$$

$$(S_2 \cdot C)/8 = (2 + 0 + 0 + 2 + 0 + 2 + 2 + 0)/8 = 1$$

$$(S_3 \cdot C)/8 = (3 + 1 + 1 - 1 + 3 + 1 + 1 - 1)/8 = 1$$

$$(S_4 \cdot C)/8 = (2 + 0 + 0 - 2 + 2 + 0 + 0 - 2)/8 = 0$$

$$(S_5 \cdot C)/8 = (1 - 1 - 1 - 3 + 1 - 1 - 1 - 3)/8 = -1$$

Thus, the receiver recovers 1 1 1 - 0 bit sequence for station C.

We assumed that all the chips are synchronized in time. In a real situation, this is impossible. The sender and receiver are synchronized by having the sender transmit a known chip sequence long enough for the receiver to lock onto. All other (unsynchronized) transmissions are then seen as random noise.

2.8 Summary

In this chapter we considered the concept of spread spectrum systems and provided the main features of the direct sequence spread spectrum system used in the IS-95 system. A key component of spread spectrum performance is the calculation of processing gain of the system, which

Table 2-7   Five Cases of One or More Stations Transmitting

Station[a] (A, B, C, D)	Transmitting	Received Chip Sequence
– – 1–	C	$S_1 = (-1+1+1-1-1+1+1-1)$
– – 1 1	C + D	$S_2 = (-2\ 0\ 0\ -2\ 0\ +2\ +2\ 0)$
1 1 1 –	A + B + C	$S_3 = (-3\ +1\ +1\ +1\ -3\ +\ 1+\ 1+1)$
1 1 – –	A + B	$S_4 = (-2\ 0\ 0\ +2\ -2\ 0\ 0\ +2)$
1 1 0 –	A + B + C	$S_5 = (-1\ -1\ -1\ +3\ -1\ -1\ -1\ +3)$

[a] Note: a dash (–) means no transmission by that station.

is the relationship between the input and output SNR of a spread spectrum receiver. We used the relationship to present some examples that evaluate the performance of a CDMA spread spectrum system.

We presented the Shannon equation for error-free communications and used it to show that error-free communication is possible (with high delays) for an energy-per-bit-to-noise-density ratio, $E_b/N_0 = -1.59$ dB. SS systems trade bandwidth for processing gain, and code division systems use a variety of orthogonal or almost orthogonal codes to allow multiple users in the same bandwidth. Thus, CDMA systems can have a higher capacity than either analog or TDMA digital systems. However, because of practical constraints on CDMA systems, it is not possible to achieve the Shannon bound in system design. The upper bound of the capacity of a CDMA system is limited by the processing gain of the system. In an actual system, the capacity is lower than the theoretical upper bound. CDMA capacity is affected by receiver modulation performance, power control accuracy, interference from other cells, voice activity, cell sectorization, and the ability to maintain synchronization of the systems. Practical CDMA systems are designed for a value of $E_b/N_0 = 6$–7 dB.

2.9 Problems

1. A total of 18 equal-power mobile users per cell are to share a frequency band through a CDMA system. Each mobile user transmits data at 19.2 kbps with a DSSS QPSK-modulated signal. Calculate the minimum chip rate of the PN code in order to maintain a bit error probability of 10^{-4}. Assume that the interference factor β from the other base stations = 0.60; power control accuracy $\alpha = 0.8$; gain from 3-sector antenna $\lambda = 2.55$. What will the chip rate be if the probability of the bit error is 10^{-6}?

2. Consider a case where 16 chips per bit are used to generate the Walsh functions. Specify the functions and sketch W_0, W_8, W_{12}, and W_{15}. Show that these Walsh functions are orthogonal.

3. For a three-stage linear shift register generator, how many maximal-length PN sequences are generated? What is the location of the modulo-2 adder for each sequence? What is the period of the maximal-length sequence?

4. Consider a four-stage linear shift register generator with the initial state of the register 1001 (see Fig. 2-15). Show that it generates a maximal-length sequence. Demonstrate the properties of maximal-length sequence. Sketch the autocorrelation function. What is the location of the modulo-2 adder for the other maximal-length sequence?

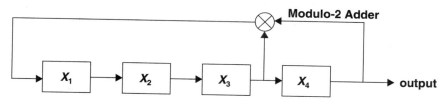

Figure 2-15   Four-Stage Register (Problem 4)

5. Estimate the number of mobile users that can be supported by a CDMA three-sector cell using an RF bandwidth of 1.23 MHz and a chip rate of 1.2288 Mcps to transmit Rate Set 2(RS2) information at 14.4 kbps. Assume $(E_b/N_0)_{reqd} = 7$ dB; the interference from neighboring cells $\beta = 0.67$; the voice activity factor $\upsilon = 0.6$; power control accuracy factor $\alpha = 0.8$; gain from three-sector antenna $\lambda = 2.55$.

6. We consider a case where 8 chips per bit are used to generate the Walsh functions. Mobile stations A, B, C, and D are assigned W_2, W_5, W_6, and W_7, respectively. The stations use the Walsh sequence to send a 1 binary bit and a negative Walsh to transmit a 0 binary bit. Assuming all stations are synchronized in time, the chip sequences begin at the same instant. When two or more stations transmit simultaneously, their bipolar signals are added linearly. Considering the following four cases when one or more stations transmit, show that the receiver recovers the bit stream of stations B and C.

Station (A, B, C, D) (−) means no transmission by that station	Transmitting Stations
− − 1 0	C + D
1 1 1 −	A + B + C
1 1 − −	A + B
1 1 0 −	A + B + C

2.10 References

1. Bhargava, V., Haccoum, D., Matyas, R., and Nuspl, P., *Digital Communications by Satellite*, John Wiley & Sons, New York, 1981.

2. Dixon, R. C., *Spread Spectrum Systems*, Ed., John Wiley & Sons, New York, 1984.

3. Feher, K., *Wireless Digital Communications Modulation and Spread Spectrum Applications,* Prentice Hall PTR, Upper Saddle River, NJ, 1995.

4. Garg, V. K., and Wilkes, J. E., *Wireless and Personal Communications Systems*, Prentice Hall, Upper Saddle River, NJ, 1996.

5. Lee, W. C. Y., *Mobile Cellular Telecommunication Systems*, McGraw-Hill, New York, 1989.

6. Pahlwan, K., and Levesque, A. H., *Wireless Information Networks,* John Wiley & Sons, New York, 1995.

7. Shannon, C. E., "Communications in the Presence of Noise," *Proceedings of the IRE*, no. 37, 1949, pp. 10–21.

8. Skalar, B., *Digital Communications—Fundamental & Applications,* Prentice Hall, Englewood Cliffs, NJ, 1988.

9. Steele, R., *Mobile Radio Communications,* IEEE Press, New York, 1992.

10. Torrien, D., *Principle of Secure Communication Systems,* Artech House, Boston, 1992.

11. Virterbi, A. J., *CDMA,* Addison-Wesley Publishing Company, Reading, MA, 1995.

12. Viterbi, A. J., and Padovani, Roberto, "Implications of Mobile Cellular CDMA," *IEEE Communication Magazine* 30(12), 1992, pp. 38–41.

Speech and Channel Coding

3.1 Introduction

In this chapter, we consider the speech and channel coding applications in the IS-95 CDMA system (the data application is discussed in chapter 14). Since speech encoding is critical to digital transmission and since it is the primary application needed for most wireless phone uses, we will focus on the speech coding algorithms used by CDMA. CDMA uses an 8-kbps (or a 13.3-kbps) data rate for voice transmission.

The wireline network is based on voice transmission using digital Pulse Code Modulation (PCM) at 64 kbps and data transmission at rates of 64 kbps or multiples of 64 kbps. Many older analog facilities still exist, especially in residential areas, and these use voice band modems at rates of up to 28.8 kbps for data and analog electrical signals for voice. At the central office the analog voice and analog data are converted to digital signals using PCM or, optionally, using modem pools for data.

It would be optimal if identical systems could be used for all wireless communications. Unfortunately error rates on radio channels are many orders of magnitude higher than those of copper or fiber-optic cables. In addition, PCM is inefficient for use over scarce and expensive radio channels.

CDMA systems use an efficient method of speech coding and extensive error recovery techniques to overcome the harsh nature of the radio channel. The CDMA system uses a Code-Excited Linear Predictor (CELP) speech coding system at 9.6 kbps (optionally at 13.3 kbps).

3.2 Speech Coding

3.2.1 Pulse Code Modulation

The simplest form of waveform coding scheme is linear PCM, in which the speech signal is band limited, compressed, sampled, quantized, and encoded (see Fig. 3-1). This approach is widely used for analog-to-digital conversion of a signal. In radio and telephone communications,

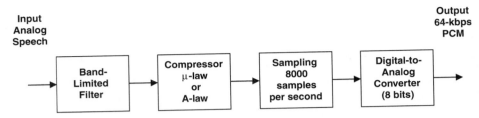

Figure 3-1 PCM Encoder

it is not necessary to send the entire 20-Hz to 20,000-Hz signal normally used for high-fidelity music. Intelligible speech communications can occur with a much narrower and therefore more efficient range of frequencies. For telephone communications, the speech signal is band limited to a frequency range of 300 to 3,300 Hz. To achieve telephone-quality speech, 12 bits per sample are required at a sampling rate of 8,000 samples per second. However, by using a logarithmic sampling system, 8 bits per sample are sufficient. Each sample is then quantized into one of 256 levels. Telephone speech uses two widely different variations of PCM to achieve quality speech (μ-law and A-law PCM). Both are based on a nonuniform quantization of the signal amplitude according to a logarithmic scale rather than a linear scale. Such coders utilize the static character-istics of amplitude nonstationary in speech to achieve good quality at a bit rate of 64 kbps. This is the basis of PCM.

The decoder for PCM (Fig. 3-2) inverts the stages of the encoding process. PCM encoding and decoding are inherently simple systems. However, they require a high bit rate for transmission.

For PCM, North America and Japan use μ-law encoding where the output digital signal, $s(t)$, is related to the input signal, $i(t)$, by

$$s(t) = \text{sgn}[i(t)]\frac{\ln(1 + \mu|i(t)|)}{\ln(1 + \mu)}, \qquad -1 \le i(t) \le 1 \tag{3.1}$$

where a typical value for $\mu = 255$ is used in the United States. In Eq. (3.1), the input signal is normalized to a range of ±1. It can be noted that, for small $i(t)$, $s(t)$ approaches a linear function; for large $i(t)$, $s(t)$ approaches a logarithmic function. The purpose of μ-law encoding is to improve the SNR for weak speech signals. The overall data rate is 64 kbps, with sampling at 8 kbps and 8 bits per sample.

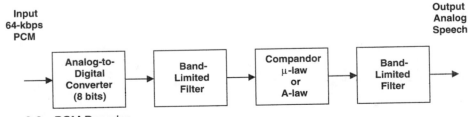

Figure 3-2 PCM Decoder

In Europe, PCM uses A-law encoding where the output digital signal, $s(t)$, is related to the input signal, $i(t)$, by

$$s(t) = \text{sgn}[i(t)] \frac{1 + \ln(A|i(t)|)}{1 + \ln A}, \qquad \frac{1}{A} \le |i(t)| \le 1$$
$$s(t) = \text{sgn}[i(t)] \frac{1 + (A|i(t)|)}{1 + \ln A}, \qquad 0 \le |i(t)| \le \frac{1}{A} \tag{3.2}$$

where a typical value of $A = 87.6$ is used in Europe. In Eq. (3.2), the input signal is also normalized to a range of ±1.

Note that $s(t)$ is logarithmic for $|i(t)| < 1/A$ and linear for $|i(t)| > 1/A$. Thus, A-law provides a somewhat flatter signal-to-distortion performance compared to μ-law when the signal is greater than $1/A$, at the expense of poorer performance at low signal levels.

Telephone communications that cross borders of continents must have conversion routines in their transmission paths if the two continents use different encoding laws.

3.2.2 Adaptive Pulse Code Modulation

High bit rates are not desirable for wireless systems since the capacity of the system is low. Higher system capacities are obtained with differential coders where compression can be applied dynamically, such as Adaptive Predictive Coding (APC) and Adaptive Differential Pulse Code Modulation (ADPCM). The reason for these coders is to achieve a better signal-to-quantization noise performance and a lower coding rate over PCM.

Differential coders generate error signals, as the difference between the input speech samples and corresponding prediction estimates. The error signals are quantized and transmitted. ADPCM and APC differential coders are often used for an intermediate bit rate—16 to 32 kbps.

ADPCM employs a short-term predictor that models the speech spectral envelope. ADPCM achieves network-quality speech (Mean Opinion Score [MOS] of 4.1 or better) at 32 kbps. This is a low-complexity coder of reasonable robustness with channel bit error rates in the range of 10^{-3} to 10^{-2}. The ADPCM coder is well suited for wireless access applications.

In an ADPCM encoder (Fig. 3-3), first analog speech is converted to PCM. If the signal is already PCM—from the network, for example—then the analog-to-PCM step is not needed. The A-law- or μ-law-encoded signal is then converted to a uniform PCM level (i.e., equal steps between levels) signal. The encoder generates a difference signal between the converted signal and an estimated signal and encodes the estimated signal using 15 levels. The resultant signal is transmitted at 32 kbps (half the rate for PCM). In the encoder, the signal estimator is generated by an inverse quantizer and an adaptive predictor. The use of differential signals and proper design of the predictor enables an overall coding efficiency improvement over PCM.

In the ADPCM decoder (Fig. 3-4), the input 32-kbps signal is processed by an inverse adaptive quantizer and an adaptive predictor. The output of the quantizer and the output of the predictor are combined to generate a reconstructed signal that is converted back to PCM. The regenerated PCM signal is then processed (in the synchronous coding adjustment stage) with signals from the input, the quantizer output, and the predictor output to generate the A-law or μ-law

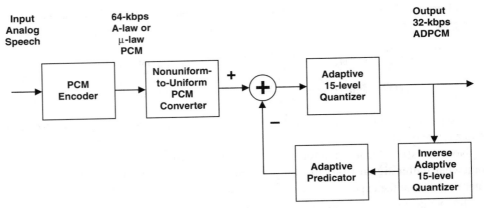

Figure 3-3 ADPCM Encoder

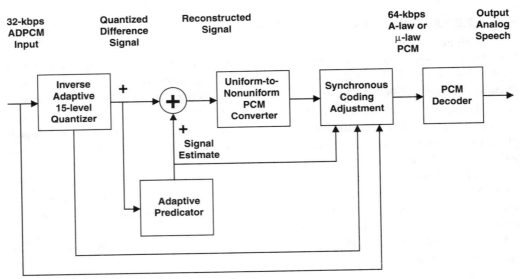

Figure 3-4 ADPCM Decoder

PCM signal. The processing in the synchronous coding adjustment stage ensures that the PCM signal is modeled correctly by converting it to uniform PCM and comparing the resulting error signals with the actual received signal. If an error occurs, it is corrected before the output PCM signal is generated. The ADPCM signal conforms to International Telecommunications Union (ITU) recommendation G.721. Finally, if analog speech is needed, the signal is processed by a PCM decoder.

The IS-665 standard supports a modified version of ADPCM that has been optimized for the PCS environment. The functional operation is the same as ADPCM; the coding algorithm has been modified. (Refer to the IS-665 standard for details.)

3.2.3 Code-Excited Linear Prediction

PCM, ADPCM, and APC operate in the time domain. No attempt is made to understand or analyze the information that is being sent. To achieve lower coding rates, the industry has successfully used redundancy removal techniques, operating in the frequency domain. Frequency domain waveform coding algorithms decompose the input speech signal into sinusoidal components with varying amplitudes and frequencies. Thus, the speech is modeled as a time-varying line spectrum. Frequency domain coders are systems of moderate complexity and operate well at a medium bit rate (16 kbps). When designed to operate in the range of 4.8 to 9.6 kbps, the complexity of the approach used to model the speech spectrum increases considerably.

The other class of speech coding techniques consists of algorithms called vocoders which attempt to describe the speech production mechanism in terms of a few independent parameters serving as the information-bearing signals. These parameters attempt to model the creation of the voice by the vocal tract, decompose the information, and send it to the receiver. The receiver attempts to model an electronic vocal tract to produce the speech output.

The model operates this way: vocoders consider that speech is produced from a source-filter arrangement. Voiced speech results from exciting the filter with a periodic pulse train (simulating the opening and closing of the vocal cords). Unvoiced speech results from exciting the filter with random noise (simulating air rushing past a constriction in the vocal tract). Vocoders operate on the input signal using an analysis process based on a particular speech production model and extract a set of source-filter parameters that are encoded and transmitted. At the receiver, they are decoded and used to control a speech synthesizer, which corresponds to the model used in the analysis process. Provided that all the perceptually significant parameters are extracted, the synthesized signal, as perceived by the human ear, resembles the original speech signal. Nonspeech signals are often not modeled well, so this method works poorly for analog modems.

Vocoders are medium-complexity systems and operate at low bit rates, typically 2.4 kbps, with synthetic-quality speech. Their poor-quality speech is due to two factors: the oversimplified source model used to drive the filter, and the assumption that the source and filter are linearly independent.

For bit rates of about 5 kbps to 16 kbps, hybrid coders that use suitable combinations of waveform coding techniques and vocoder techniques produce the best speech quality. Residual-Excited Linear Prediction (RELP) coding is a simple hybrid coding scheme for telephone-quality speech with a few integrated digital signal processors. RELP belongs to a class of coders known as analysis-synthesis coders based on Linear Predictive Coding (LPC).

RELP systems employ short-term (and, in certain cases, long-term) linear prediction to formulate a difference signal (residual) in a feed-forward manner. RELP systems are capable of producing communication-quality speech at 8 kbps. These systems utilize either pitch-aligned high-frequency regeneration procedures or full-band pitch prediction in time domain to remove the pitch information from the residual signal prior to band limitation/decimation. At bit rates of less than 9.6 kbps, the quality of the recovered speech signal can be improved significantly by using an Analysis-by-Synthesis (AbS) optimization procedure to define the excitation signal. In these systems both the filter and the excitation are defined on a short-term basis using a closed-loop optimization process that minimizes a perceptually weighted error measure formed between the input and decoded speech signals.

CDMA uses a variation of RELP called Code-Excited Linear Prediction (CELP). With this technique, the CELP decoder (Fig. 3-5) uses a codebook to generate inputs to a synthesis filter. The codebook is characterized by its codebook index (I) and gain (G). The spectral filter is characterized by three sets of parameters: the pitch spectral lines (a), the lag generated by the AbS process (L), and the pitch gain (b). The output of the filter is processed by a postfilter and gain adjustment.

CDMA implements a rate 1 encoder at 8.55 kbps and supports rates of 4, 2, and 0.8 kbps (rates 1/2, 1/4, and 1/8, respectively). Each of the rates uses successively fewer bits for encoding the values of I, G, L, b, and a. At rate 1/8 (Fig. 3-6), insufficient bits are available to send the codebook index (I), and a pseudorandom code generator (synchronized at both ends) is used and seeded by a random seed of value *CBSEED* (codebook seed).

The basic frame for CDMA is 20 milliseconds (ms). At rate set 1, 160 bits, plus an 11-bit parity check field, are sent for encoding the data. Lower numbers of bits are used at lower data rates (see Table 3-1).

Implementation of the CELP speech encoder requires three steps. First the Line Spectral Pairs (LSP), i, values are determined. Then the LSP values are used in an AbS process to determine the values for the pitch lag (L) and pitch gain (b). Finally, the values of i, L, and b are used in a second AbS step to determine the codebook indices (I) and gains (G). These steps are described in more detail in the following paragraphs.

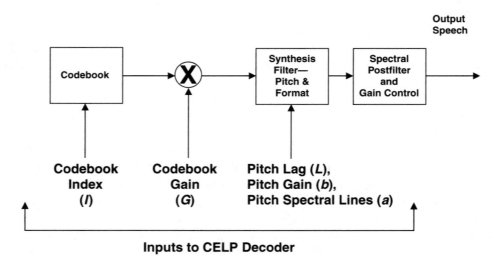

Figure 3-5   CELP Decoder for Rates 1, 1/2, and 1/4

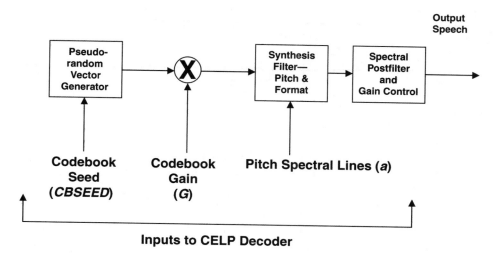

Figure 3-6   CELP Decoder for Rate 1/8

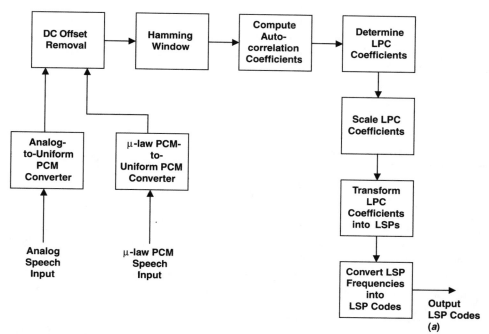

Figure 3-7   CELP Encoder for LSP Codes

Table 3-1 CELP Parameters for Various Coding Rates

Rate CELP Parameters	Rate 1	Rate 1/2	Rate 1/4	Rate 1/8
Line spectral pairs (i) bits	40	20	10	10
i updates per frame	1	1	1	1
Total i bits per frame	40	20	10	10
Pitch lag (L) bits	7	7	7	0
L updates per frame	4	2	1	0
Total L bits per frame	28	14	7	0
Pitch gain (b) bits	3	3	3	0
b updates per frame	4	2	1	0
Total b bits per frame	12	6	3	0
Codebook index (I) bits	7	7	7	0
I updates per frame	8	4	2	—
Total I bits per frame	56	28	14	0
Codebook gain (G) bits	3	3	3	2
G updates per frame	8	4	2	1
Total G bits per frame	24	12	6	2
Codebook seed ($CBSEED$) bits	0	0	0	4
$CBSEED$ updates per frame	—	—	—	1
Total $CBSEED$ bits	—	—	—	4
Parity check bits per frame	11	0	0	0
Total number of bits per frame	171	80	40	16

LSP determination (Fig. 3-7). The encoder for the LSP codes first converts the speech to uniform PCM with at least 14 bits. If the encoder is in a base station, then the received speech is most likely µ-law PCM; if the encoder is in a mobile station, then the received speech is analog. After the speech is converted to PCM, it is processed to remove the DC component and filtered by a Hamming window. The autocorrelation of the sampled output is then computed and used to determine the coefficients for the LPC. The LPC coefficients are then scaled, transformed into the frequency components, and converted into the values for the i bits of the coder output.

The pitch lag and gain bits (Fig. 3-8). These are computed by a recursive process where the output of the PCM encoder is combined with the LSP codes previously calculated and with all possible values of pitch and gain. For each value of pitch and gain, an error function is computed and the transmitted values for pitch and gain are chosen to minimize the error.

The codebook index and gain (Fig. 3-9). These are computed in a recursive process similar to the pitch lag and gain bits using the uniform PCM signal; the computed values for frequency, pitch, and gain; and all possible codebook values and gains.

For the rate 1/8 system, codebook indices are not computed—this type of system uses a random vector generated at both sides.

For every frame or subframe the excitation waveform is selected from a codebook consisting of a large number of candidate waveform vectors. The codebook vector chosen to excite the speech coder filters minimizes the weighted error between the original and synthesized speech.

3.2.4 Enhanced Variable-Rate Codec

The technique used by the Enhanced Variable-Rate Codec (EVRC) [5] to reduce the number of bits required for linear predictor coefficients and pitch synthesis enables the algebraic codebook to generate excitation. As a result, EVRC has higher voice quality.

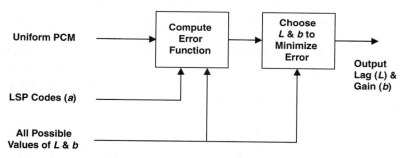

Figure 3-8 CELP Encoder for Pitch Parameters

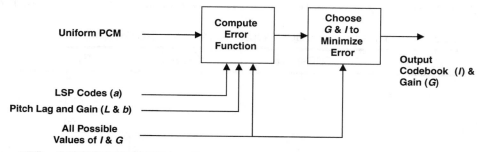

Figure 3-9 CELP Encoder for Codebook Values

Unlike conventional CELP encoders, EVRC does not attempt to match the original speech signal exactly. Instead, EVRC matches a time-wrapped version of the residual that conforms to a simplified pitch contour. The contour is obtained by estimating the pitch delay in each frame and linearly interpolating the pitch from frame to frame. While this adds to computational complexity, the result is higher voice quality per each bit transmitted.

The simplified pitch representation also leaves more bits available in each packet for stochastic excitation and the channel impairment protection than would be possible if a traditional fractional pitch approach were used. The result is enhanced error performance without degraded speech quality at the small cost of added processing requirements.

EVRC also enhances call quality by suppressing background noise. The IS-127 [5] standard recommends a noise suppressor algorithm, but allows system designers to define their own. This is an important factor in choosing a processing platform, making programmable DSP a desirable choice.

The EVRC algorithm is based upon the CELP algorithm. It uses the Relaxed Code-Excited Linear Prediction (RCELP) algorithm and thus does not match the original residual signal but rather a time-wrapped version of the original residual signal that conforms to a simplified pitch contour. This approach reduces the number of bits per frame that are dedicated to pitch representation, allowing additional bits to be dedicated to stochastic excitation and to channel impairment protection. The EVRC algorithm categorizes speech into full-rate (8.55-kbps), 1/2-rate (4-kbps), and 1/8-rate (0.8-kbps) frames that are formed every 20 ms. The EVRC algorithm offers a significant performance improvement over the IS-96A speech codec. Table 3-2 shows the performance of the CDMA Development Group's 13-kbps (CDG-13kbps) speech codec along with IS-96A and EVRC codecs. The CDG-13kbps offers high voice quality but results in a decrease in channel capacity of about 40%. Fig. 3-10 shows a functional diagram of the EVRC algorithm.

Table 3-3 provides the bit allocations by packet type.

3.3 Channel Coding

Channel coding is used to provide the coding gain that is defined as reduction in the required E_b/N_0 (in dB) to achieve a specified error probability of the encoded system over an uncoded system with the same modulation and channel characteristics. The channel coding process usually falls into two classes: *block codes* and *convolutional codes*. There

Table 3-2 Comparisons of CDG-13kbps, IS-96A, and EVRC Vocoders in MOS

FER[a] %	CDG-13 kbps	IS-96A	EVRC
0	4.00	3.29	3.95
1	3.95	3.17	3.83
2	3.88	2.77	3.66
3	3.67	2.55	3.50

[a] FER = frame error rate.

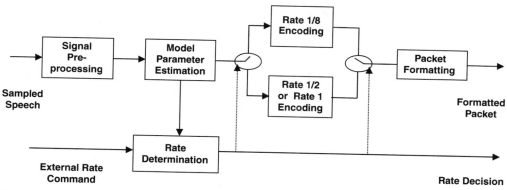

Figure 3-10 Functional Diagram of EVRC

Table 3-3 Bit Allocations by Packet Type in EVRC

Field	Packet Type			
	Rate 1	Rate 1/2	Rate 1/8	Blank
Spectral Transition Indicator	1			
LSP[a]	28	22	8	
Pitch delay	7	7		
Delta delay	5			
ACB gain[b]	9	9		
FCB shape[c]	105	30		
FCB gain[c]	15	12		
Frame energy			8	
Unused	1			
Total encoded bits	171	80	16	
Mixed mode (MM) bit	1			
Frame quality indicator (CRC)[d] (F)	12	8		
Encoder tail bits (T)	8	8	8	8
Total bits	192	96	24	8
Rate (kbps)	9.6	4.8	1.2	0.4

[a] A representation of digital filter coefficient in a pseudofrequency domain; this representation has good quantization and interpolation properties.
[b] Adaptive codebook.
[c] Fixed codebook.
[d] Cyclic redundancy check.

are many subclasses of block codes, including linear block codes, binary cyclic codes, and Bose-Chadhusi-Hocquenghem (BCH) codes. Binary cyclic codes are also called *Cyclic Redundancy Check (CRC)* codes. A BCH code is a special CRC code. BCH codes are represented by (n,k,q) where k bits are mapped into n output bits $(n > k)$, and q is error correction capability. For example a $(15,7,2)$ BCH code transmits 7 information bits by using a 15-bit code word, and it can correct any random errors up to two errors in the code word.

IS-95 systems use convolutional code based on the Viterbi algorithm.

3.3.1 Convolutional Code

Convolutional encoders can be thought of as finite-state machines that change states as the function of the input sequence. A convolutional code [6,7] is generated by passing an information sequence through a finite-state shift register. The shift register contains K stages and m linear algebraic function generators based on generator polynomials. The number of output bits for each k-bit input data sequence is n bits. The code rate is $r = k/n$. The parameter K is called the *constraint length* and indicates the number of input data bits upon which the current output is dependent (see Fig. 3-11).

The error-correcting capability of a convolutional coding scheme increases as rate r decreases. However, the channel bandwidth and decoder complexity both increase with K. The advantage of lower code rates when using convolutional code with coherent phase-shift keying (PSK) is that the required E_b/N_0 is decreased, permitting the transmission of higher data rates for a given amount of power or permitting reduced power for a given data rate. Simulation studies

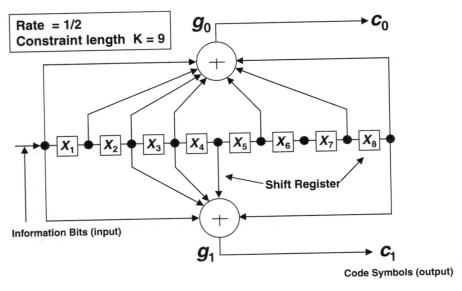

Figure 3-11   Convolutional Encoder

have indicated that, for a fixed constraint length, a decrease in the code rate from 1/2 to 1/3 results in a reduction of the required E_b/N_0 of about 0.4 dB. However, the corresponding increase in decoder complexity is about 17%. For a smaller value of code rate, the improvement in performance relative to the increased decoding complexity diminishes rapidly. Eventually, a point is reached where a further decrease in code rate is characterized by a reduction in coding gain.

The Viterbi algorithm performs maximum-likelihood decoding. It reduces the computational load by taking advantage of the special structure in coder trellis. The complexity of a Viterbi decoder is not a function of the number of symbols in the code word sequence. The Viterbi algorithm removes from consideration those trellis paths that could not possibly be candidates for the maximum-likelihood choice. When two paths enter the same state, the one having the best metric is selected; this path is called the *surviving path*. This selection of surviving paths is performed for all the states. The decoder continues in this way to advance deeper into the trellis, making decisions by eliminating the least likely paths. The major drawback of the Viterbi algorithm in that, while error probability decreases exponentially with constraint length, the number of code states—and consequently decoder complexity—grows exponentially with constraint length.

In IS-95, traffic data frames on uplink and downlink are fed to convolutional encoders. Both uplink and downlink encoders use an 8-bit shift register with a constraint length of 9. The rate of the uplink coder is 1/3—it outputs 3 bits for every input bit. At a rate below 9.6 kbps, output bits are repeated to bring the number of bits in a 20-ms block to 576, for a gross rate of 28.8 kbps. The rate of the downlink encoder is 1/2—it outputs 2 bits for every input bit. At a rate below 9.6 kbps, output bits are repeated to bring the number of bits in a 20-ms block to 384, for a gross rate of 19.2 kbps.

EXAMPLE 3.1

Compare the error probability for a CDMA link with and without the use of error correction coding. Assume that the uncoded transmission characteristics are: QPSK modulation, Gaussian noise, E_b/N_0 = 10 dB, data rate = 9.6 kbps. For the coded case, assume the use of a (15,11) error-correcting code, which is capable of correcting any single error pattern within a block of 15 bits. Also assume the demodulator makes hard decisions and feeds the demodulated code bit directly to the decoder, which in turn outputs an estimate of the original message.

Uncoded Condition

$$p_{bu} = Q\left[\sqrt{\frac{2E_b}{N_0}}\right] = Q[\sqrt{20}] = 4.05 \times 10^{-6}$$

The probability that the uncoded message block will be received in error is

$$P^u_M = 1 - (1 - p_{bu})^k = 1 - (1 - 4.05 \times 10^{-6})^{11} = 4.455 \times 10^{-5}$$

This is the probability that at least 1 bit out of 11 is in error.

Coded Condition

$$R_c = \frac{9600 \times 15}{11} = 13,091 \ bps$$

$$\frac{E_b}{N_0} = \frac{10}{(13.091/9.6)} = 7.333(8.653 \text{ dB})$$

$$p_c = Q\left[\sqrt{\frac{2E_b}{N_0}}\right] = Q[\sqrt{14.666}] = 6.83 \times 10^{-5}$$

$$P^c{}_M = \sum_2^{15}\left(\begin{array}{c}15\\j\end{array}\right)(p_c)^j(1-p_c)^{15-j}$$

The summation starts with $j = 2$ since the code corrects all single errors within a block of 15 bits.

$$P^c_M \approx \left(\begin{array}{c}15\\2\end{array}\right)(p_c)^2(1-p_c)^{13} = 1.96 \times 10^{-6}$$

The probability of message error has improved by $\dfrac{4.455 \times 10^{-5}}{1.96 \times 10^{-6}} = 22.73$ due to error-correcting code.

3.4 Summary

This chapter discussed the digital voice encoding systems used for both CDMA and wideband CDMA. Conventional wireline systems transmit voice by digitizing the voice signal using PCM at a rate of 64 kbps. While it is possible to use PCM in wireless systems, the capacity of the wireless system is lower compared to using other digitizing methods for voice. The CDMA system uses different approaches to digitizing the voice signal—it uses a CELP at 8 or 13.3 kbps to digitize voice. CELP systems model the operation of the human vocal tract to efficiently code speech. We described the operation of the CELP encoder and decoder in detail and provided a high-level description of EVRC. Throughout the descriptions on voice coding, the goal has been to explain the coding systems at a high level so that you can understand the operation of the system and read the standards with some understanding of the motivation for them. If you need to design systems or want additional information, we encourage you to read the standards.

3.5 Problems

1. Repeat Example 3.1 for $E_b/N_0 = 6.8$ dB and a data rate of 14.4 kbps.
2. Determine the resulting output from a one-half convolutional encoder for the input 1 0 1 1 0 0 ... 0 0. The left-most bit is transmitted first.

$$g_0 = 1 + D + D^2 + D^3 + D^5 + D^7 + D^8, \text{ and}$$

$$g_1 = 1 + D^2 + D^3 + D^4 + D^8$$

Assume that the encoder is initially reset (all registers are 0).

3.6 References

1. ITU Recommendation G.711.

2. Recommendation G162, CCITT Plenary Assembly, Geneva, May–June 1964, Blue Book, Vol. 111, p. 52.

3. Skalar, B., *Digital Communication—Fundamentals and Applications*, Prentice Hall, Englewood Cliffs, NJ, 1988.

4. TIA-IS-96A, "Speech Service Option Standard for Wideband Spread Spectrum Digital Cellular System."

5. TIA-IS-127, "Enhanced Variable Rate Codec (EVRC) 8.5 kbps Speech Coder."

6. TIA-IS-665, "W-CDMA (Wideband Code Division Multiple Access) Air Interface Compatibility Standard for 1.85–1.99 GHz PCS Applications."

7. Ziemer, R. E., and Peterson, R. L., *Introduction to Digital Communication,* Macmillan Publishing Co., New York, 1992.

Diversity, Combining, and Antennas

4.1 Introduction

In this chapter, we discuss the concepts of diversity reception where multiple signals are combined to improve the SNR of the system. Time diversity is used to improve system performance for IS-95 CDMA systems; therefore we explore that system in more detail. We then describe various combining schemes that are used to combine the signals. Finally, we consider some practical antennas used in the cellular telephone industry today.

4.2 Diversity Reception

Buildings and other obstacles in built-up areas scatter the signal. Furthermore, because of the interaction between the several incoming waves, the resultant signal at the antenna is subject to rapid and deep fading. Average signal strength can be 40 to 50 dB below the free-space path loss. Fading is most severe in heavily built-up areas in an urban environment. In these areas, the signal envelope follows a Rayleigh distribution over short distances and a log-normal distribution over large distances [10].

Diversity reception techniques are used to reduce the effects of fading and improve the reliability of communication without increasing either the transmitter's power or the channel bandwidth.

The basic idea of diversity reception is that, if two or more independent samples of a signal are taken, these samples will fade in an uncorrelated manner. This means that the probability of all the samples being simultaneously below a given level is much lower than the probability of any individual sample being below that level. The probability of M samples all being simultaneously below a certain level is p^M, where p is the probability that a single sample is below that level. Thus, we can see that a signal composed of a suitable combination of the various samples will have much less severe fading properties than any individual sample.

In principle, diversity reception techniques can be applied either at the base station or at the mobile station, although each type of application has different problems that must be addressed. Typically, the diversity receiver is used in the base station instead of the mobile station. The cost of the diversity combiner can be high, especially if multiple receivers are required. Also the power output of the mobile station is limited by its battery life. The base station, however, can increase its power output or antenna height to improve coverage to a mobile station. Most diversity systems are implemented in the receiver instead of the transmitter since no extra transmitter power is needed to install the receiver diversity system. Since the path between the mobile station and the base station is assumed to be reciprocal, diversity systems implemented in a mobile station work similarly to those in a base station.

4.3 Types of Diversity

There are different ways by which diversity can be achieved: time, frequency, space, angle, multipath, and polarization [1]. In order to gain complete advantage of diversity, combining must be performed at the receiving end. Combiners are designed so that input signal levels, after phase and time delay corrections for the multipath effects, add vectorially while noise outputs are added randomly. Thus, on the average, the combined output SNR will be greater than that present at the input of a single receiver.

4.3.1 Macroscopic Diversity

Macroscopic diversity is used to reduce large-scale fading caused by shadowing. The local mean signal strength varies because of variations in the terrain between the mobile station transmitter and the base station receiver. If only one antenna site is used, the traveling mobile unit may not be able to transmit a signal to the base station at certain geographical locations because of terrain variations such as hills or mountains. Therefore, two separated antenna sites can be used to receive two signals and to combine them to reduce long-term fading. The selective combining technique discussed later in this chapter works best in the macroscopic diversity scheme since other methods require coherent combining which is difficult to achieve when the receivers are some distance apart. Macroscopic diversity is often used in shortwave radio systems to reduce the effects of fading from the ionosphere. Cellular and PCS systems achieve the same effect by handoffs to nearby cell sites when the signal strength becomes weak.

With CDMA systems, macroscopic diversity (i.e., soft handoff) is essential in order to achieve reasonable system performance because of frequency reuse of unity and fast power control. If the mobile station is not connected to the base station that has the lowest attenuation, unnecessary interference is generated in adjacent cells. In the reverse direction (MS to BS), the macroscopic diversity is beneficial because the more base stations try to detect the signals, the higher the probability is for at least one to succeed. In the reverse direction, the detection process itself does not utilize the information from the other base stations receiving the same signal, but the diversity is selection diversity in which the best frame is utilized in the network based on the frame error rate (FER) indication from a CRC.

In the forward direction, macroscopic diversity is different because the transmission originates from several sources and diversity reception is handled by one receiver in the mobile station. All extra transmissions contribute to interference. Capacity improvement is based on a
principle similar to a RAKE receiver in a multipath channel, in which the received power-level
fluctuations tend to decrease as separable paths increase. With forward-link macroscopic diversity, the RAKE receiver capability to gain from extra diversity depends also on the number of
available RAKE fingers. If the RAKE receiver is not able to collect enough energy from transmissions from two or, in some cases, three base stations due to a limited number of RAKE fingers, the extra transmissions to the mobile station can have a negative effect on total system
capacity due to increased interference. This is most likely in the macroscopic cellular environment because the typical number of RAKE fingers considered adequate to capture the channel
energy in most cases is four. If all connections offered that amount of diversity, then the receiver
would have only one or two branches to allocate for each connection.

4.3.2 Microscopic Diversity

Microscopic diversity uses two or more antennas that are at the same site (co-located) but
that are designed to exploit differences in arriving signals from the receiver. Microscopic diversity techniques are used to prevent deep fades from occurring. Once the diversity branches are
created, any of the combining schemes (e.g., selective, maximal-ratio, or equal-gain) can be
used. The following methods are used to obtain uncorrelated signals for combining:

Space diversity. Several transmission paths are used. Two antennas separated physically
by a short distance d can provide two signals with low correlation between their fades. The
separation d in general varies with antenna height h and with frequency. The higher the
frequency, the closer the two antennas can be to each other. Typically a separation of a few
wavelengths is enough to obtain uncorrelated signals.

Frequency diversity. Signals received on two frequencies, separated by the coherence
bandwidth, B_c, are uncorrelated. To use frequency diversity in an urban or suburban environment for cellular and PCS frequencies, the frequency separation must be 300 kilohertz
(kHz) or more. The use of frequency hopping typical of TDMA system (such as GSM)
provides frequency diversity.

Polarization diversity. Horizontally or vertically polarized carrier waves are used. The
horizontal and vertical polarization components, E_x and E_y, respectively, transmitted by
two polarized antennas at the base station and received by two polarized antennas at the
mobile unit, can provide two uncorrelated fading signals. Polarization diversity results in a
3-dB power reduction at the transmitting site since the power must be split between two
polarized antennas.

Angle diversity. When the operating frequency is ≥ 10 gigahertz (GHz), the scattering of
the signals from transmitter to receiver generates received signals from different directions
that are uncorrelated with each other. Thus, two or more directional antennas can be

pointed in different directions at the receiving site and can provide signals for a combiner. This scheme is more effective at the mobile unit than at the base station since the scattering is from local buildings and vegetation and is more pronounced at street level than at the height of base station antennas.

Time diversity. The transmission of a symbol is spread out over time. If the identical signal is transmitted in different time slots, the received signals will be uncorrelated. This system will work for an environment where the fading occurs independent of the movement of the receiver. In a mobile radio environment, the mobile unit may be at a standstill at a location having a weak local mean or caught in deep fade. Although fading still occurs even when the mobile is still, the time-delayed signals are correlated and time diversity will not reduce the fades.

Time diversity is achieved by coding, interleaving, and retransmitting. Channel coding is applied to achieve lower power levels and required signal quality in terms of bit error rate (BER)/FER. Interleaving and channel coding processes are used to correct errors due to channel fades and interference peaks.

4.3.3 RAKE Receiver

In 1958, Price and Green [9] proposed a method of resolving multipath problems using wideband pseudorandom sequences modulated onto a transmitter using other modulation methods (AM or FM). The pseudorandom sequence has the property that time-shifted versions of itself are almost uncorrelated. Thus, a signal that propagates from transmitter to receiver over multiple paths (hence multiple different time delays) can be resolved into separately fading signals by cross-correlating the received signal with multiple time-shifted versions of the pseudorandom sequence. Fig. 4-1 shows a block diagram of a typical system. In the receiver, the outputs are time shifted and, therefore, must be sent through a delay line before entering the

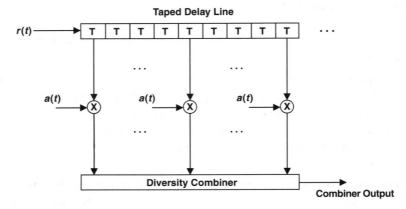

Figure 4-1 RAKE Receiver

diversity combiner. The receiver is called a *RAKE receiver* since the block diagram looks like a garden rake.

When the CDMA systems were designed for cellular systems, the inherent wide-bandwidth signals with their orthogonal Walsh functions were natural for implementing a RAKE receiver. In addition, the RAKE receiver mitigates the effects of fading and is in part responsible for the claimed 10:1 spectral efficiency improvement of CDMA over analog cellular.

In the CDMA system, the bandwidth (1.25 to 15 MHz) is wider than the coherence bandwidth of the cellular or PCS channel. Thus, when the multipath components are resolved in the receiver, the signals from each tap on the delay line are uncorrelated with each other. The receiver can then combine them using any of the combining schemes. The CDMA system then uses the multipath characteristics of the channel to its advantage to improve the operation of the system (see Fig. 4-2).

The combining scheme used governs the performance of the RAKE receiver. An important factor in the receiver design is synchronizing the signals in the receiver to match that of the transmitted signal. Since adjacent cells are also on the same frequency with different time delays on the Walsh codes, the entire CDMA system must be tightly synchronized.

A RAKE receiver uses multiple correlators to separately detect the M strongest multipath components. The relative amplitudes and phases of the multipath components are found by correlating the received waveform with delayed versions of the signal or vice versa. The energy in the multipath components can be recovered effectively by combining the (delay-compensated) multipath components in proportion to their strengths. This combining is a form of diversity and can help to reduce fading. Multipath components with relative delays of less than $\Delta t = 1/B_w$ can-

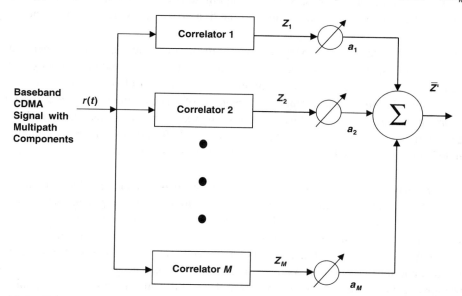

Figure 4-2 RAKE Receiver Correlator

not be resolved and, if present, contribute to fading; in such cases forward error-correction coding and power control schemes play the dominant role in mitigating the effects of fading.

The outputs of the M correlators are denoted as $Z_1, Z_2, \ldots,$ and Z_M. The weights of the outputs are $a_1, a_2, \ldots,$ and a_M, respectively [10] (see Fig. 4-2). The weighting coefficients are based on the power or the SNR from each correlator output. If the power or SNR is small from a particular correlator, it is assigned a small weighting factor. The composite signal, $\bar{Z}$, is given by

$$\bar{Z} = \sum_{k=1}^{M} a_k \cdot Z_k \qquad (4.1)$$

The weighting coefficients, a_k, are normalized to the output signal power of the correlator in such a way that the coefficients sum to unity, as shown in Eq. (4.2).

$$a_k = \frac{Z_k^2}{\sum\limits_{k=1}^{M} Z_k^2} \qquad (4.2)$$

In CDMA cellular/PCS systems, the forward link (BS to MS) uses a *three-finger* RAKE receiver, and the reverse link (MS to BS) uses a *four-finger* RAKE receiver [5]. In the IS-95 CDMA system, the detection and measurement of multipath parameters are performed by a *searcher receiver*, which is programmed to compare incoming signals with portions of I- and Q-channel PN codes. Multipath arrivals at the receiver unit manifest themselves as correlation peaks that occur at different times. A peak's magnitude is proportional to the envelope of the path signal. The time of each peak, relative to the first arrival, provides a measurement of the path's delay.

The PN chip rate of 1.2288 Mcps allows for resolution of multipath components at time intervals of 0.814 µs. Because all of the base stations use the same I and Q PN codes, differing only in code phase offset, not only multipath components but also other base stations are detected by correlation (in a different *search window* of arrival times) with the portion of the codes corresponding to the selected base stations. The searcher receiver maintains a table of the stronger multipath components and/or base station signals for possible diversity combining or for handoff purposes. The table includes time of arrival, signal strength, and the corresponding PN code offset.

On the reverse link, the base station's receiver assigned to track a particular mobile transmitter uses the I- and Q-code times of arrival to identify mobile signals from users affiliated with that base station. Of the mobile signals using the same I- and Q-code offsets, the searcher receiver at the base station can distinguish the desired mobile signal by means of its unique scrambling, long-PN-code offset, acquired before voice transmission begins on the link using a special preamble for that purpose. As the call proceeds, the searcher receiver is able to monitor the strengths of the multipath components from the mobile unit to the base station and to use more than one path through diversity combining.

4.4 Basic Combining Methods

After obtaining the necessary signal samples, we need to consider the question of processing these samples to obtain the best results. For most communication systems, the process can be broadly classified as the *linear combination* of the samples. In the combining process, the various signal inputs are individually weighted and added together as [8]

$$r(t) = a_1 r_1(t) + a_2 r_2(t) + \ldots + a_M r_M(t) \tag{4.3a}$$

$$r(t) = \sum_{i=1}^{M} a_i r_i(t) \tag{4.3b}$$

where $r_i(t)$ = the envelope of the ith signal, and
$\quad a_i$ = the weight factor applied to the ith signal.

We make the following assumptions in the analysis of a combiner:

1. The noise in each branch is independent of the signal and is additive.
2. The signal amplitudes change because of fading, but the fading rate is much smaller than the lowest modulation frequency present in the signal.
3. The noise components are locally incoherent and have 0 mean, with a constant local mean-square value (i.e., constant noise power).
4. The local mean-square values (powers) of the signals are statistically independent.

Since the goal of the combiner is to improve the noise performance of the system, the analysis of combiners is generally performed in terms of SNR. We will examine several different types of combiners and compare their SNR improvements over no diversity.

4.4.1 Selection Combiner

The selection combiner is the simplest of all the schemes. An ideal selection combiner chooses the signal with the highest instantaneous SNR, so the output SNR is equal to that of the best incoming signal. In practice, the system cannot function on an instantaneous basis; to be successful, it is essential that the internal time constants of a selection system are substantially shorter than the reciprocal of the signal fading rate.

We assume that the signal received by each branch is statistically independent of the signals in other branches and is Rayleigh distributed with equal mean signal power P_0. The probability density function of the signal envelope, on branch i, is given by

$$p(r_i) = \frac{r_i}{P_0} e^{-r_i^2/2P_0} \tag{4.4}$$

where $2P_0$ = mean-square signal power per branch = $<r_i^2>$, and
$\quad r_i^2$ = instantaneous power in the ith branch.

Let $\xi_i = r_i^2/2N_i$ and $\xi_0 = 2P_0/2N_i$, where N_i is the noise power in the ith branch.

$$\therefore \frac{\xi_i}{\xi_0} = r_i^2/2P_0 \tag{4.5}$$

The probability density function for ξ_i is given by

$$p(\xi_i) = \frac{1}{\xi_0}e^{(-\xi_i/\xi_0)} \tag{4.6}$$

We assume that the signal in each branch has a constant mean; thus, the probability that the SNR on any one branch is less than or equal to any given value ξ_g is

$$P[\xi_i \le \xi_g] = \int_0^{\xi_g} p(\xi_i)d\xi_i = 1 - e^{(-\xi_g/\xi_0)} \tag{4.7}$$

Therefore, the probability that the SNRs in all branches are simultaneously less than or equal to ξ_g is given by

$$P_M(\xi_g) = P[\xi_1, \xi_2, ..., \xi_M \le \xi_g] = [1 - e^{(-\xi_g/\xi_0)}]^M \tag{4.8}$$

The probability that at least one branch will exceed the threshold SNR value of ξ_g is given by

$$P \text{ (at least one branch } \ge \xi_g) = 1 - P_M(\xi_g) \tag{4.9}$$

The percentage of time the instantaneous output SNR ξ_M is below or equal to the threshold value, ξ_g, is equal to $P(\xi_M \le \xi_g)$. We plot results for $M = 1, 2,$ and 4 in Fig. 4-3. Note that the largest gain occurs for the two-branch combiner. By differentiating Eq. (4.8) we get the probability density function

$$p_M(\xi_g) = (M/\xi_0)[1 - e^{(-\xi_g/\xi_0)}]^{M-1}e^{(-\xi_g/\xi_0)} \tag{4.10}$$

The mean value of the SNR can be given as

$$\overline{\xi_M} = \int_0^{\infty} M\left(\frac{\xi_g}{\xi_0}\right)[1 - e^{(-\xi_g/\xi_0)}]^{M-1}e^{(-\xi_g/\xi_0)}d\xi_g \tag{4.11}$$

Let $x = \xi_g/\xi_0$ and $dx = d\xi_g/\xi_0$

$$\therefore \frac{\overline{\xi_M}}{\xi_0} = M\int_0^{\infty} x[1 - e^{-x}]^{M-1}e^{-x}dx \tag{4.12}$$

Substituting $y = 1 - e^{-x}$ or $x = -\ln(1 - y)$; then $dy = e^{-x}dx$

$$\therefore \frac{\overline{\xi_M}}{\xi_0} = M\int_0^1 [-\ln(1 - y)]y^{M-1}dy = M\sum_{K=1}^{\infty} \int_0^1 \frac{1}{K}y^{M+K-1}dy \tag{4.13}$$

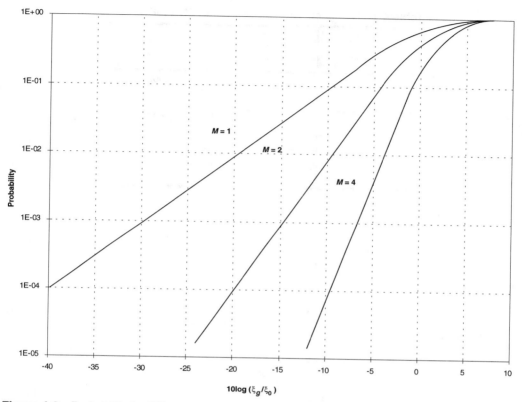

Figure 4-3 Probability for Different Values of *M*-Selection Combiner

$$\therefore \frac{\overline{\xi_M}}{\xi_0} = \sum_{K=1}^{M} \frac{1}{K} \tag{4.14}$$

Table 4-1 shows that the mean SNR increases slowly with M.

4.4.2 Maximal-Ratio Combiner

Maximal-ratio combining was first proposed by Kahn [3]. The M signals are weighted proportional to their signal voltage-to-noise power ratios and then summed.

$$r_M = \sum_{i=1}^{M} a_i r_i(t) \tag{4.15}$$

Since noise in each branch is weighted according to noise power

$$\overline{n_i^2(t)} = \sum_{j=1}^{M} \sum_{i=1}^{M} a_i a_j \overline{n_i(t) n_j(t)} \tag{4.16}$$

Table 4-1 Number of Branches vs. Mean SNR (dB)

M	$\dfrac{\overline{\xi_M}}{\overline{\xi_0}}$	$10\log\dfrac{\overline{\xi_M}}{\overline{\xi_0}}$
1	1.000	0.000
2	1.500	1.761
3	1.833	2.632
4	2.083	3.187
5	2.283	3.585
6	2.450	3.892

The average noise power

$$N_T = \sum_{i=1}^{M} a_i^2 \overline{n_i^2(t)} = 2\sum_{i=1}^{M} |a_i|^2 N_i \tag{4.17}$$

where

$$\overline{n_i^2(t)} = 2N_i \tag{4.18}$$

The SNR at the output is given as

$$\xi_M = \frac{1}{2} \frac{\left|\displaystyle\sum_{i=1}^{M} a_i r_i(t)\right|^2}{\displaystyle\sum_{i=1}^{M} |a_i|^2 N_i} \tag{4.19}$$

We want to maximize ξ_M. This can be done by using the Schwartz inequality.

$$\left|\sum_{i=1}^{M} a_i r_i\right|^2 \leq \left[\sum_{i=1}^{M} |r_i^2|\right]\left[\sum_{i=1}^{M} |a_i|^2\right] \tag{4.20}$$

If $a_i = r_i / \sqrt{N_i}$, then

$$\xi_M = \frac{1}{2} \frac{\displaystyle\sum_{i=1}^{M} r_i^2 \sum_{i=1}^{M} \frac{r_i^2}{N_i}}{\displaystyle\sum_{i=1}^{M} r_i^2} \tag{4.21}$$

$$\therefore \xi_M = \frac{1}{2}\sum_{i=1}^{M} \frac{r_i^2}{N_i} = \sum_{i=1}^{M} \xi_i \tag{4.22}$$

Thus, the SNR at the combiner output equals the sum of the SNR of the branches.

$$\overline{\xi_M} = \sum_{i=1}^{M} \overline{\xi_i} = \sum_{i=1}^{M} \xi_0 = M\xi_0 \qquad (4.23)$$

$$\therefore \frac{\overline{\xi_M}}{\xi_0} = M \qquad (4.24)$$

The probability density function of the combiner output SNR is given by

$$p(\xi_M) = \frac{\xi_M^{M-1} e^{-\frac{\xi_M}{\xi_0}}}{\xi_0^M (M-1)!} \quad , \quad \xi_M \geq 0 \qquad (4.25)$$

The probability that $\xi_M \leq \xi_g$ is given by

$$P(\xi_M \leq \xi_g) = 1 - e^{-\frac{\xi_g}{\xi_0}} \sum_{K=1}^{M} \frac{\left(\frac{\xi_g}{\xi_0}\right)^{K-1}}{(K-1)!} \qquad (4.26)$$

$$P(\xi_M > \xi_g) = e^{-\frac{\xi_g}{\xi_0}} \sum_{K=1}^{M} \frac{\left(\frac{\xi_g}{\xi_0}\right)^{K-1}}{(K-1)!} \qquad (4.27)$$

The plot of P for $M = 1$, 2, and 4 is shown in Fig. 4-4.

4.4.3 Equal-Gain Combining

Equal-gain combining is similar to maximal-ratio combining, but there is no attempt to weight the signal before addition; thus $a_i = 1$ [2]. The envelope of the output signal is given by Eq. (4.3b) with all $a_i = 1$

$$r = \sum_{i=1}^{M} r_i \qquad (4.28)$$

and the mean output SNR is given as

$$\overline{\xi_M} = \frac{1}{2} \frac{\left[\sum_{i=1}^{M} r_i\right]^2}{\sum_{i=1}^{M} \overline{N_i}} \qquad (4.29)$$

If we assume that the mean noise power in each branch is the same (i.e., N), then Eq. (4.29) becomes

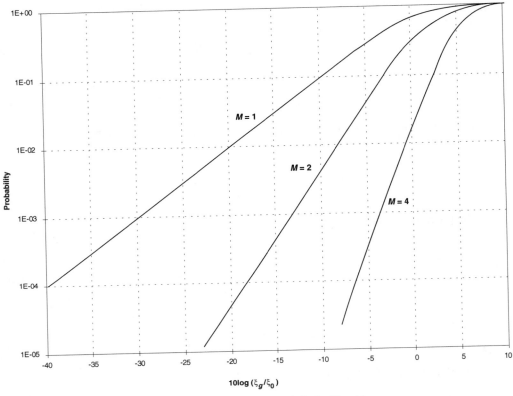

Figure 4-4   Probability for Different Values of Maximal-Ratio Combiner

$$\overline{\xi_M} = \frac{1}{2NM}\overline{\left[\sum_{i=1}^{M} r_i\right]^2} = \frac{1}{2NM}\sum_{j,\,i=1}^{M}\overline{r_j r_i} \tag{4.30}$$

but $\overline{r_i^2} = 2P_0$; and $\overline{r_i} = \sqrt{\dfrac{\pi P_0}{2}}$.

Since the various branch signals are uncorrelated, $\overline{r_j r_i} = \overline{r_i r_j} = \overline{r_i}\,\overline{r_j}$, for i not equal to j. Therefore Eq. (4.30) will be

$$\overline{\xi_M} = \frac{1}{2NM}\left[2MP_0 + M(M-1)\frac{\pi P_0}{2}\right] = \xi_0\left[1 + (M-1)\frac{\pi}{4}\right] \tag{4.31}$$

$$\frac{\overline{\xi_M}}{\xi_0} = 1 + (M-1)\frac{\pi}{4} \tag{4.32}$$

For $M = 2$, the probability P can be written in closed form as

$$P(\xi_M \le \xi_g) = 1 - e^{-\left(\frac{2\xi_g}{\xi_0}\right)} - \sqrt{\pi\left(\frac{\xi_g}{\xi_0}\right)} \, e^{-\frac{\xi_g}{\xi_0}} \cdot \text{erf}\sqrt{\frac{\xi_g}{\xi_0}} \qquad (4.33)$$

For $M > 2$, the probability can be obtained by numerical integration techniques. The plot of probability $P(\xi_M \le \xi_g)$ is shown in Fig. 4-5.

Table 4-2 shows M versus SNR at 1% probability for the selection, maximal-ratio, and equal-gain combiners. Table 4-3 shows SNR improvement for $M = 2$, 4, and 6 at 1% probability for the selection, maximal-ratio, and equal-gain combiners. Note that the selection diversity scheme has the poorest performance and the maximal-ratio combiner has the best. The performance of equal-gain combining is only marginally inferior to maximum-ratio. The implementation complexity for equal-gain combining is significantly less than maximal-ratio combining because of the requirement of correct weighting factors. The data is compared in Fig. 4-6.

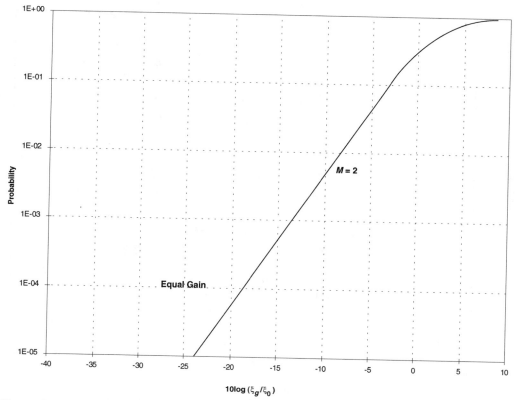

Figure 4-5 Probability for $M = 2$ Equal Gain Combiner

Table 4-2   SNR (dB)

M	Selection	Maximal-Ratio	Equal-Gain
1	−20.0	−20.0	−20.0
2	−10.0	−8.5	−9.2
4	−4.0	−1.0	−2.0
6	−2.0	2.0	1.5

Table 4-3   SNR Improvement (dB)

M	Selection	Maximal-Ratio	Equal-Gain
2	10.0	11.5	10.8
4	16.0	19.0	18.0
6	18.0	22.0	21.5

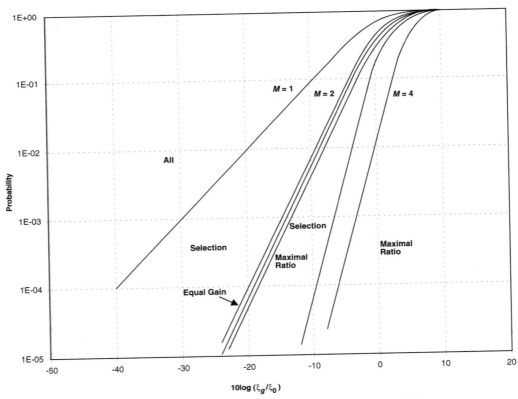

Figure 4-6   Performance Improvement Comparison of Various Combining Schemes

4.5 BPSK Modulation and Diversity

As we discussed earlier, to prevent deep fades from occurring, microscopic diversity techniques can exploit the rapidly changing signal. Macroscopic diversity can be used to reduce large-scale fading caused by shadowing due to variations in the terrain between the transmitter and receiver. Macroscopic diversity is useful at the base station receiver. By using base station antennas sufficiently separated in space, the base station improves the reverse link performance choosing the antenna with strongest signal from the mobile.

For a BPSK system, the bit error probability is given by

$$p_{\text{BPSK}} = Q\left(\sqrt{\frac{2E_b}{N_0}}\right) \tag{4.34}$$

The bit error probabilities over M-branch diversity with selection, equal-gain, and maximal-ratio combining are given as [5]

- **Selection combining**

$$p_{\text{BPSK}, M} = \frac{M}{2} \sum_{k=0}^{M-1} \binom{M-1}{K} \cdot \frac{(-1)^k}{k+1} \cdot \left[1 - \sqrt{\frac{\rho_c}{k+1+\rho_c}}\right] \tag{4.35}$$

- **Equal-gain combining**

$$p_{\text{BPSK}, M} = (\bar{p}_{\text{BPSK}})^M \cdot \sum_{k=0}^{M-1} \binom{M+k-1}{k} \cdot (1 - \bar{p}_{\text{BPSK}})^k \tag{4.36}$$

- **Maximal-ratio combining**

$$p_{\text{BPSK}, M} \approx \binom{2M-1}{M} \cdot \left(\frac{1}{4\rho_c}\right)^M \tag{4.37}$$

where $\rho_c = \dfrac{E_b}{N_0 \cdot M}$ = average SNR per diversity branch

$\bar{p}_{\text{BPSK}} = \frac{1}{2} \cdot \left(1 - \sqrt{\frac{\rho_c}{1+\rho_c}}\right)$

EXAMPLE 4.1

Compare the bit error performance of BPSK modulation having SNR =10 dB with 2-branch diversity using selection, equal-gain, and maximal-ratio combining.

- **Selection combining**

$$p_{\text{BPSK}, 2} = \sum_{k=0}^{1} \binom{1}{k} \cdot \frac{(-1)^k}{k+1} \cdot \left[1 - \sqrt{\frac{10}{k+1+10}}\right] = [1 - 0.95346] - \frac{1}{2}[1 - 0.91287] = 0.002976$$

Table 4-4 Bit Error Performance Comparison

Combining Type	$P_{\text{BPSK},2}$	Performance with Respect to Maximal-Ratio Combining
Maximal-ratio	0.001875	1.0
Equal-gain	0.002153	1.148
Selection	0.002976	1.587

• **Equal-gain combining**

$$\bar{p}_{\text{BPSK}} = \frac{1}{2} \cdot \left(1 - \sqrt{\frac{10}{11}}\right) = 0.02327$$

$$p_{\text{BPSK},2} = (0.02327)^2 \cdot \left[\sum_{k=0}^{1} \binom{1+k}{k} \cdot (1 - 0.02327)^k\right] = 0.002153$$

• **Maximal-ratio combining**

$$p_{\text{BPSK},2} \approx \binom{3}{2} \cdot \left(\frac{1}{4 \times 10}\right)^2 = 0.001875$$

The bit error performance of equal-gain and selection combining is about 15% and 59% worse than the maximal-ratio combining, respectively (see Table 4-4).

4.6 Examples of Base Station and Mobile Antennas

While the simple dipole antenna (Fig. 4-7) is the reference example for antenna specifications, most practical antenna designs aim to improve on the gain of the dipole antenna. Practical antenna design must consider the following issues:

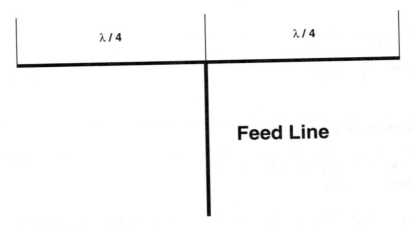

Figure 4-7 Dipole Antenna

- **Antenna pattern.** Closely related to the gain of the antenna is the antenna pattern. As gain is increased, the beam width is decreased. This can be an advantage or a disadvantage depending on the antenna orientation and the needs of the system design.
- **Bandwidth.** The antenna must operate over the full range of frequencies in use for the cellular or PCS system. If the antenna bandwidth is small, channels at the edge of the band may not receive signals as well as those near the band center.
- **Gain.** The higher the gain of the antenna, the lower the power that is necessary at the transmitter. Since the antenna is purchased once and the transmitter power is purchased continuously, high-gain antennas save money by using less electricity, thus conserving the natural resources used to create the electricity.
- **Ground plane.** Some antennas must be mounted above a reflecting surface to function correctly. For example, a quarter-wave antenna is one-half of a dipole and requires that the other half of the dipole be developed by a mirror image below a ground plane. This can be used to advantage in designing antennas for vehicles, but it is a disadvantage when base station antennas, which are high above the surface of the earth, are designed.
- **Height.** The higher the antenna, the better the coverage of the system. However, if the coverage of the system is *too* good, interference from other cells may become troublesome. In an interference-limited system, all levels scale equally so that, at the first order, there will not be a problem. However, since radio wave propagation is statistical, there may be locations where good propagation exists from a point far removed from a base station. The higher the base station antenna, the more likely that these anomalous events will occur.
- **Input impedance.** Most cables used as feed line from the transmitter/receiver to the antenna are either 50 ohms or 72/75 ohms. If the input impedance of the antenna is far removed from either of these values, it will be difficult to get the antenna to accept the power delivered to it and its efficiency η will be low.
- **Mechanical rigidity.** If the antenna flexes in the wind, it will introduce an additional fading component to the received signal. Ultimately, the continuous flexing will cause metal fatigue and mechanical failure of the antenna.
- **Polarization.** For wireless cellular and PCS communications, a vertical antenna is the easiest to mount on a vehicle; therefore vertical polarization has been standardized. In general, horizontal or vertical polarization works equally well.

With this background, we will examine some simple antennas that are used for base and mobile operation.

4.6.1 Quarter-Wave Vertical Antenna

The simplest antenna for a vehicle is the quarter-wave vertical (see Fig. 4-8). A length of wire a quarter of a wavelength long is mounted on the roof of the vehicle. With metal vehicles, the other half of the dipole is developed in the image in the ground plane. Since a vertical dipole antenna has an omnidirectional pattern [6], the quarter-wave vertical has an omnidirectional pattern. The gain of the antenna is the same as that of a dipole (0 dB dipole [dBd] or 2.1 isotropic dipole [dBi]). The impedance of a quarter-wave vertical is 36.5 ohms and requires a matching transformer for proper feeding of the antenna.

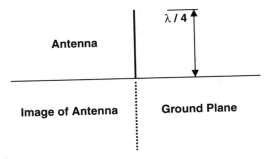

Figure 4-8 Quarter-Wave Vertical Antennas

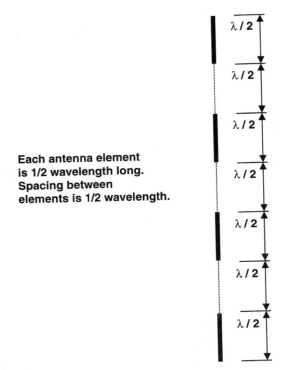

Figure 4-9 Stacked Vertical Dipoles

4.6.2 Stacked Dipoles

Since a vertical dipole has an omnidirectional pattern, three or more dipoles can be stacked vertically to produce increased gain and maintain the omni pattern. A typical base station antenna will have four half-wave dipoles spaced apart by one wavelength vertically (see Fig. 4-9). All the antenna elements are fed in phase with a signal from the transmitter. The resultant pattern is omnidirectional in the horizontal plane and has an 8.6-dBi gain on the horizon (0-degree elevation) with a vertical beamwidth of ±6.5 degrees. Stacked dipoles have an impedance of 63 ohms; thus a matching transformer is necessary, but it is easier to build than the one for the quarter-wave vertical.

A variation of the base station antenna for vehicles uses a half-wave dipole above a quarter-wave vertical (see Fig. 4-10). The other half of the antenna is in the ground plane image. The two elements are decoupled from each other by a quarter-wavelength-long decoupling coil. This is the most common cellular antenna for vehicles. It has a gain of 7–10 dBi.

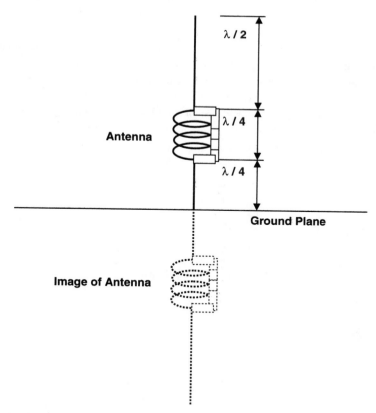

Figure 4-10 Half-Wave over Quarter-Wave Antenna Mounted on Vehicle

4.6.3   Corner Reflectors

The antennas we have discussed so far have omnidirectional patterns. A directional antenna at a base station can improve the SNR of the system and thus improve its spectral efficiency. In 1939 Kraus [4] designed the corner reflector antenna consisting of a vertical dipole and two sheets of metal at a 45, 60, or 90 degree angle (see Fig. 4-11). The impedance and gain of the antenna depend on the angle of the corner and the spacing from the corner to the dipole antenna. The gain of the antenna varies from 7 to 13 dBd, and the impedance varies from 0 to 150 ohms. A colinear antenna can be used in place of the dipole for additional gain.

4.6.4   Smart Antenna

Smart antennas include either a Switched Beam System (SBS) or an adaptive antenna system [7] (see Figs. 4-12 and 4-13). The SBS uses multiple fixed beams in a sector and a switch to select the best beam for receiving a particular signal. In an adaptive antenna system, the received signals by multiple antennas are weighted and combined to maximize the SNR using either the minimum mean-square error (MMSE) or least squares (LS) criterion. The advantage of the adaptive antenna system over the SBS is that, in addition to the M-fold antenna gain, it provides an M-fold diversity gain.

Smart antennas offer a broad range of methods to improve system performance. They provide enhanced coverage through range expansion, hole filling, and better building penetration.

Given the same transmitter power output at the base station and the mobile unit, smart antennas can increase range by increasing the gain of the base station antenna. The M-fold antenna gain increases the range by a factor of $M^{1/\gamma}$ where γ is the path loss exponent; it reduces the number of base stations to cover a given area by $M^{2/\gamma}$ [7]. An SBS with M beams can increase the system capacity by a factor of M by reducing the number of interferers. An adaptive antenna

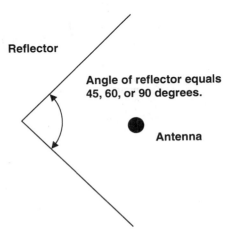

Reflector

Angle of reflector equals 45, 60, or 90 degrees.

Antenna

Figure 4-11   Corner Reflector

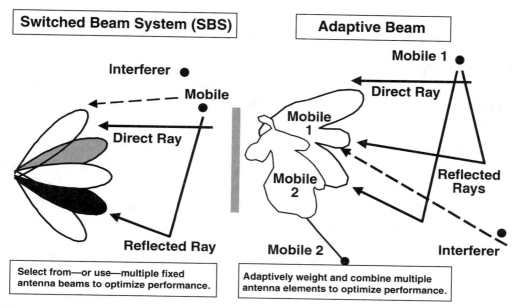

Figure 4-12   Smart Antennas

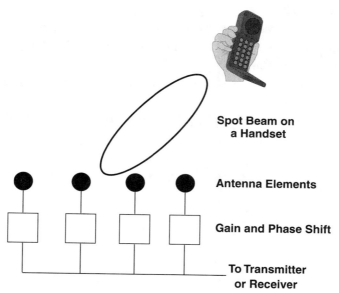

Figure 4-13   Adaptive Array Smart Antenna

system can provide some additional gain by suppressing interferers further. However, since there are so many interferers, the additional gain may not be worth the complexity.

4.7 Summary

In this chapter we considered the role of antennas in the wireless system. We also presented the concepts of diversity reception where multiple signals are combined to improve the SNR of the system.

4.8 References

1. Garg, V. K., and Wilkes, J. E., *Wireless and Personal Communications Systems*, Prentice Hall, Upper Saddle River, NJ, 1996.

2. Halpern, S. W., "The Theory of Operation of an Equal-Gain Predication Regenerative Diversity Combiner with Rayleigh Fading Channel," *IEEE Transactions on Communication Technology*, COM-22 (8), August 1974, pp. 1099–106.

3. Kahn, L. R., "Radio Squarer," *Proceedings of the IRE* 42, November 1954, p. 1704.

4. Kraus, J. D., *Antenna*, McGraw-Hill, New York, 1988.

5. Lee, J. S., and Miller, L. E., *CDMA Systems Engineering Handbook*, Artech House, Boston, 1998.

6. Lee, W. C. Y., "Antenna Spacing Requirements for a Mobile Radio Base Station Diversity," *Bell System Technical Journal* 50(6), July–August 1971.

7. Liberti, J. C. Jr., and Rappaport, T. S., *Smart Antennas for Wireless Communications*, Prentice Hall, Upper Saddle River, NJ, 1999.

8. Mahrotra, A., *Cellular Radio Performance Engineering,* Artech House, Boston, 1994.

9. Price, R., and Green, P. E. Jr., "A Communication Technique for Multipath Channels," *Proceedings of the IRE* 46, March 1958, pp. 555–70.

10. Rappaport, T. S., *Wireless Communications: Principle and Practice*, Prentice Hall, Upper Saddle River, NJ, 1996.

11. Special Issue on Smart Antennas, *IEEE Personal Communications* 5(1), February 1998.

IS-95 System Architecture

5.1 Introduction

A wireless system—whether for cellular operation (Band Class 0), i.e., 850-MHz band, or for PCS operation (Band Class 1), i.e., 1.8-GHz band—must support communication with the mobile station and interact with the Public-Switched Telephone Network (PSTN). As the mobile station changes its location during a call, the wireless system must insure that the connection between the mobile station and the PSTN is maintained.

A wireless system consists of discrete logical components. These components may be discrete physical entities or they may be physically located with another logical entity. These functional entities must interact in order to coordinate operation. Such interaction is achieved by messaging over interfaces between two entities. If two functional entities are physically separate and if the interface is standardized, the service provider can purchase products from different manufacturers. However, successful operation is not guaranteed since the associated standard often does not cover all facets of operation. Manufacturers need to cooperate to eliminate differences that jeopardize proper interaction.

This chapter first will cover the functional entities of the wireless network and the interfaces between the entities that have been standardized by the wireless communication industry. Then it examines the activities of the International Telecommunication Union (ITU) to add Intelligent Network (IN) to wireless systems.

5.2 TR-45/TR-46 Reference Model

Key to the North American systems is the use of a common reference model from the cellular standards group TR-45. When work started on PCS, the TR-46 standards group adopted the TR-45 reference model for PCS, but with some minor changes in the names of the elements (see Fig. 5-1). A second reference model was proposed by T1P1, but it is similar to the TR45/46 model.

The names of each of the network elements are similar, and some of the functionality is partitioned differently between the models. The main difference between the two reference models is how mobility is handled. Mobility is the capability for users to place and receive calls in systems

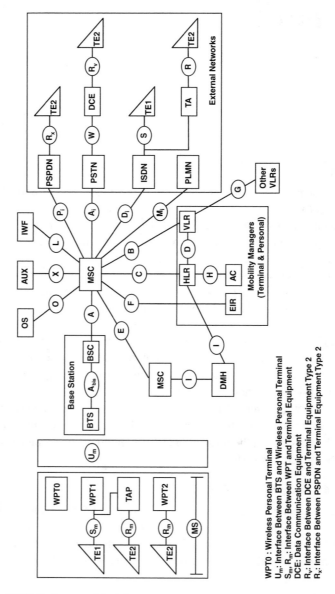

Figure 5-1   TR-45/46 Reference Model

other than their home system. In the T1P1 reference model, the user data and the terminal data are separate; thus, users can communicate with the network via different mobile stations. In the TR-45/46 reference model, only terminal mobility is supported. A user can place or receive calls at only one terminal (the one the network has identified as owned by the user). The T1P1 functionality is migrating toward independent terminals and user mobility, but all aspects of it are not currently supported. We also discuss a Wireless Intelligent Network (WIN) architecture and reference model that overcomes some of the mobility problems in the current architecture.

The main elements of the reference model are

- **Mobile Station (MS).** The MS terminates the radio path on the user side and enables the user to gain access to services from the network. The MS can be a stand-alone device or can have other devices (e.g., personal computers, fax machines) connected to it.
- **Base Station (BS).** The BS terminates the radio path and connects to the Mobile Switching Center (MSC). The BS is often segmented into the BTS and the BSC.
    - **Base Transceiver System (BTS).** The BTS consists of one or more transceivers placed at a single location and terminates the radio path on the network side. The BTS may be co-located with a BSC or may be independently located.
    - **Base Station Controller (BSC).** The BSC is the control and management system for one or more BTSs. The BSC exchanges messages with both the BTS and the MSC. Some signaling messages may pass through the BSC transparently.

- **Mobile Switching Center (MSC).** The MSC is an automatic system that interfaces the user traffic from the wireless network with the wireline network or other wireless networks. The MSC functions as one or more of the following:
    - *Anchor MSC:* first MSC providing radio contact in a call
    - *Border MSC:* an MSC controlling BTSs adjacent to the location of a mobile station
    - *Candidate MSC:* an MSC that could possibly accept a call or a handoff
    - *Originating MSC:* the MSC directing an incoming call toward a mobile station
    - *Remote MSC:* the MSC at the other end of an intersystem handoff trunk
    - *Serving MSC:* the MSC currently providing service to a call
    - *Tandem MSC:* an MSC providing only trunk connections for a call in which a handoff has occurred
    - *Target MSC:* the MSC selected for a handoff
    - *Visited MSC:* an MSC providing service to the mobile station

- **Home Location Register (HLR).** The HLR is the functional unit that manages mobile subscribers by maintaining all subscriber information (e.g., electronic serial number, directory number, international mobile station identification, user profiles, current location). The HLR may be co-located with an MSC as an integral part of the MSC or may be independent of the MSC. One HLR can serve multiple MSCs, or an HLR may be distributed over multiple locations.

- **Data Message Handler (DMH).** The DMH is used to collect billing data.
- **Visited Location Register (VLR).** The VLR is linked to one or more MSCs and is the functional unit that dynamically stores subscriber information (e.g., the user's electronic serial number [ESN], directory number, user profile information) obtained from the user's HLR when the subscriber is located in the area covered by the VLR. When a roaming MS enters a new service area covered by an MSC, the MSC informs the associated VLR about the MS by querying the HLR after the MS goes through a registration procedure.
- **Authentication Center (AC).** The AC manages the authentication or encryption information associated with an individual subscriber. The AC may be located within an HLR or MSC or may be located independently of both.
- **Equipment Identity Register (EIR).** The EIR provides information about the mobile station for record purposes. The EIR may be located within an MSC or may be located independently of it.
- **Operations System (OS).** The OS is responsible for overall management of the wireless network.
- **Interworking Function (IWF).** The IWF enables the MSC to communicate with other networks.
- **External Networks.** These are other communications networks—the Public-Switched Telephone Network (PSTN), the Integrated Services Digital Network (ISDN), the Public Land Mobile Network (PLMN), and the Public-Switched Packet Data Network (PSPDN).

The following interfaces are defined between the various elements of the system:

- **BS to MSC (A-Interface).** This interface between the base station and the MSC supports signaling and traffic (both voice and data). A-Interface protocols have been defined using SS7, ISDN BRI/PRI, and frame relay transport (TIA IS-634).
- **BTS-to-BSC Interface (A_{bis}).** If the base station is segmented into a BTS and BSC, this internal interface is defined.
- **MSC-to-PSTN Interface (A_i).** This interface is defined as an analog interface using either Dual-Tone Multifrequency (DTMF) signaling or Multifrequency (MF) signaling.
- **MSC to VLR (B-Interface).** This interface is defined in the TIA IS-41 protocol specification [8].
- **MSC to HLR (C-Interface).** This interface is defined in the TIA IS-41 protocol specification [8].
- **HLR to VLR (D-Interface).** This interface is the signaling interface between an HLR and a VLR and is based on SS7. It is currently defined in the TIA IS-41 protocol specification [8].
- **MSC to ISDN (D_i-Interface).** This is the digital interface to the PSTN and is a T1 interface (24 channels of 64 kbps) and uses Q.931 signaling.

- **MSC-MSC (E-Interface).** This interface is the traffic and signaling interface between wireless networks. It is currently defined in the TIA IS-41 protocol specification.
- **MSC to EIR (F-Interface).** Since the EIR is not yet defined, the protocol for this interface is not defined.
- **VLR to VLR (G-Interface).** When communication is needed between VLRs, this interface is used. It is defined by TIA IS-41.
- **HLR to AC (H-Interface).** The protocol for this interface is not defined.
- **DMH to MSC (I-Interface).** The protocol for this interface is defined in the IS-124 [9].
- **MSC to IWF (L-Interface).** This interface is defined by the interworking function.
- **MSC to PLMN (M_i-Interface).** This interface is to another wireless network.
- **MSC to OS (O-Interface).** This is the interface to the OS. It is currently being defined in ATIS standard body T1M1.
- **MSC to PSPDN (P_i-Interface).** This interface is defined by the packet network that is connected to the MSC.
- **Terminal Adapter (TA) to Terminal Equipment (TE) (R-Interface).** These interfaces are specific for each type of terminal connected to an MS.
- **ISDN to TE (S-Interface).** This interface is outside the scope of PCS and is defined within the ISDN system.
- **BS to MS (U_m-Interface).** This is the air interface.
- **PSTN to DCE (W-Interface).** This interface is outside the scope of PCS and is defined within the PSTN system.
- **MSC to AUX (X-Interface).** This interface depends on the auxiliary equipment connected to the MSC.
- **BS Management Application Part (BSMAP).** Message sent between BS and MSC.
- **Direct Transfer Application Part (DTAP).** Message sent between MS and MSC.

5.3 Functional Model Based on Reference Model

Fig. 5-2 shows the functional model derived from the reference model. Several physical scenarios can be developed using the functional entities shown in Fig. 5-2. Fig. 5-3 shows the Functional Entity (FE) grouping in which the physical interface between the radio system (RS) and the switching system platform (SSP) carries both the call control and mobility management messages.

- **Radio Terminal Function (RTF) FE.** This is the subscriber unit (SU). The only physical interface is to the Radio System (RS) through the air interface.
- **Radio Control Function (RCF) FE and Radio Access Control Function (RACF) FE.** These are included in the RS. Combining these FEs onto the same platform allows air-interface-specific functions (such as those that would impact handoff) to be isolated from the other interfaces. OS information, including performance data and accounting records, is generated, collected, and formatted on this platform. There is only one physical interface to SSP to carry both the call control and mobility management signaling.

•   **Service Switching Function (SSF)/ Call Control Function (CCF) FE.** The SSF/CCF
    FE is contained in SSP and provides interfaces to operator services, E911, international
    calls, and network repair/maintenance centers. Physical interfaces for this collection
    include: to the RS, to the mobility management platform, to the internal peripheral (IP),
    and to other SSPs and external networks.

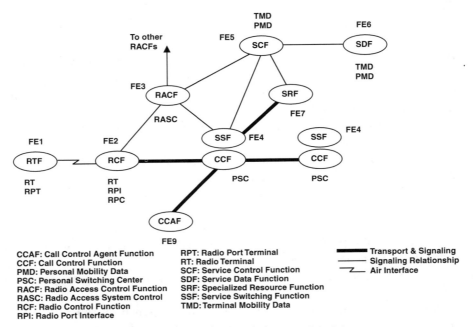

CCAF: Call Control Agent Function
CCF: Call Control Function
PMD: Personal Mobility Data
PSC: Personal Switching Center
RACF: Radio Access Control Function
RASC: Radio Access System Control
RCF: Radio Control Function
RPI: Radio Port Interface

RPT: Radio Port Terminal
RT: Radio Terminal
SCF: Service Control Function
SDF: Service Data Function
SRF: Specialized Resource Function
SSF: Service Switching Function
TMD: Terminal Mobility Data

▬▬▬ Transport & Signaling
───── Signaling Relationship
──ᵼ── Air Interface

Figure 5-2   Functional Model Derived from the Reference Model

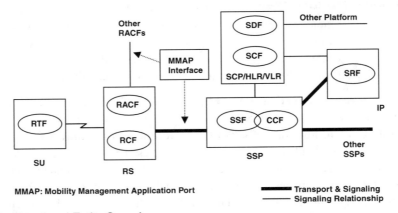

MMAP: Mobility Management Application Port

▬▬▬ Transport & Signaling
───── Signaling Relationship

Figure 5-3   Functional Entity Groupings

- **Specialized Resource Function (SRF) FE and Data Interworking Function.** These are contained in the IP. Physical interfaces for this collection include one to the SSP and another to the mobility management platform.

Individually the SSF/CCF FE and CCF FE represent interswitch and internetwork functional entity collections and physical interfaces.

As shown in Fig. 5-3, the only interface to the RS is from the SSP. There is no direct physical path between the RS and the switch control point (SCP)/VLR. All operations to or from the RS pass through the SSP, whether or not the SSP consumes or produces the operation.

5.4 Wireless Intelligent Network

For the wireline network, BellCore has defined a set of protocols called Intelligent Network (IN) [4] that enable a rich set of new telephony capabilities to be generated without additional software development in the central switch. As IN has grown in popularity for the wireline network, the wireless network has also embraced the concept. IN improves the ability to locate and efficiently direct calls to roaming mobile stations and provides other advanced features. In this section we examine the work of the ITU to add IN to the wireless systems.

An examination of the needs of intelligent networking and the mobility aspects of wireless communications shows several similarities. The wireline user should have one phone number anywhere in the world. Anyone dialing that number would have their calls routed to the owner's destination independent of number's owner's location. The services that a user has available in the home or office should also be available in a home or office where the user is visiting as well as at public phones.

The functions of the HLR/VLR in a wireless network are similar to the functions of an SCP in a wireline network.

The ITU has adapted IN for use in wireless networks. Question 8 of Study Group (SG) 11 [7] has defined a communications control plane (Fig. 5-4) and a radio resource control plane (Fig. 5-5) using IN concepts.

On the network side of the communications control plane, the following functional elements are defined:

- The **Bearer Control Function (BCF)** provides those bearer functions needed to process handoffs. A common example of this is a conference bridge to support soft handoffs.
- The **Bearer Control Function for the radio bearer (BCFr)** provides the functions necessary to select bearer functions and radio resources. It also detects and responds to pages from the network and performs handoff processing. Some example bearer functions are PCM voice, ADPCM voice, packet data, and circuit-switched data.
- The **Call Control Function-Enhanced (CCF')** provides the call and connection control in the network. For example, it establishes, maintains, and releases call instances requested by the CCAF'; provides IN triggers to the SSF; controls bearer connection elements in the network.

- The **Service Access Control Function (SACF)** provides the network side of mobility management functions. Some examples are registration and location of the mobile station.
- The **Service Control Function (SCF)** contains the service and mobility control logic and call processing to support the functions of a mobile terminal.

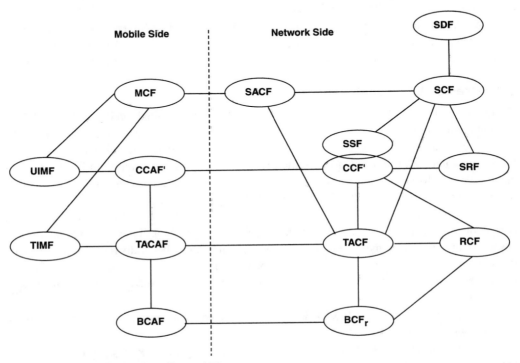

Figure 5-4 Communications Control Plane

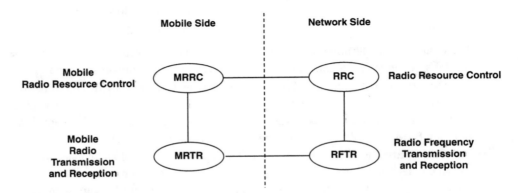

Figure 5-5 Radio Resource Control Plane

- The **Service Data Function (SDF)** provides data storage and data access in support of mobility management and security data for the network.
- The **Specialized Resource Function (SRF)** provides the specialized functions needed to support execution of IN services. Some examples are dialed digit receivers, conference bridges, and announcement generators.
- The **Service Switching Function (SSF)** provides the functions required for interaction between the CCF' and the SCF. It supports extensions of the CCF' logic to recognize IN triggers and interact with the SCF. It manages the signaling between the CCF' and the SCF and modifies functions in the CCF' to process IN services under control of the SCF.
- The **Terminal Access Control Function (TACF)** provides control of the connection between the mobile station and the network. It provides paging of mobile stations, page response handling, handoff decision, and completion. It also provides trigger access to IN functionality.

On the mobile side, the following functional elements are defined:

- The **Bearer Control Agent Function (BCAF)** establishes, maintains, modifies, and releases bearer connections between the mobile station and the network.
- The **Call Control Agent Function-Enhanced (CCAF')** supports the call processing functions of the mobile station.
- The **Mobile Control Function (MCF)** supports the mobility management functions of the mobile station.
- The **Terminal Access Control Agent Function (TACAF)** provides the functions necessary to select bearer functions and radio resources. It also detects and responds to pages from the network and performs handoff processing.
- The **Terminal Identification Management Function (TIMF)** stores the terminal-related security information. It provides terminal identification to other functional elements and provides the terminal authentication and cryptographic calculations.
- The **User Identification Management Function (UIMF)** provides user-related security information similar to the TIMF.

Both the TIMF and the UIMF can be stored in either the mobile station or a separate security module often implemented in a smart card.

The radio resource control plane (Fig. 5-5) is responsible for assigning and supervising radio resources. Four function entities (two on the mobile side and two on the network side) perform the functions of the radio access subsystem.

- The **Radio Resource Control (RRC)** provides functionality in the network to select radio resources (channels, spreading codes, etc.), make handoff decisions, control the RF power of the mobile station and provide system information broadcasting.
- The **Radio Frequency Transmission and Reception (RFTR)** provides the network side of the radio channel. It provides the radio channel encryption and decryption (if

used) and channel quality estimation (data error rates for digital channels), sets the RF power of the mobile station, and detects when the mobile station accesses the system.

- The **Mobile Radio Resource Control (MRRC)** processes the mobile side of the radio resource selection. It provides base station selection during start-up, mobile-assisted handoff control, and system access control.
- The **Mobile Radio Transmission and Reception (MRTR)** provides the mobile side of the radio channel and performs functions similar to RFTR.

The two control planes interact to provide services to the mobile station and the network.

The TR-46 PCS network reference model working group has generated a simplified version of the ITU model and has explicitly shown the operations functions on the model. The IN Function Reference Model for PCS (Fig. 5-6) has many of the functional elements with the same name as those of the ITU model. The differences are that

- The **Radio Terminal Function (RTF)** contains all functionality of the mobile side of the reference model.
- The **Radio Access Control Function (RACF)** is similar to the SACF and provides mobility management functions.

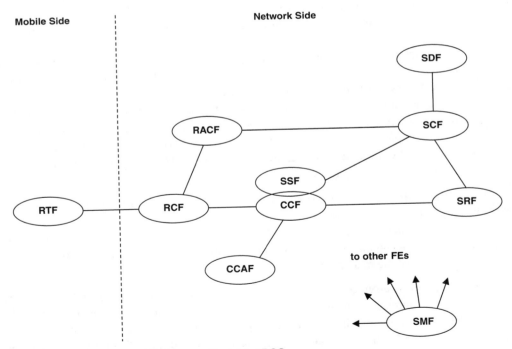

Figure 5-6 IN Functional Reference Model for PCS

- The **Radio Control Function (RCF)** provides the capabilities of the TACF, the BCFr, and the BCF in the ITU model. It provides the radio ports and the radio port controller capabilities in the PCS network.
- The **Call Control Agent Function (CCAF)** provides access to the wireless network by wireline users.
- The **Service Management Function (SMF)** provides the network management functions for each functional element.

With the ITU reference model, the functionality to support all of the features and capabilities for a wireless network can be partitioned into different functional elements and still meet the variety of national and worldwide standards. Therefore no exact partitioning of the functions of the network reference models can be made. We encourage you to examine the standards and implementations of the various manufacturers for partitioning examples.

Around the world, two common mobile application parts (MAP) are used. In GSM systems deployed worldwide, the GSM MAP is used. In the systems that were originally deployed in the United States, the IS-41 MAP is used. The use of two different MAPs makes roaming between the two systems difficult. The use of the IN application part with enhancements will offer the opportunity for a common MAP for all mobile and wireline services.

5.5 Summary

This chapter presented the TR-45/46 reference model, which is used as the basis by standards committees. The main elements of this model are the Mobile Station (MS), Base Station (BS), Mobile Switching Center (MSC), Home Location Register (HLR), and Visited Location Register (VLR). Next we discussed the MSC-BS interface (TIA IS-634), which standardizes the messaging between the base station and the MSC. Messages between the BS and the MSC are categorized into two types: Base Station Application Part (BSAP) and Direct Transfer Application Part (DTAP). Messages can be associated with the functions of call processing and supplementary services, radio resource management, mobility management, and terrestrial facility management. The effects of a CDMA system upon the architecture are emphasized. Finally, we focused on the concepts of wireless intelligent networking. Current systems have difficulty with roaming because of the support of two MAPs in the world (GSM and IS-41). With wireless intelligent networking and extensions to the IN MAP, true worldwide roaming may become possible once a common frequency band is available.

5.6 References

1. American National Standards Institute, Inc., "Signaling System No. 7 (SS7)—Message Transfer Part (MTP)," ANSI T1.111-1992, June 1992.
2. American National Standards Institute, Inc., "Signaling System No. 7 (SS7)—Signaling Connection Control Part (SCCP)," ANSI T1.112-1992, June 1992.

3. American National Standards Institute, Inc., "Synchronization Interface Standards for Digital Networks," 1087, ANSI T1.101-1987, March 1987.

4. Bellcore, "Advanced Intelligent Network (AIN) Release 1 Switching System Generic Requirements," TA-NWT-001123, Issue 1, May 1991.

5. Committee T1, "Stage 2 Service Description for Circuit Mode Switched Bearer Services," Draft T1.704.

6. Committee T1—Telecommunications, "A Technical Report on Network Capabilities, Architectures, and Interfaces for Personal Communications," T1 Technical Report #34, May 1994.

7. ITU Study Group 11, "Version 1.1.0 of Draft New Recommendation Q.FNA, Network Functional Model for FPLMTS," Document Q8/TYO-50, September 15, 1995.

8. TIA Interim Standard, IS-41 C, "Cellular Radiotelecommunications Intersystem Operations," January 1996.

9. TIA IS-124, "Cellular Radio Telecommunications Intersystem Non-Signaling Data Message Handlers (DMH)," 1994.

10. TIA IS-634, "MSC-BS Interface for Public 800 MHz," Revision 0, 1995.

11. TIA SP-2977, "Cellular Features Description," Prepublication Version, March 14, 1995.

12. TIA TR-45 Reference Model, 1990.

13. TIA TR-46 Reference Model, 1991.

IS-95 CDMA Air Interface

6.1 Introduction

This chapter provides a high-level description of the IS-95 CDMA air interface. The intent of this chapter is to lay a foundation for the system details given in chapters 7–11. We describe important aspects of the forward link (base station to mobile) and reverse link (mobile to base station) and include modulation parameters for the channels.

6.2 TIA IS-95 CDMA System

The TIA IS-95 CDMA system operates on the same frequency band as the Advanced Mobile Phone System (AMPS) using Frequency Division Duplex (FDD) with 25 MHz in each direction.[*] For Band Class 0 the reverse link (mobile to base station) uses frequencies from 869 to 894 MHz, and forward link (base station to mobile) uses frequencies from 824 to 849 MHz. The mobile station supports CDMA operations on AMPS channel numbers 1013 through 1023, 1 through 311, 356 through 644, 689 through 694, and 739 through 777 inclusive. The PCS version of the specification J-STD-008 for use at 1800-MHz (Band Class 1) has been also standardized. The CDMA channels are defined in terms of an RF frequency and code sequence. Sixty-four Walsh functions are used to identify the forward (down) link channels, whereas 64 long PN codes are used for the identification of the reverse (up) link channels. Figs. 6-1 and 6-2 show the CDMA forward (down) link and reverse (up) link channels structure. The modulation and coding features of the IS-95 CDMA system are listed in Table 6-1. We will discuss the system in detail in chapters 7–11.

---

[*] The frequency spectrum for the A-System cellular service provider is split such that the spectrum is not divisible by 1.25 MHz. Thus, the A-System cellular provider cannot partition the spectrum into 10 1.25-MHz CDMA channels. This restriction is not imposed for the B-System, however.

Modulation and coding details for the reverse link and forward link channels differ. Pilot signals are transmitted by each cell to assist the mobile radio in acquiring and tracking the cell site forward link signals. The strong coding helps these radios to operate effectively at E_b/N_0 ratio of 5- to 7-dB range.

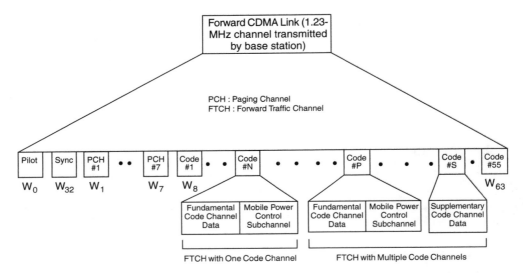

Figure 6-1   Forward Link Channels Structure

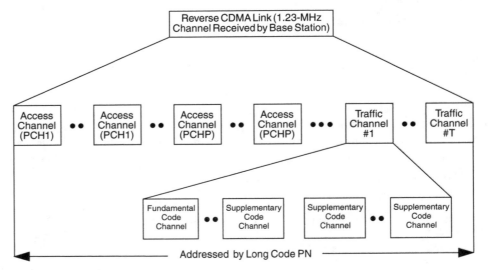

Figure 6-2   Reverse Link Channels Structure

Table 6-1 Modulation and Coding Features of the IS-95 CDMA System

Modulation	Quadrature Phase-Shift Keying (QPSK)
Chip rate	1.2288 Mcps
Nominal data rate (RS1)	9600 bps
Filtered bandwidth	1.23 MHz
Coding	convolution with Viterbi decoding
Interleaving	with 20-ms span

The CDMA system uses power control and voice activation to minimize mutual interference. Voice activation is provided by using a variable-rate vocoder that operates at a maximum rate of 8 kbps to a minimum rate of 1 kbps for Rate Set 1 (RS1), depending on the level of voice activity. With the decreased data rate, the power control circuits reduce the transmitter power to achieve the same bit error rate. A precise power control, along with voice activation circuits, is critical to avoid the excessive transmitter signal power responsible for contributing the overall interference in the system. A coding algorithm at 13.3 kbps for RS2 is also supported.

A time interleaver with a 20-ms span is used with error-control coding to overcome rapid multipath fading and shadowing. The time span used is the same as the time frame of the voice compression algorithm.

The CDMA radio uses a RAKE receiver to take advantage of a multipath delay greater than 1 μs, which occurs commonly in cellular/PCS networks in urban and suburban areas.

6.2.1 Forward Link

In this section, we summarize the operation of the forward link. For more details refer to chapter 7. The forward link channels (see Fig. 6-1) include one *pilot channel*, one *synchronization* (sync) *channel*, up to seven *paging channels*, and a number of *forward traffic channels* (if multiple carriers are implemented, the pilot channel and sync channels do not need to be duplicated). Each forward traffic channel contains one *forward fundamental code channel* and may contain one to seven *forward supplemental code channels*. The information on each channel is modulated by the appropriate Walsh function and then modulated by a quadrature pair of PN sequences at a fixed chip rate of 1.2288 Mcps. The pilot channel is always assigned to code channel number 0. If the sync channel is present, it is given the code channel number 32. Whenever paging channels are present, they are assigned the code channels number 1 through 7 (inclusive), in sequence. The remaining code channels are used by forward traffic channels.

The sync channel always operates at a fixed data rate of 1200 bps and is convolutionally encoded to 2400 bps, repeated to 4800 bps, and interleaved over the period of the pilot pseudorandom binary sequence. Each of the interleaved symbols uses four Walsh symbols.

The forward traffic channels are grouped into sets. RS1 has four elements—9600, 4800, 2400, and 1200 bps. RS2 contains four elements—14,400, 7200, 3600, and 1800 bps. All radio systems support RS1 on the forward traffic channels. RS2 is optionally supported on the forward traffic channels. When a radio system supports a rate set, it supports all four elements of the set.

Speech is encoded using a variable-rate vocoder to generate forward-traffic-channel data depending on voice activity. Since the frame duration is fixed at 20 ms, the number of bits per frame varies according to the traffic rate. Half-rate convolutional encoding is used, which doubles the traffic rate to give rates from 2400 to 19,200 symbols per second. Interleaving is performed over 20 ms. A long code of $2^{42} - 1$ ($= 4.4 \times 10^{12}$) is generated containing the user's ESN embedded in the MS long code mask (with voice privacy, the MS long-code mask does not use the ESN). The scrambled data is multiplexed with power control information that steals bits from the scrambled data. The multiplexed signal remains at 19,200 bps and is changed to 1.2288 Mcps by the Walsh code W_i assigned to the ith user traffic channel (TCH). The signal is spread at 1.2288 Mcps by pilot quadrature pseudorandom binary sequence signals, and the resulting quadrature signals are then weighted. The power level of the traffic channel depends on its data transmission rate.

The paging channels (PCH) provide the mobile stations with system information and instructions, in addition to acknowledging messages following access requests on the mobile stations' access channels. The paging channel data is processed in a manner similar to the traffic channel data. However, there is no variation in the power level on a per-frame basis. The 42-bit mask is used to generate the long code. The paging channel operates at a data rate of 9600 or 4800 bps.

All 64 channels are combined to give single I and Q channels. The signals are applied to quadrature modulators and the resulting signals are summed to form a QPSK signal, which is linearly amplified.

The pilot CDMA signal transmitted by a base station provides a reference for all mobile stations. It is used in the demodulation process. The pilot signal level for all base stations is about 4 to 6 dB higher than the traffic channel with a constant value. The pilot signals are quadrature pseudorandom binary sequence signals with a period of 32,768 chips. Since the chip rate is 1.2288 Mcps, the pilot pseudorandom binary sequence corresponds to a period of 26.66 ms, which is equivalent to 75 pilot channel code repetitions every 2 seconds. The pilot signals from all base stations use the same pseudorandom binary sequence, but each base station is identified by a unique time offset of its pseudorandom binary sequence. These offsets are in increments of 64 chips, providing 511 unique offsets relative to 0 offset code. These large numbers of offsets ensure that unique base station identification can be obtained, even in dense microcellular environments. Tables 6-2 through 6-5 provide modulation parameters for the forward link channels.

A mobile station processes the pilot channel to find the strongest signal components. The processed pilot signal provides an accurate estimation of time delay and the phase and magnitude of the three multipath components. These components are tracked in the presence of fast fading, and coherent reception with combining is used. The chip rate on the pilot channel and on all channels is locked to precise system time, e.g., by using the Global Positioning System (GPS). Once the mobile station identifies the strongest pilot offset by processing the multipath components from the pilot channel correlator, it examines the signal on its sync channel which is

Table 6-2 Sync Channel Modulation Parameters

Parameter	Data Rate 1200 bps	Units
PN chip rate	1.2288	Mcps
Code rate	1/2	bits per code symbol
Code symbol repetition	2	modulation symbols per code symbol
Modulation symbol rate	4800	symbols per second (sps)
PN chips per modulation symbol	256	PN chips per modulation symbol
PN chips per bit	1024	PN chips per bit

Table 6-3 Paging Channel Modulation Parameters

Parameter	Data Rate (bps)		Units
	9600	4800	
PN chip rate	1.2288	1.2288	Mcps
Code rate	1/2	1/2	bits per code symbol
Code symbol repetition	1	2	modulation symbols per code symbol
Modulation symbol rate	19,200	19,200	sps
PN chips per modulation symbol	64	64	PN chips per modulation symbol
PN chips per bit	128	256	PN chips per bit

Table 6-4 Forward Traffic Channel Modulation Parameters for RS1

Parameters	Data Rate (bps)				Units
	9600	4800	2400	1200	
PN chip rate	1.2288	1.2288	1.2288	1.2288	Mcps
Code rate	1/2	1/2	1/2	1/2	bits per code symbol
Code symbol repetition	1	2	4	8	repeated symbols per code symbol
Modulation symbols rate	19,200	19,200	19,200	19,200	sps
PN chips per modulation symbol	64	64	64	64	PN chips per modulation symbol
PN chips per bit	128	256	512	1024	PN chips per bit

Table 6-5   Forward Traffic Channel Modulation Parameters for RS2

Parameters	Data Rate (bps)				Units
	14,400	7200	3600	1800	
PN chip rate	1.2288	1.2288	1.2288	1.2288	Mcps
Code rate	1/2	1/2	1/2	1/2	bits per code symbol
Code symbol repetition	1	2	4	8	repeated symbols per code symbol
Puncturing rate	4/6	4/6	4/6	4/6	modulation symbols per repeated symbol
Effective code rate	3/4	3/4	3/4	3/4	
Modulation symbol rate	19,200	19,200	19,200	19,200	sps
PN chips per modulation symbol	64	64	64	64	PN chips per modulation symbol
PN chips per bit	85.33	170.67	341.33	682.67	PN chips per bit

locked to the pseudorandom binary sequence signal on the pilot channel. Since the sync channel is time aligned with its base station's pilot channel, the mobile station finds the information pertinent to this particular base station on the sync channel. The sync channel message contains time of day and long-code synchronization to ensure that long-code generators at the base station and mobile station are aligned and identical. The mobile station now attempts to access the paging channel and listens for system information. The mobile station enters the idle state when it has completed acquisition and synchronization. It listens to the assigned paging channel and is able to receive and initiate the calls. When informed by the paging channel that voice traffic is available on a particular channel, the mobile station recovers the speech data by applying the inverse of the spreading procedures.

6.2.2   Reverse Link

In this section, we summarize the operation of the reverse link. For a more detailed discussion see chapter 7. The reverse link is separated from the forward link by 45 MHz at cellular frequencies and 80 MHz at PCS frequencies. The reverse link uses the same 32,768-chip code as that used on the forward link. The reverse link channels are either *access channels* or *reverse traffic channels*. A reverse traffic channel is further subdivided into a single *fundamental code channel* and 0 through 7 *supplemental code channels* (see Fig. 6-2). There are 62 traffic channels and up to 32 access channels. The access channel is used by the mobile station to communicate nontraffic information, such as originating calls and responding to paging. The access rate is fixed at 4800 bps. All mobile stations accessing a radio system share the same frequency assignment. Each access channel is identified by a distinct access-channel long-code sequence having an access number, a paging channel number associated with the access channel, and other system data. Each mobile station uses a different PN code; therefore the radio system can correctly decode the information from an individual mobile station. Data transmitted on the reverse channel is grouped into 20-ms frames. All data on the reverse channel is convolutionally encoded, block interleaved, and modulated by modulation symbols transmitted for each 6 code symbols. The modulation symbol is one of the 64 mutually orthogonal waveforms that are generated using Walsh functions.

The reverse traffic channel may use either 9600-, 4800-, 2400-, or 1200-bps data rates for transmission. The duty cycle for transmission varies proportionally with data rate, being 100% at 9600 bps and dropping to 12.5% at 1200 bps. An optional second rate set is also supported in the PCS version of CDMA and new versions of cellular CDMA (see chapter 7 for details). The actual burst transmission rate is fixed at 28,800 code symbols per second. Since 6 code symbols are modulated as one of 64 modulation symbols for transmission, the modulation symbol transmission rate is fixed at 4800 modulation symbols per second. This results in a fixed Walsh chip rate of 307.2 kilo-chips per second (kcps). The rate of spreading PN sequence is fixed at 1.2288 Mcps, so that each Walsh chip is spread by 4 PN chips. Table 6-6 provides the signal rates and

Table 6-6 CDMA Reverse Traffic Channel Modulation Parameters for RS1

Parameter	9600 bps	4800 bps	2400 bps	1200 bps	units
PN chip rate	1.2288	1.2288	1.2288	1.2288	Mcps
Code rate	1/3	1/3	1/3	1/3	bits per code sym
Transmitting duty cycle	100	50	25	12.5	%
Code symbol repetition	1	2	4	8	repeated code symbol per code symbol
Code symbol rate	$3 \times 9600 = 28,800$	28,800	28,800	28,800	sps
Modulation	6	6	6	6	code symbol per mod. symbol
Modulation symbol rate	$28,800/6 = 4800$	4800	4800	4800	sps
Walsh chip rate	$64 \times 4800 = 307.2$	307.2	307.2	307.2	kcps
Mod. symbol duration	$1/4800 = 208.33$	208.33	208.33	208.33	μs
PN chips per code symbol	$12,288/288 = 42.67$	42.67	42.67	42.67	PN chips per code symbol
PN chips per mod. symbol	$1,228,800/4800 = 256$	256	256	256	PN chips per mod. symbol
PN chips per Walsh chip	4	4	4	4	PN chips per Walsh chip

their relationship for the various transmission rates on the reverse traffic channel for RS1. Reverse traffic channel modulation parameters for RS2 are given in Table 6-7.

Following the orthogonal spreading, the reverse traffic channel and access channel are spread in quadrature. Zero-offset I and Q pilot PN sequences are used for spreading. These sequences are periodic with 2^{15} chips (32,768 PN chips in length) and are based on characteristic polynomials $g_I(x)$ and $g_Q(x)$.

$$g_I(x) = x^{15} + x^{13} + x^9 + x^8 + x^7 + x^5 + 1 \tag{6.1}$$

$$g_Q(x) = x^{15} + x^{12} + x^{11} + x^{10} + x^6 + x^5 + x^4 + x^3 + 1 \tag{6.2}$$

The maximum-length linear feedback register sequences I(n) and Q(n), based on these polynomials, have a period $2^{15} - 1$ and are generated by using the following recursions:

$$I(n) = I(n-15) \oplus I(n-10) \oplus I(n-8) \oplus I(n-7) \oplus I(n-6) \oplus I(n-2) \tag{6.3}$$

Table 6-7   CDMA Reverse Traffic Channel Modulation Parameters for RS2

Parameter	14,400 bps	7200 bps	3600 bps	1800 bps	units
PN chip rate	1.2288	1.2288	1.2288	1.2288	Mcps
Code rate	1/2	1/2	1/2	1/2	bits per code symbol
Transmitting duty cycle	100	50	25	12.5	%
Code symbol repetition	1	2	4	8	repeated code symbols per code symbol
Code symbol rate	28,800	28,800	28,800	28,800	sps
Modulation	6	6	6	6	repeated code symbol per modulation symbol
Modulation symbol rate	4800	4800	4800	4800	sps
Walsh chip rate	307.2	307.2	307.2	307.2	kcps
Mod. symbol duration	208.33	208.33	208.33	208.33	μs
PN chips per code symbol	42.67	42.67	42.67	42.67	PN chips per repeated code symbol
PN chips per mod. symbol	256	256	256	256	PN chips per mod. symbol
PN chips per Walsh chip	4	4	4	4	PN chips per Walsh chip

based on $g_I(x)$ as the characteristic polynomial, and

$$Q(n) = Q(n-15) \oplus Q(n-12) \oplus Q(n-11) \oplus \\ Q(n-10) \oplus Q(n-9) \oplus Q(n-5) \oplus \\ Q(n-4) \oplus Q(n-3) \tag{6.4}$$

based on $g_Q(x)$ as the characteristic polynomial, where $I(n)$ and $Q(n)$ are binary numbers (0 and 1) and the additions are modulo-2. To obtain the I and Q pilot PN sequences (of period 2^{15}), a 0 is inserted in $I(n)$ and $Q(n)$ after 14 consecutive 0 outputs (this occurs only once in each period). Therefore, the pilot PN sequences have one run of 15 consecutive 0 outputs instead of 14.

The chip rate for the pilot PN sequence is 1.2288 Mcps and its period is 26.666 ms. There are exactly 75 pilot PN repetitions every 2 seconds. The spreading modulation is Offset Quadrature Phase-Shift Keying (OQPSK). The data spread by Q pilot PN sequence is delayed by half a chip time (406.901 nanoseconds [ns]) with respect to the data spread by I pilot PN sequence (see chapter 7). Fig. 6-3 and Table 6-8 describe the characteristics of OQPSK.

Table 6-8 CDMA Reverse Traffic Channel Modulation Parameters for RS2

I	Q	Phase
0	0	$\pi/4$
1	0	$(3\pi)/4$
1	1	$-(3\pi)/4$
0	1	$-\pi/4$

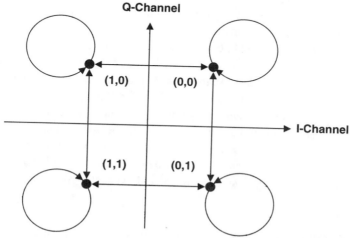

Figure 6-3 Signal Constellation and Phase Transition of OQPSK Used on Reverse CDMA Channel

Table 6-9   CDMA Access Channel Modulation Parameters

Parameter	Data Rate 4800 bps	Units
PN chip rate	1.2288	Mcps
Code rate	1/3	bits per code symbol
Code symbol repetition	2	symbols per code symbol
Transmit duty cycle	100	%
Code symbol rate	28,800	sps
Modulation	6	code symbol per modulation symbol
Modulation symbol rate	4800	sps
Walsh chip rate	307.2	kcps
Modulation symbol duration	208.33	μs
PN chips per code symbol	42.67	PN chips per code symbol
PN chips per modulation symbol	256	PN chips per modulation symbol
PN chips per Walsh chip	4	PN chips per Walsh chip

Each base station transmits a pilot signal of constant power on the same frequency. The received power level of the received pilot signal enables the mobile station to estimate the path loss between the base station and the mobile station. Knowing the path loss, the mobile station adjusts its transmitted power such that the base station will receive the signal at the requisite power level. The base station measures the mobile station's received power and informs the mobile station to make the necessary adjustment to its transmitter power. One command every 1.25 ms adjusts the transmitted power from the mobile station in ±0.5 dB steps. The base station uses frame errors reported by the mobile station to increase or decrease the transmitted power.

CDMA access channel modulation parameters are listed in Table 6-9.

CDMA provides a soft handoff. As the mobile station moves to the edge of its single cell, the adjacent base station assigns a modem to the call; meanwhile, the current base station continues to handle the call. The call is handled by both base stations on a make-before-break basis. Handoff diversity occurs with both base stations handling the call until the mobile station moves sufficiently close to one of the base stations for it to then exclusively handle the call. This handoff procedure is different from conventional break-before-make or hard-handoff procedures. The soft-handoff procedure will be discussed in more detail in chapter 10.

6.3 Summary

A CDMA system operates with a low E_b/N_0 ratio, exploits voice activity, and uses sectorization of cells. Each sector has 64 CDMA channels. It is a synchronized system with three receivers to provide path diversity at the mobile station and four receivers at the cell site.

This chapter provided a high-level description of the IS-95 CDMA air interface. The forward link channels are pilot, sync, paging, and forward traffic (fundamental and supplementary) channels. The reverse link channels include access and reverse traffic (fundamental and supplementary) channels.

6.4 References

1. Garg, V. K., Smolik, K. F., and Wilkes, J. E., *Applications of CDMA in Wireless Communications,* Prentice Hall PTR, Upper Saddle River, NJ, 1997.

2. TIA/EIA IS-95, "Mobile Station–Base Station Compatibility Standard for Dual-Mode Wideband Spread Spectrum Cellular Systems," 1998.

Physical and Logical Channels of IS-95 CDMA

7.1 Introduction

This chapter first covers how to introduce a CDMA carrier in an existing AMPS or TDMA system, establishing the number of AMPS or TDMA channels that should be removed in order to introduce the first and second CDMA carrier of 1.23 MHz without interfering with the remaining AMPS or TDMA carriers. After this, we briefly describe modulation schemes, bit repetition, block interleaving, and channel coding that are used in processing logical channels on the IS-95 CDMA forward and reverse links. Details about information processing, message types, and message framing are then presented for the pilot, sync, paging, and traffic channels on the forward link. This chapter also provides similar details for the access and traffic channels on the reverse link.

7.2 Physical Channels

In IS-95, physical channels are defined in terms of an RF frequency and a code sequence. There are 64 Walsh codes available for the forward link (BS to MS), providing 64 logical channels. On the reverse link, channels are identified by long PN code (see chapter 6) sequences.

A CDMA system is implemented using N wideband frequency carriers, each capable of supporting M circuits that can be accessed by any user. A unique circuit is defined by a different code sequence for each user. Frequency assignment remains under control of the system in both forward link (BS to MS) and reverse link (MS to BS) directions. Fig. 7-1 shows N RF carriers for each direction. This is a Frequency Division Duplex (FDD) arrangement and is called the *CDMA/FDD* system.

In IS-95, CDMA carrier band center frequencies are denoted in terms of AMPS channel numbers. We refer to Fig. 7-2 where the AMPS channel number 283 is the center of a CDMA carrier band. To introduce one CDMA carrier, we need 41 30-kHz AMPS channels to provide a

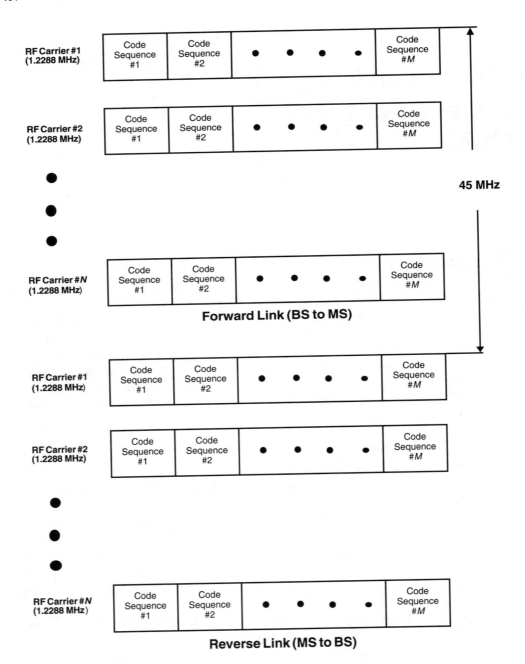

Figure 7-1   CDMA/FDD System

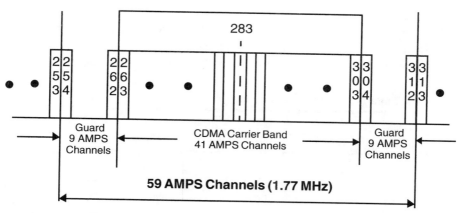

Figure 7-2 Introducing One CDMA Carrier

CDMA carrier bandwidth of 1.23 MHz. The recommended guard band between the CDMA carrier band edge and an AMPS or a TDMA carrier is 0.27 MHz—that is equal to 9 AMPS or TDMA channels. Thus to introduce the first CDMA carrier without interfering with the remaining AMPS or TDMA channels, it is necessary that 59 AMPS channels be removed. In order to introduce the second CDMA carrier, we should remove only 41 additional AMPS channels. We see from Fig. 7-3 that, to introduce two CDMA carriers, we should remove 100 AMPS channels, a total of 3 MHz.

The *primary* and *secondary* CDMA carriers are the preassigned frequencies (AMPS channel numbers) that allow the mobile to acquire the CDMA system. A base station can support primary, secondary, or both types of channels.

The 1.23-MHz bandwidth for a CDMA carrier suggests that the minimum center frequency separation between two carrier frequencies is 1.23 MHz. The MS and BS frequencies

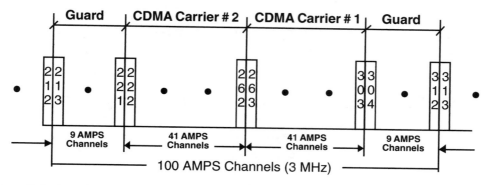

Figure 7-3 Introducing Two CDMA Carriers

for cellular band (Band Class 0) are specified in Table 7-1. The channel spacings, CDMA channel designations, and transmit center frequencies of Band Class 0 are given in Table 7-2. The valid CDMA carrier frequencies are on AMPS channel numbers 1013–1023, 1–311, 356–644, 689–694, and 739–777 (see Table 7-3). Only the primary and secondary CDMA carrier

Table 7-1 Band Class 0 System Frequencies

| System | Transmit Frequency Band (MHz) | |
	Mobile Station	Base Station
A	824.025–835.005	869.025–880.005
	844.995–846.495	889.995–891.495
B	835.005–844.995	880.005–889.995
	846.495–848.985	891.495–893.985

Table 7-2 CDMA Channel Number to CDMA Frequency Assignment for Band Class 0

Transmitter	CDMA Channel Number	CDMA Frequency Assignment (MHz)
Mobile station	$1 \leq N \leq 777$	$0.030 \, N + 825.000$
	$1013 \leq N \leq 1023$	$0.030 \, [N - 1023] + 825.000$
Base station	$1 \leq N \leq 777$	$0.030 \, N + 870.000$
	$1013 \leq N \leq 1023$	$0.030 \, [N - 1023] + 870.000$

Table 7-3 Channel Numbers for Band Class 0

Frequency Block	Valid CDMA Frequency Assignment	CDMA Channel Number
A" (1 MHz)	not valid	991–1012
	valid	1013–1023
A (10 MHz)	valid	1–311
	not valid	312–333
B (10 MHz)	not valid	334–355
	valid	356–644
	not valid	645–666
A' (1.5 MHz)	not valid	667–688
	valid	689–694
	not valid	695–716
B' (2.5 MHz)	not valid	717–738
	valid	739–777
	not valid	778–799

center frequencies are specified in IS-95 standard. Other center frequencies are selected by each system operator.

Table 7-4 shows the CDMA center frequency (AMPS channel number) assignments for systems A and B with 41 AMPS channel separation.

The mobile station and base station frequencies for the PCS band (Band Class 1) are specified in Table 7-5. The channel spacings, CDMA channel designations, and transmit center frequencies of Band Class 1 are given in Table 7-6. Mobile stations in Band Class 1 support operations on channel numbers 25 through 1175 as shown in Table 7-7. Note that certain channel assignments are not valid and that others are conditionally valid. Transmission on conditionally valid channels is allowed if the adjacent block is allocated to the licensee or if other valid authorization has been obtained.

A preferred set of CDMA frequency assignments for Band Class 1 is given in Table 7-8.

Table 7-4   CDMA Center Frequency Assignment for Systems A and B for Cellular Band Class 0

CDMA Channel Type	CDMA Frequency System A	AMPS Channels System B
Primary	283	384
	242	425
	201	466
	160	507
	119	548
	78	589
	37	630
	1019	
Secondary	691	777
Total number of CDMA channels	9	8

Table 7-5   Band Class 1 System Frequencies

Block	Transmit Frequency Band (MHz)	
	Mobile Station	Base Station
A	1850–1865	1930–1945
D	1865–1870	1945–1950
B	1870–1885	1950–1965
E	1885–1890	1965–1970
F	1890–1895	1970–1975
C	1895–1910	1975–1990

Table 7-6   CDMA Channel Number to CDMA Frequency Assignment for Band Class 1

Transmitter	CDMA Channel Number	Center Frequency of CDMA Channel (MHz)
Mobile station	$0 \leq N \leq 1199$	$1850.000 + 0.050\,N$
Base station	$0 \leq N \leq 1199$	$1930.000 + 0.050\,N$

Table 7-7   CDMA Channel Numbers for Band Class 1

Block	Valid CDMA Frequency Assignment	CDMA Channel Number
A (15 MHz)	not valid	0–24
	valid	25–275
	conditionally valid	276–299
D (5 MHz)	conditionally valid	300–324
	valid	325–375
	conditionally valid	376–399
B (15 MHz)	conditionally valid	400–424
	valid	425–675
	conditionally valid	676–699
E (5 MHz)	conditionally valid	700–724
	valid	725–775
	conditionally valid	776–799
F (5 MHz)	conditionally valid	800–824
	valid	825–875
	conditionally valid	876–899
C (15 MHz)	conditionally valid	900–924
	valid	925–1175
	not valid	1176–1199

Table 7-8   CDMA Preferred Set of Frequency Assignments for Band Class 1

Block	Preferred Set of Channel Numbers
A	25, 50, 75, 100, 125, 150, 175, 200, 225, 250, 275
D	325, 350, 375
B	425, 450, 475, 500, 525, 550, 575, 600, 625, 650, 675
E	725, 750, 775
F	825, 850, 875
C	925, 950, 975, 1000, 1025, 1050, 1075, 1100, 1125, 1150, 1175

7.3 Modulation

The signals from each channel (pilot, sync, paging, and traffic) are modulo-2 added to I and Q PN short-code sequences. The I and Q spread signals are baseband filtered and sent to a linear adder with gain control (see Fig. 7-4). The gain control allows individual channels to have different power levels assigned to them. The CDMA system assigns power levels to different channels depending on the quality of the received signal at a mobile station. The algorithms for determining the power levels are proprietary to each equipment manufacturer. The I and Q baseband signals are modulated by I and Q carrier signals, combined together, amplified, and sent to the base station antenna. The net signal from the CDMA modulator is a complex quadrature signal that looks like noise.

The same PN short-code sequences are used on all channels (i.e., pilot, sync, paging, and traffic) of the forward link. All base stations in a system are synchronized using the GPS satellite. Different base stations use time-shifted versions of these PN sequences to allow mobile stations to select the appropriate base station.

Unlike the forward direction, the CDMA system uses a different modulation scheme to generate the signal in the reverse direction. The net signal from modulator is a 4-phase quadrature signal. The output from either the access channel or the traffic channel is sent to two modulo-2 adders—one for the in-phase channel and the other for the quadrature channel. Two different short-code PN sequences are modulo-2 added to the data and filtered by a baseband filter. For a quadrature channel, a delay of 1/2 PN symbol (406.9 ns) is added before the filter. Thus, the reverse channel uses OQPSK (see Fig. 7-5). No pilot signal is used on the reverse link.

7.4 Bit Repetition

The nominal RS1 data rate on the forward and reverse traffic channels is 9600 bps. If data is transmitted at a lower rate (4800, 2400, or 1200 bps), then the data bits are repeated n times to increase the rate to 9600 bps. Likewise, the nominal RS2 data rate on the forward and reverse

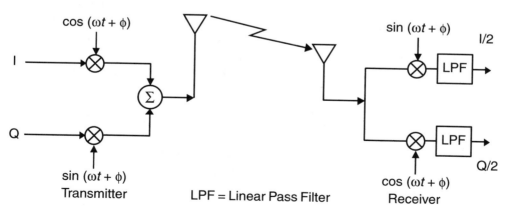

Figure 7-4 QPSK Modulator

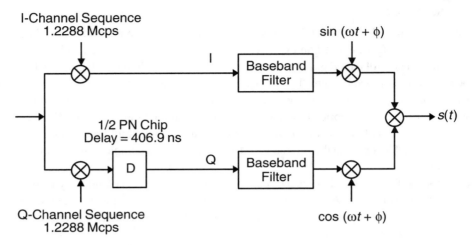

Figure 7-5 OQPSK Modulation

traffic channels is 14,400 bps. If the data is transmitted at the lower rates (7200, 3600, or 1800 bps), then the data bits are repeated either 2, 4, or 8 times to increase the rate to 14,400 bps.

7.5 Block Interleaving

Communications over a radio channel are characterized by deep fades that can cause large numbers of consecutive errors. Most coding schemes perform better on random data errors than on blocks of errors. By interleaving the data, no two adjacent bits are transmitted near to each other, and the data errors are randomized.

The interleaver spans a 20-ms frame. In the reverse direction, the output of the interleaver is 28.8 kbps if the data rate is 9.6 kbps (RS1). The resultant signal transmits with 100% duty cycle. If the data rate is lower (4800, 2400, or 1200 bps), the interleaver plus randomizer deletes redundant bits and transmits with a lower duty cycle (50%, 25%, 12.5%). Thus, bits are not repeated on the reverse CDMA traffic channel. On the access channel, the data bits are repeated. In the forward direction, the nominal data rate for RS1 is 19.2 kbps; lower data rates use a lower duty cycle.

On the reverse traffic channel, the output of the interleaver is processed by a data randomizer. The randomizer removes redundant data blocks generated by the code repetition. It uses a masking pattern determined by the data rate and the last 14 bits of the long code. For a 20-ms block (192 bits at 9600 bps), the data randomizer segments the block into 16 blocks of 1.25 ms each. At a data rate of 9600 bps (RS1), all blocks are filled with data. At a data rate of 4800 bps, 8 out of 16 blocks are filled with data in a random manner. Similarly, for 2400 and 1200 bps, 4 of 16 and 2 of 16 blocks, respectively, are randomly filled with data. Thus, no redundant data are transmitted over the reverse channel.

7.6 Channel Coding

In IS-95, traffic data frames on uplink and downlink are fed to convolutional encoders. Both uplink and downlink encoders use an eight-stage shift register with constraint length of 9. The rate of the uplink coder is 1/3—it outputs 3 bits for every input bit. At rates below 9.6 kbps, output bits are repeated to bring the number of bits in a 20-ms block to 576, for a gross rate of 28.8 kbps (RS1). The rate of the downlink encoder is 1/2—it outputs 2 bits for every input bit. At rates below 9.6 kbps (RS1), output bits are repeated to bring the number of bits in a 20-ms block to 384, for a gross rate of 19.2 kbps.

7.7 Logical Channels

Logical channels in CDMA are the *control* and *traffic* channels. The control channels are the *pilot channel* (downlink), the *paging channels* (downlink), the *sync channels* (downlink), and the *access channels* (uplink) (see Fig. 7-6)

The traffic channels are used to carry user information (speech or data), along with signaling traffic, between the base station and the mobile station. Four different rates are used—1, 1/2,

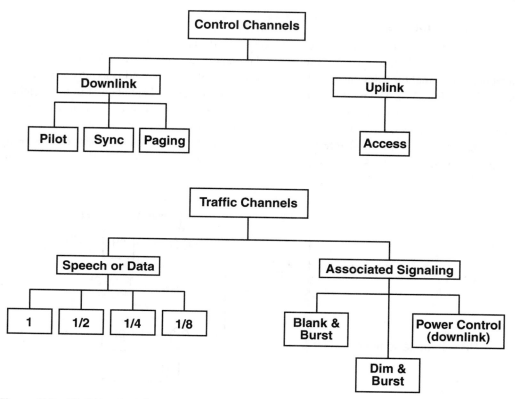

Figure 7-6 IS-95 Logical Channels

1/4, and 1/8. The downlink traffic channel is called the *forward link*, whereas the uplink traffic channel is referred to as the *reverse link*.

When user speech or data is replaced by associated signaling, it is called *blank-and-burst signaling*. When part of the speech is replaced by signaling information, it is called *dim-and-burst signaling*. All associated signaling is sent in Rate 1. In addition, on the downlink there is a power control subchannel that allows the mobile to adjust its transmitted power by ±1 dB every 1.25 ms (i.e., 16 times during a 20-ms speech frame).

In the forward direction there are the pilot channel, the sync channel, up to seven paging channels, and a number of forward traffic channels that all share the same center frequency. Out of the 64 Walsh-coded channels available for use (W_0, W_1, W_2, ..., W_{63}), the pilot channel on W_0 is always required. There can be one sync channel (W_{32}) and seven paging channels (W_1 to W_7— the maximum allowed); the remaining channels are traffic channels. A traffic channel is further subdivided into a *fundamental code channel* and a *supplementary code channel* (see Figs. 6-1 and 6-2). The primary paging channel is always assigned the Walsh code W_1. The mobile examines the number of paging channel parameters in the *system parameter message*. If this value is not 1, a *hashing algorithm* is invoked to determine the correct paging channel number.

7.7.1 Pilot Channel

The pilot channel is used by a base station to provide a reference for all mobile stations. It provides a phase reference for coherent demodulation at the mobile receiver to enable coherent detection. Note that the pilot channel does not carry any information and is assigned the Walsh code W_0 (see Fig. 7-7). The pilot signal level for all base stations is kept about 4 to 6 dB higher than a traffic channel with a constant signal power. The pilot signal is used for comparisons of signal strength between different base stations to decide when to perform handoff.

The pilot channel is needed to lock onto other logical channels on the same RF carrier. It carries an unmodulated DSSS signal that is transmitted continuously by each base station. The

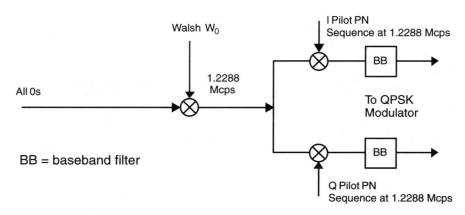

Figure 7-7 Pilot Channel Processing

pilot signals are quadrature pseudorandom (PN) binary sequence signals with a period of 32,768 (2^{15}) chips. Since the chip rate is 1.2288 Mcps, the pilot PN sequence corresponds to a period of 26.667 ms. This is equivalent to 75 pilot channel code repetitions every 2 seconds. The pilot signals from all base stations use the same PN sequences, but each base station is identified by a unique time offset. These offsets are in increments of 64 chips to provide 512 unique offsets (see Fig. 7-8). The high number of offsets ensures that unique base station identification can be obtained even in dense microcellular environments.

7.7.2 Sync Channel

The sync channel is an encoded, interleaved, and modulated spread spectrum signal that may be used with the pilot channel to acquire initial time synchronization. It is assigned the Walsh code W_{32}. The sync channel always operates at a fixed rate of 1200 bps and is convolutionally encoded to 2400 bps, repeated to 4800, and interleaved. (see Fig. 7-9). The sync channel is used with the pilot channel to acquire initial time synchronization. Only the Sync Channel message is transmitted over this channel.

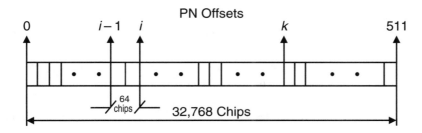

Figure 7-8   Short PN Offsets

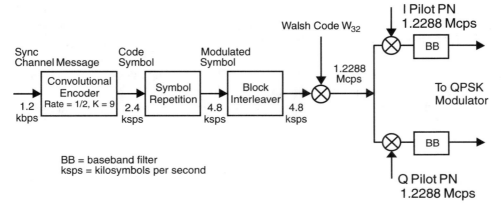

Figure 7-9   Sync Channel Processing

The Sync Channel message parameters are

- **System Identification (SID):** identifier number for the system
- **Network Identification (NID):** subidentifier for the system
- **Pilot short PN sequence offset (PILOT_PN) index:** offset index, in units of 64 chips, for the base station or sector
- **Long-code state (LC_STATE):** long code at the time specified in system time parameter
- **System time (SYS_TIME)**
- **Leap seconds (LP_SEC):** number of leap seconds that have occurred since the start of system
- **Offset of local time:** offset from the system time
- **Daylight saving time indicator**
- **Paging channel data rate (PRAT):** 4.8 or 9.6 kbps

The Sync Channel message itself is long and may occupy more than one *sync channel frame*.The Sync Channel message is organized in a *Sync Channel message capsule*. A Sync Channel message capsule contains the Sync Channel message and padding. When the Sync Channel message occupies more than one sync channel frame, padding is used to fill the bit positions up to the beginning of the next *sync channel superframe*, where the next Sync Channel message starts.

The Sync Channel message has an 8-bit message-length header, a message body of a minimum of 2 bits and a maximum of 1146 bits, and a cyclic redundancy check (CRC) code of 30 bits (see Fig. 7-10). The message length includes the header, body, and CRC, but not the padding. The CRC is computed on the message-length header and the message body using the following polynomial:

$$g(x) = x^{30} + x^{29} + x^{21} + x^{20} + x^{15} + x^{13} + x^{12} + x^{11} + x^8 + x^7 + x^6 + x^2 + x + 1 \qquad (7.1)$$

After a message is formed, it is segmented into 31-bit groups and sent in a sync frame (see Fig. 7-11) consisting of a 1-bit start of message (SOM) field and 31 of the sync channel frame bodies. The value of 1 for SOM indicates that the frame is the start of the Sync Channel message, whereas a value of 0 for SOM indicates that the frame is a continuation of a Sync Channel message or padding. The sync channel frames are transmitted in groups of *sync channel superframes*. Three sync frames are combined to form a superframe. Each superframe carries 96 bits and lasts for 80 ms (see Fig. 7-12). The entire Sync Channel message is then sent in N super-

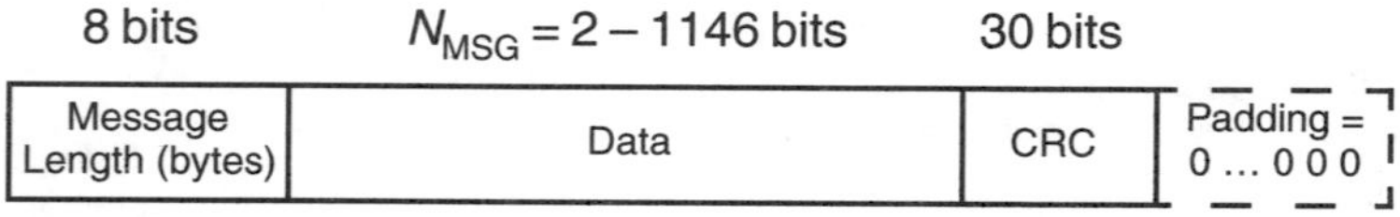

8 bits	$N_{MSG} = 2 - 1146$ bits	30 bits	
Message Length (bytes)	Data	CRC	Padding = 0 ... 0 0 0

Figure 7-10 Message Framing on Sync Channel and Paging Channel

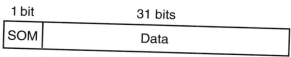

Note: SOM = 1 for first body of Sync Channel message
SOM = 0 for all other bodies in Sync Channel message

Figure 7-11   Sync Channel Frame

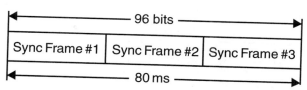

Figure 7-12   Sync Channel Superframe

frames. The padding bits are used so that the Start message always starts at 1 bit after the beginning bit of the superframe.

7.7.3 Paging Channel

The paging channel is used to transmit control information to the mobile station. When a mobile station is to receive a call, it will receive a page from the base station on an assigned paging channel. The paging channel provides the mobile station with system information and instructions, in addition to acknowledging messages following access requests on the mobile station's access channels. Paging channel data are processed in a manner similar to the processing of traffic channel data. However, there is no power control on a per-frame basis. The 42-bit mask is used to generate the long code (see Fig. 7-13). The paging channel operates at a data rate of 4800 or 9600 bps.

The Paging Channel message is similar in form to the Sync Channel message. It has an 8-bit message-length header, a message body of a minimum of 2 bits and a maximum of 1146 bits, and a CRC code of 30 bits. The message length includes the header, body, and CRC, but not the padding. The CRC is computed on the message-length header and the message body using the same code as the sync channel.

A Paging Channel message can use synchronized capsules that end on a half-frame boundary or unsynchronized capsules that can end anywhere within a half-frame. If synchronized Paging Channel messages are less than an integer multiple of 47 bits for 4800-bps transmission (or 95 bits for 9600-bps transmission), they are padded with 0 bits at the end of the message. Unsynchronized messages are not padded.

After a message is formed, it is segmented into 47- or 95-bit chunks and sent in a paging channel half-frame (see Fig. 7-14) that consists of a 1-bit synchronized capsule indicator (SCI) field and 47 or 95 bits of the paging channel frame body. A value of 1 for SCI indicates that the

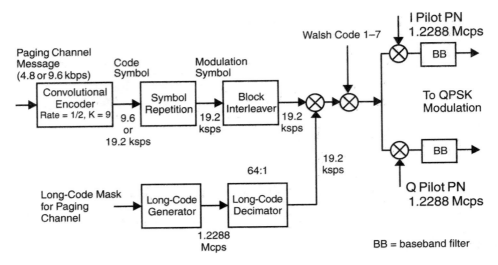

Figure 7-13   Paging Channel Processing

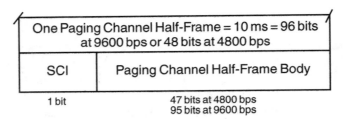

Note: SCI = 1 for first new capsule of synchronized Paging Channel message
       SCI = 0 for all other capsules in Paging Channel message

Figure 7-14   Paging Channel Half-Frame

frame is the start of a Paging Channel message (either synchronized or unsynchronized). Messages can also start in the middle of a frame and immediately after the end of an unsynchronized message (with no padding bits). A value of 0 for SCI indicates that the frame is not the start of a message and that it can include a message (with or without padding), padding only, or the end of one message and start of another.

Eight paging channel half-frames are combined to form a *paging channel slot* (see Fig. 7-15) of length 80 ms (384 bits at 4800 bps and 768 bits at 9600 bps). The entire Paging Channel message is then sent in *N* slots. The maximum number of slots that a message can use is 2048. The base station always starts a slot with an asynchronized message capsule that starts at 1 bit after the beginning of a slot. The first bit in a slot is SCI = 1.

The paging channel carries information to allow the network to

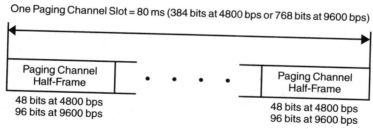

One Paging Channel Slot = 80 ms (384 bits at 4800 bps or 768 bits at 9600 bps)

Figure 7-15 Paging Channel Slot Structure

- Supply display information to be displayed by the mobile station
- Identify the called party's number
- Identify the calling party's number
- Convey information to the mobile station by means of tones or other alerting signals
- Indicate the number of messages waiting

The messages carried by the paging channel include

1. **System Parameter message.** Provides overhead information such as the pilot PN sequence offset index, i, in PN-I-i (t) and PN-Q-i (t), base station identifier, the number of paging channels, PAGE_CHAN, and other system information. PAGE_CHAN and a hashing algorithm (IS-95) are used by the mobile station to determine the correct paging channel number. Initially, the mobile station expects a PAGE_CHAN equal to 1— the primary paging channel. A different value invokes the hashing algorithm.

2. **Access Parameters message.** Defines parameters required by a mobile station to transmit on an access channel.

3. **Neighbor List message.** Provides information about neighbor base station parameters, e.g., the neighbor pilot PN sequence offset index, i. If the neighbor does not have a paging channel on the current CDMA carrier frequency, the *CDMA Channel List message* contains this information.

4. **CDMA Channel List message.** Provides the list of CDMA carriers.

5. **Slotted Page or Page message.** Provides data used to inform the mobile that it can receive a call. The mobile station monitors for its identification number in the mobile identification number (MIN) field. With the slotted page message, the mobile needs to monitor only specific time slots of the Paging Channel message.

6. **Page message.** Provides a page to the mobile station. The mobile monitors every time slot of the Paging Channel message.

7. **Typical Order message.** Several order messages can be carried on the paging channel, such as abbreviated alert, base station challenge confirmation, reorder, audit, intercept, base station acknowledgment, lock until power cycled, maintenance required, unlock, release (with or without reason), registration accepted, registration request, registration rejected, and local control.

8. **Channel Assignment message.** Message to inform the mobile station to tune to a new frequency.

9. **Data Burst message.** Data message sent by base station to the mobile station.

10. **Authentication Challenge message.** Allows the base station to validate the mobile identity. The unique mobile authentication keys and/or Shared Secret Data (SSD) for each mobile registered in the system will be used to perform the authentication calculations. These are then sent back to the base station in an Authentication Challenge Response message.

11. **SSD Update message.** Request by the base station for the mobile station to update the SSD.

12. **Feature Notification message.** Contains information records to allow the network to supply information to be displayed by the mobile, to identify the called party's number, to identify the calling party's number, to convey information to the mobile by means of tones or other alerting signals, and to indicate the number of messages waiting.

The paging channel is divided into 80-ms slots called the *paging channel slots.* IS-95 allows two modes of paging—*slotted* and *nonslotted.* In the slotted mode, a mobile listens for pages only at certain times (i.e., during its page slot). This feature allows the mobile to turn off its receiver for most of the time, saving the battery power and increasing time between battery charging. In the nonslotted mode of operation, the mobile is required to monitor all paging slots.

In the slotted mode, a mobile generally monitors the paging channel for one or two slots per slot cycle. The mobile can specify its preferred slot using the SLOT_CYCLE_INDEX field in the Registration message, in the Origination message, in the Page Response message, in traffic signaling to the base station. The length of the slot cycle, T, in units of 1.28 seconds is given by $T = 2^i$, where i is the selected slot cycle index.

There are $16T$ slots in a cycle for a particular mobile using some value of i, and four 20-ms full frames in an 80-ms slot. A value of $i = 0$ means that the mobile listens to every 16th paging slot; $i = 1$ implies that the mobile monitors every 32nd slot; for $i = 2$, the mobile monitors every 64th slot. The $i = 0$ value ensures that the pages are not missed by the mobile, but it is a drain on mobile's battery power. The value of $i = 1$ is suggested. *PGSLOT* is a randomly calculated number that specifies the slot out of the $16T$ slots to be monitored by the mobile. This number is fixed for each mobile.

7.7.4 Access Channel

The access channel is used by the mobile station to transmit control information to the base station. The access channel allows the mobile station to communicate nontraffic information (e.g., call origination and response to page). The access rate is fixed at 4800 bps. All mobile stations accessing a system share the same frequency assignment. Each access channel is identified by a distinct access channel long-code sequence having an access number, a paging channel number associated with the access channel, and other system data. There are many messages that can be carried on the access channel. When a mobile places a call it uses the access channel to inform the base station. This channel is also used to respond to a page.

The messages carried by access channel include

1. **Registration message.** The mobile station sends this message to inform the base station about its location, status, identification, and other parameters required to register with the system. This is necessary so that the base station can page the mobile whenever a call is to be delivered to the mobile.

2. **Order message.** Typical order messages include base station challenge, SSD update confirmation, SSD update registration, mobile station acknowledgment, local control response, mobile station reject (with or without reason).

3. **Data Burst message.** This is a user-generated data message sent by the mobile station to base station.

4. **Origination message.** This message allows the mobile station to place a call—sending dialed digits.

5. **Page Response message.** This message is used by the mobile to respond to a page or slotted page in continuation of the process of receiving a call.

6. **Authentication Challenge Response message.** This message contains necessary information to validate the mobile station's identity.

Fig. 7-16 shows the processing of the access channel. The baseband information is error protected using a convolutional encoder of rate $r = 1/3$. The lower encoding rate on the reverse link is used to make error protection more robust. The symbol repetition repeats the symbol, yielding a code symbol rate of 28.8 kilosymbols per second (ksps). The data is then interleaved to combat fading, after which 64-ary orthogonal modulation is used. In this case, for each exited

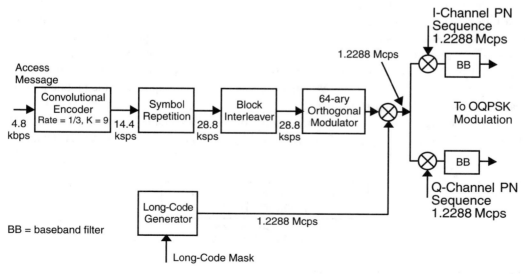

Figure 7-16 Access Channel Processing

6 input symbols to the modulator, one output Walsh function is generated. The reason for using orthogonal modulation of the symbol is the noncoherent nature of the reverse link. Since each group of 6 symbols is represented by a unique Walsh function, it is much easier for the base station to detect 6 symbols at a time by deciding which 64-bit Walsh function was transmitted during that period. An exited of 6 coded symbols is used as 6 binary symbols corresponding to a decimal value between 1 and 64. The pattern of the 6-symbol group determines the choice of the particular Walsh function transmitted.

The Walsh function is defined by

$$W_i = c_0 + 2c_1 + 4c_2 + 8c_3 + 16c_4 + 32c_5 \tag{7.2}$$

where c_5 is the most recent and c_0 the oldest of the 6 symbols to be transmitted, and W_i is selected from 1 of 64 orthogonal Walsh functions.

As an example suppose an exited of 6 symbols $(-1, 1, 1, -1, -1, -1)$ is an input. The corresponding bit values are $(1\ 0\ 0\ 1\ 1\ 1)$. The output of the modulator in terms of the Walsh function will be

$$W_{39} = 1 + 2 \times 1 + 4 \times 1 + 8 \times 0 + 16 \times 0 + 32 \times 1 = 39$$

The 64-ary modulated data at 4.8 ksps (modulation symbols) or at 307.2 ksps (code symbols) is spread using the long PN sequence at 1.2288 Mcps. The long code has a length of $2^{42} - 1$ bits and is generated by the following polynomial:

$$L(x) = \begin{aligned} &x^{42} + x^{35} + x^{31} + x^{27} + x^{26} + x^{25} + x^{22} + x^{21} + x^{19} + x^{18} + \\ &x^{17} + x^{16} + x^{10} + x^7 + x^6 + x^5 + x^3 + x^2 + x + 1 \end{aligned} \tag{7.3}$$

The output of the long-code generator is modulo-2 added with a shared 42-bit long-code mask to generate a long code. The long PN sequence is used to distinguish the access channel from all other channels occupying the reverse link. The data is scrambled in the I and Q paths by the short PN sequence. Since the reverse link uses OQPSK modulation, the data in the Q path is delayed by 1/2 a PN chip. The primary reason for this delay is to prevent the collapse of the QPSK signal envelope to 0. This is essential because the power amplifier of the mobile is typically small and limited in performance.

The message on the access channel consists of an access preamble of multiple frames of 96 0 bits with a length of 1_PAM_SZ frames (Fig. 7-17), followed by an Access Channel

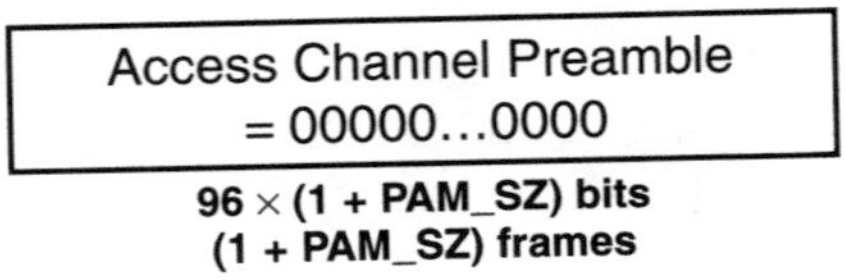

```
┌─────────────────────────────────┐
│   Access Channel Preamble        │
│     = 00000…0000                 │
└─────────────────────────────────┘
     96 × (1 + PAM_SZ) bits
     (1 + PAM_SZ) frames
```

Figure 7-17 Access Channel Preamble

message capsule of length 3 + MAX_CAP_SZ frames. The message capsule also consists of frames 96 bits in length. Since the data rate on the access channel is 4800 bps, each frame has a duration of 20 ms.

The entire access channel transmission, therefore, occurs in an access channel slot that has a length of

$$4 + MAX\_CAP\_SZ + PAM\_SZ \text{ frames} \tag{7.4}$$

where the values of MAX_CAP_SZ and PAM_SZ are received on the paging channel.

An access channel slot nominally begins at a frame where

$$t \bmod (4 + MAX\_CP\_SZ + PAM\_SZ) = 0 \tag{7.5}$$

where t is the system time in frames.

The actual start of transmission on the access channel is randomized to minimize collisions between multiple mobiles accessing the channel at the same time.

All access channels corresponding to a paging channel have the same slot length. Different base stations may have different slot lengths.

The Access Channel message (Fig. 7-18) is similar in form to the Sync Channel message and has an 8-bit message-length header, a message body of a minimum of 2 bits and a maximum of 842 bits, and a CRC code of 30 bits. Following the message are padding bits to make the message end on a frame boundary. The message length includes the header, body, and CRC, but not the padding bits. The CRC is computed on the message-length header and message body using the same code as the sync channel (Eq. 7.1).

Each access channel frame contains either preamble bits (all 0s) or message bits. Frames containing message bits (Fig. 7-19) have 88 message bits and 8 encoder tail bits (set to all 0s). Multiple frames are combined with an access channel preamble to form an *access channel slot* (Fig. 7-20).

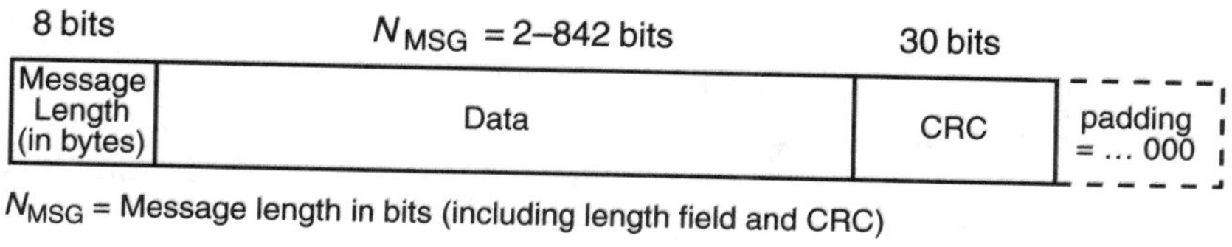

8 bits	N_{MSG} = 2–842 bits	30 bits	
Message Length (in bytes)	Data	CRC	padding = ... 000

N_{MSG} = Message length in bits (including length field and CRC)

Figure 7-18 Message Framing on Access Channel

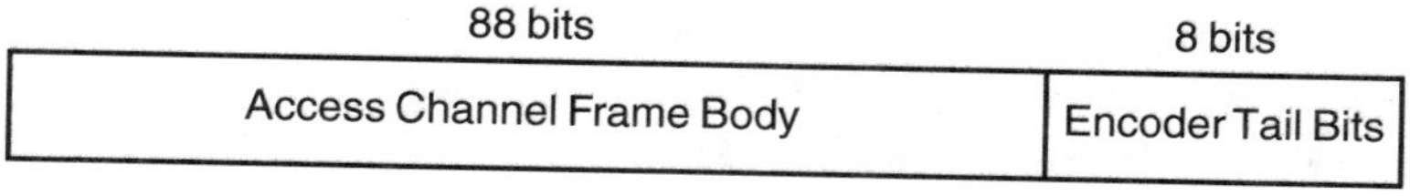

88 bits	8 bits
Access Channel Frame Body	Encoder Tail Bits

Figure 7-19 Access Channel Framing

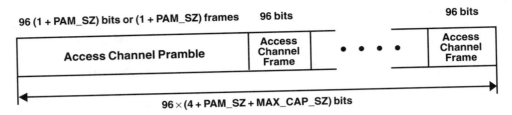

Figure 7-20   Access Channel Slot

7.7.5   Forward Traffic Channels

The forward traffic channels are grouped into rate sets. RS1 has four elements: 9600, 4800, 2400, and 1200 bps. RS2 uses four elements: 14,400, 7200, 3600, and 1800 bps. All systems support RS1 on the forward traffic channels. RS2 is optionally supported on the forward traffic channels. When a system supports a rate set, it supports all four elements. Walsh codes that can be assigned to forward traffic channels are available at a cell or sector (W_2 through W_{31} and W_{33} through W_{63}). After the first paging channel, each additional paging channel consumes one traffic channel Walsh code. So, if all seven paging channels are used, only 55 Walsh codes will be available for the forward traffic channels.

Each forward traffic channel contains one *fundamental code channel* and may contain one to seven *forward supplemental code channels*. Each of these code channels is orthogonally spread by the appropriate Walsh function and is then spread by a quadrature pair of PN sequences at a fixed chip rate of 1.2288 Mcps. Multiple forward CDMA channels may be used within a base station in a frequency-division-multiplexed manner.

The speech is encoded using a variable-rate vocoder to generate the forward traffic data depending on voice activity. Since frame duration is fixed at 20 ms, the number of bits per frame varies according to the traffic rate. Since half-rate convolutional encoding is used, it doubles the traffic rate to provide rates from 2400 to 19,200 symbols per second. Interleaving is performed over 20 ms. A long code of $2^{42} - 1 = 4.4 \times 10^{12}$ is generated containing the user's ESN embedded in the mobile station long-code mask (with voice privacy, the mobile station long-code mask does not use the ESN). The scrambled data are multiplexed with power control information that steals bits from the scrambled data. The multiplexed signal remains at 19,200 bps and is changed to 1.2288 Mcps by the Walsh code W_i assigned to the ith user traffic channel. The signal is spread at 1.2288 Mcps by quadrature PN binary sequence signals, and the resulting quadrature signals are then weighted. The power level of the traffic channel depends on its data transmission rate.

A *power control subchannel* is continuously transmitted on the forward traffic channel. A 0 specifies that the mobile increases its mean output power level by 1 dB (nominal), and a 1 indicates a decrease in mean output power level by 1 dB (nominal).

The power control bits puncture the modulated data symbols at a rate of 800 bps; a single power control bit replaces 2 data symbols. The location of a power control bit in a power control group is determined by bits 20–23 of the 1/64 long code in the previous 1.25-ms control group.

The 19.2-ksps long code is decimated to 800 bps (see Fig. 7-21) to establish the location of the bits in the power control subchannel. This is done to randomize the location of the power control bits to avoid any spikes due to periodic repetition. Fig. 7-22 shows forward traffic channel processing for RS2.

Each power control group contains 24 scrambled bits. The 24-bits position in a 1.25-ms period is numbered 0, 1, 2, ..., 23 (see Fig. 7-23). The power control bits are always transmitted at full power.

Channels not used for paging or sync can be used for traffic. Thus, the total number of traffic channels at a base station is 63 minus the number of paging plus sync channels in operation at that base station. Information on the forward traffic channels includes the primary traffic (voice or data), secondary traffic (data), and signaling in frames 20 ms in length.

When the data rate on the forward traffic channel is 9600 bps (RS1), each frame of 192 bits carries 172 information bits, 12 frame-quality bits, and 8 encoder tail bits (set of all 0s). At 4800 bps, there are 80 information bits, 8 frame-quality bits, and 8 tail bits for a total of 96 bits. At 2400 and 1200 bps, there are 40 and 16 information bits and 8 tail bits, for a total of 48 and 24 bits, respectively. The base station can select the data transmission rate on a frame-by-frame basis. A data rate of 9600 bps can support multiplexed traffic and signaling. Data rates of 1200, 2400, and 4800 bps can support only primary traffic information.

The frame-quality indicator is a CRC on the information bits in the frame. At 9600 bps the generator polynomial is

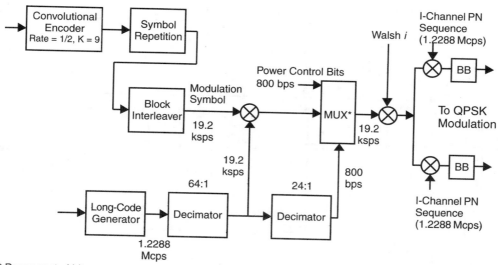

* Power control bits are not multiplexed in for supplemental code channel of the forward traffic channels.
BB = baseband filter
MUX = multiplexer

Figure 7-21 Forward Traffic Channel Processing for RS1

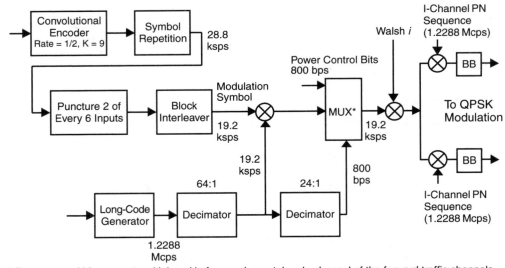

* Power control bits are not multiplexed in for supplemental code channel of the forward traffic channels.
BB = baseband filter
MUX = multiplexer

Figure 7-22   Forward Traffic Channel Processing for RS2

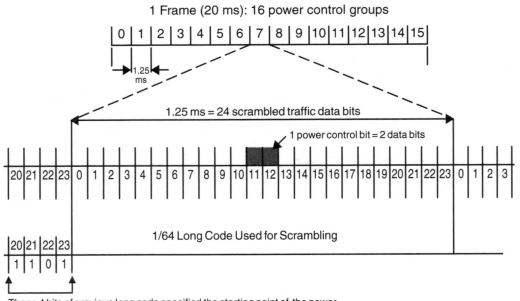

These 4 bits of previous long code specified the starting point of the power control bits value: 1 1 0 1 = 11; the power control bit starts at position 11.

Figure 7-23   Position of the Power Control Bits

$$g(x) = x^{12} + x^{11} + x^{10} + x^9 + x^8 + x^4 + x + 1 \tag{7.6}$$

At 4800 bps, the generator polynomial is

$$g(x) = x^8 + x^7 + x^4 + x^3 + x + 1 \tag{7.7}$$

The generator polynomials for the 10-bit and 6-bit frame-quality indicators are as follows:

- 10-bit frame-quality indicator

$$g(x) = x^8 + x^7 + x^4 + x^3 + x + 1 \tag{7.8}$$

- 6-bit frame-quality indicator

$$g(x) = x^6 + x^2 + x + 1 \tag{7.9}$$

When the data rate on the forward traffic channel is 14,400 bps (RS2), each frame of 288 bits carries 267 information bits, 12 frame-quality bits, and 8 encoder tail bits. Table 7-9 provides a summary of the forward traffic channel frame structure for RS1 and RS2. Figs. 7-24 and 7.25 show the forward traffic channel frame structure for RS1 and RS2, respectively.

At 9600 bps, the 172 information bits consist of 1 or 4 format bits and 171 or 168 traffic bits. A variety of different multiplexing options are supported. The entire 171 information bits can be used for primary traffic, or the 168 bits can be used for 80 primary traffic bits and 88 signaling traffic bits or 88 secondary traffic bits. Other options use 40 and 128 or 16 and 152 bits for primary and signaling/secondary traffic. Alternatively, the entire 168 bits can be used for signaling or secondary traffic.

Table 7-9 Forward Traffic Channel Frame Structure Summary

Rate Set	Transmission Rate (bps)	Number of Bits per Frame				
		Total	Reserved/ Flag	Information	Frame-Quality Indicator	Encoder Tail Bits
1	9600	192	0	172	12	8
	4800[a]	96	0	80	8	8
	2400[a]	48	0	40	0	8
	1200[a]	24	0	16	0	8
2	14,400	288	1	267	12	8
	7200[a]	144	1	125	10	8
	3600[a]	72	1	55	8	8
	1800[a]	36	1	21	6	8

[a] Applicable to forward fundamental code channel only; not allowed on forward supplemental code channels.

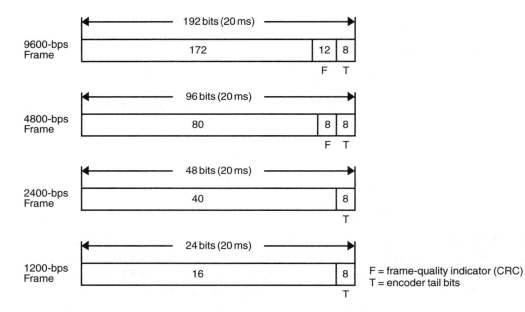

Figure 7-24 Forward Traffic Channel Frame Structure for RS1

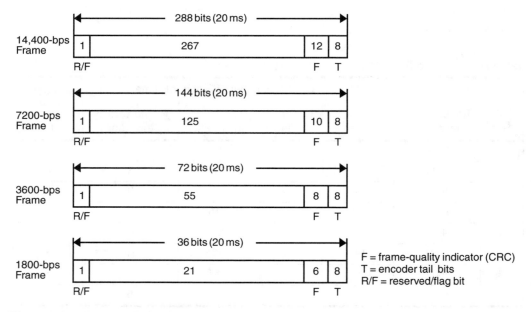

Figure 7-25 Forward Traffic Channel Frame Structure for RS2

When the forward traffic channel is used for signaling, the message is similar in form to the paging channel (see Fig. 7-10) and has an 8-bit message-length header, a message body of a minimum of 16 bits and a maximum of 1160 bits, and a CRC code of 16 bits. Following the message are padding bits to make the message end on a frame boundary. The message length includes the header, body, and CRC, but not the padding. The CRC is computed on the message-length header and the message body using

$$g(x) = x^{16} + x^{12} + x^5 + 1 \tag{7.10}$$

When the forward traffic channel is used for signaling, the following are typical messages that can be sent:

1. **Order message.** Similar to the Order message on the paging channel.
2. **Authentication Challenge message.** When the base station suspects the validity of the mobile, it can challenge the mobile to prove its identity.
3. **Alert with Information message.** Allows the base station to validate the mobile identity.
4. **Data Burst message.** A data message sent by the base station to the mobile.
5. **Handoff Direction message.** Provides the mobile with information needed to begin the handoff process.
6. **Analog Handoff Direction message.** Tells the mobile to switch to the analog mode and begin the handoff process.
7. **In-Traffic System Parameters message.** Updates some of the parameters set by the System Parameters message in the paging channel.
8. **Neighbor List Update message.** Updates the neighbor base station parameters set by the Neighbor List message on the paging channel.
9. **Power Control message.** Tells the mobile how long the period is or what threshold is to be used in measuring frame-error statistics that will be sent in the mobile's power measurement report message.
10. **Send Burst Dual-Tone Multifrequency (DTMF) message.** When the base station needs dialed digits, it can request them in this message. For example, this message would be used for digits for a three-way call.
11. **Retrieve Parameters message.** Requests the mobile to report on any of the retrievable and settleable parameters (refer to IS-95 appendix E).
12. **Set Parameter message.** Informs the mobile to adjust any of the retrievable and settleable parameters (refer to IS-95 appendix E).
13. **SSD Update message.** A request from the base station for the mobile to update the shared secret data.
14. **Flash with Information message.** Contains information records to allow the network to supply display information to be displayed by the mobile, to identify the responding party's number (the connected number), to convey information to the mobile by means of tones or other alerting signals, and to indicate the number of messages waiting.

15. **Mobile Registration message.** Informs the mobile that it is registered and supplies the necessary system parameters.

16. **Extended Handoff Direction message.** One of several handoff messages sent by the base station.

7.7.6 Reverse Traffic Channels

For the RS1, the reverse traffic channel may use either 9600-, 4800-, 2400-, or 1200-bps data rates for transmission. The duty cycle for transmission varies proportionally with the data rate, from 100% at 9600 to 12.5% at 1200 bps. An optional RS2 is also supported. The actual burst transmission rate is fixed at 28,800 code symbols per second. Reverse traffic channel processing is similar to that of the access channel. The major difference is that the reverse traffic channel (fundamental code) uses a data burst randomizer (see Fig. 7-26). The reverse CDMA channel structure for the supplementary code channel is shown in Fig. 7-27.

Since hextets (6 code symbols) are modulated as one of 64-ary modulation symbols for transmission, the modulation symbol transmission rate is fixed at 4800 modulation symbols per second. This results in a fixed Walsh chip rate of 307.2 kcps. The data from the 64-ary modulator is fed into the data burst randomizer. The data burst randomizer takes advantage of the voice activity on the reverse link. This is used to reduce reverse link power during a quiet period of speech by pseudorandom masking out redundant symbols produced by symbol repetition. This is achieved by the data burst randomizer. The data burst randomizer generates a masking pattern of 0s and 1s to randomly mask out redundant data. The masking pattern depends on the vocoder

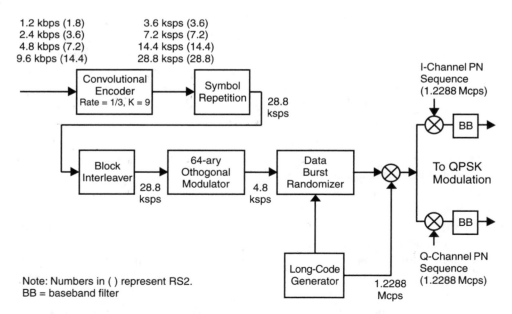

Figure 7-26 Reverse Traffic Channel Processing for Fundamental Code Channel

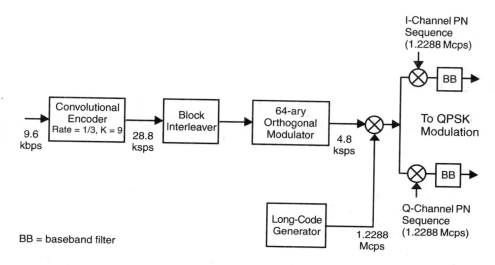

Figure 7-27 Reverse CDMA Channel Structure for Supplementary Code Channel

rate. For a vocoder operating at 9.6 kbps, no data is masked out, whereas, if the vocoder is operating at 1.2 kbps, then the symbols are repeated seven times and the data burst randomizer masks out seven out of eight groups of symbols.

Each 20-ms traffic channel frame is divided into 16 power control groups of 1.25 ms each (as discussed earlier). The data burst randomizer pseudorandomly masks out individual power control groups (PCG). With 9.6 kbps, no PCG is masked out; with 4.8 kbps, 8 PCGs are masked out in a frame; with 2.4 kbps, 12 PCGs are masked; with 1.2 kbps, 14 PCGs are masked. An example of this operation with a vocoder operating at 2.4 kbps is shown in Fig. 7-28.

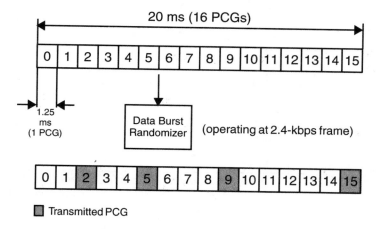

Figure 7-28 Data Burst Randomizer Operation at 2.4-kbps Frame

The reverse channel structure for RS2 is similar. The RS2 vocoder supports 14.4, 7.2, 3.6, and 1.8 kbps data rates. For RS2, the convolutional encoder is 1/2 instead of 1/3 as in RS1.

The system can multiplex primary (voice) and secondary (data) or signaling traffic on the same traffic channel. Multiplex option 1 is used to transmit primary and secondary traffic. This option is also used to transmit primary (voice) and signaling (messaging) traffic. Multiplex option 1 uses the following methods to simultaneously transmit primary and secondary traffic.

- **Blank and burst.** The entire traffic channel frame is used to send only secondary data *or* only signaling data. The secondary or signaling data blank out the primary data (see Fig. 7-29).
- **Dim and burst.** The traffic frame is used to send both primary and secondary data *or* both primary and signaling data (see Fig. 7-30). Fig. 7-31 shows the traffic channel frame structure for the reverse link.

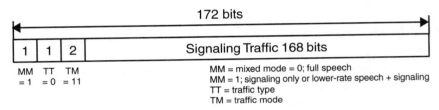

Figure 7-29 Blank and Burst for Speech RS1

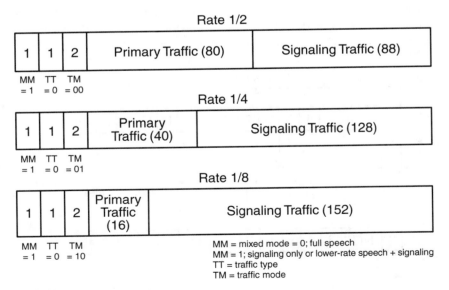

Figure 7-30 Dim and Burst for Speech RS1

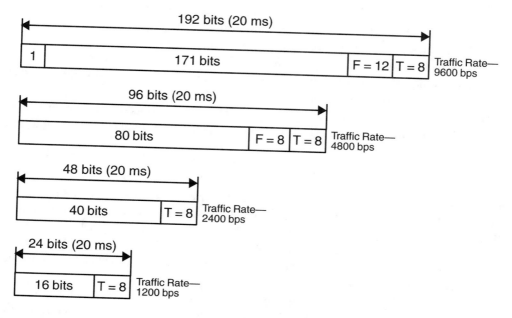

Figure 7-31   Traffic Channel Frame Structure for Reverse Link

The reverse traffic channel carries the following typical messages:

1. **Order messages** on the reverse traffic channel typically include base station challenge, SSD update confirmation, SSD update rejection, parameter update confirmation, mobile station acknowledgment, service option request, service option response, release (normal with power-down indication), long-code transition request (public and private), connect, continuous DTMF tone (start and stop), service option control, mobile station reject (with and without a reason), and local control.

2. **Authentication Challenge Response message** contains the information to validate the mobile's identity.

3. **Flash with Information message** contains information records from the mobile station concerning mobile features, mobile key pay facility, called party number, calling party number, and the connected number (i.e., the responding party).

4. **Data Burst message** is a user-generated data message sent by the mobile to the base station.

5. **Pilot Strength Measurement message** sends information about the strength of other pilot signals that are not associated with the serving base station.

6. **Power Measurement Report message** sends FER statistics to the base station. The report is generated at specified intervals or when a threshold is reached.

7. **Send Burst DTMF message** uses two tones DTMF—one low- and one high-frequency tone to represent a dialed digit—and transmits dialed digits to the base station.

8. **Status message** contains information records from the mobile about mobile identification, mobile call mode, mobile terminal information, and security status.

9. **Origination Continuation message** is a continuation of the origination message that was sent on the access channel if additional dialed digits need to be sent.

10. **Handoff Completion message** is the mobile response to a Handoff Direction message.

11. **Parameter Response message** is the mobile response to the base station for a Retrieve Parameters message.

7.8 Summary

This chapter discussed the implementation of CDMA carriers in an existing AMPS or TDMA system, establishing that 59 AMPS carrier channels must be removed to introduce the first CDMA carrier without interfering with the remaining AMPS or TDMA channels. The chapter provided details of information processing and message framing for the pilot, sync, paging, and traffic channels on the forward link. The pilot channel (Walsh code 0) is unmodulated; it consists of only short-code-spreading sequences.

The pilot channel is used by all mobiles attached to a cell as a coherent phase reference and also provides a unique identifier for different base stations. The sync channel (Walsh code 32) transmits system timing information to allow mobiles to synchronize themselves with base stations. The paging channel(s) (Walsh codes 1–7) are the digital control channel(s) for the CDMA forward link. One base station can have up to 7 paging channels. The first paging channel is always assigned Walsh code 1. The traffic channels (Walsh 8–31 and 33–63) carry digitized voice data.

On the reverse link, the access channel is used by the mobile to register with the system, to access the system before assignment of a traffic channel, to originate a call, or to respond to a page. Several access channels can be used per paging channel. The reverse traffic channels are used to deliver encoded voice and reverse link signaling from mobile to base station.

We also presented typical messages that are carried over the logical channels of the forward and reverse links.

7.9 References

1. Garg, V. K., Smolik, K., and Wilkes, J. E., *Applications of CDMA to Wireless/Personal Communications,* Prentice Hall PTR, Upper Saddle River, NJ, 1997.

2. Garg, V. K., and Wilkes, J. E., *Wireless and Personal Communications Systems,* Prentice Hall PTR, Upper Saddle River, NJ, 1996.

3. Rappaport, T. S., *Wireless Communications*, Prentice Hall PTR, Upper Saddle River, NJ, 1996.

4. TIA/EIA/SP-3693, "Mobile Station–Base Station Compatibility Standard for Dual-Mode Wideband Spread Spectrum Cellular Systems," November 1997.

CHAPTER 8

IS-95 CDMA Call Processing

8.1 Introduction

This chapter discusses IS-95 CDMA call processing states that a mobile station goes through in getting to a traffic channel. These include the system initialization state, the system idle state, the system access state, and the traffic channel state. Each of the call processing states has several substates that are also discussed, in addition to idle handoff, slotted paging operation, CDMA registration, and authentication procedures. The chapter covers messages used to exchange data in different call processing states and concludes by providing call flows for CDMA call origination, call termination, call release, and authentication.

8.2 CDMA Call Processing State

In getting to a traffic channel, a mobile station in CDMA goes through several states: system initialization state, system idle state, system access state, and traffic channel state (see Fig. 8-1).

In the *system initialization state* the mobile acquires a pilot channel by searching all the PN-I and PN-Q possibilities and selecting the strongest pilot signal. Once the pilot is acquired, the synchronization (sync) channel is acquired using the W_{32} Walsh function and the detected time offset of the pilot channel. Then the mobile obtains the system configuration and timing information.

Next the mobile enters the *system idle state* where it monitors the paging channel. The mobile can receive messages from the base station containing necessary parameters to initiate or to receive a call.

If a call is being placed or received, the mobile enters the *system access state* where the necessary parameters are exchanged. The mobile transmits its response on the access channel and the base station transmits its response on the paging channel.

When the access attempt is successful, the mobile enters the *traffic channel state*. In this state, speech communication takes place, with associated control messages replacing digital speech by either of two methods:

- **Blank-and-burst signaling.** The case where the complete speech packet is replaced with signaling.
- **Dim-and-burst signaling.** The case where part of the speech packet is replaced with signaling.

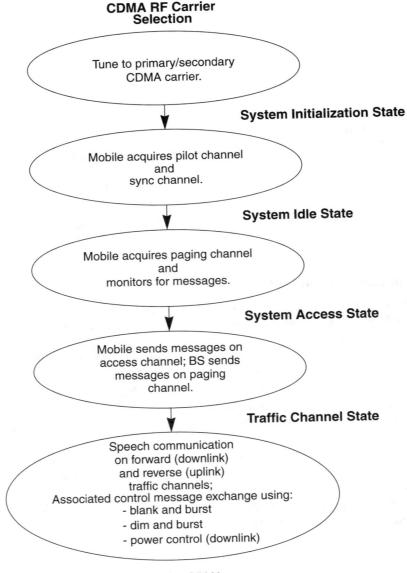

Figure 8-1 Getting to a Traffic Channel in CDMA

In addition, power control messages are sent by a method called *bit puncturing* on the forward link channel. In bit puncturing two gross-data bits are replaced by a single power control bit.

8.2.1 System Initialization State

In the system initialization state, the mobile selects a system to use and then proceeds to acquire and then synchronize to a CDMA carrier. The system initialization state consists of the following substates (Fig. 8-2):

- **System determination substate.** In this substate, the mobile selects the system to use (analog or CDMA) if it is a dual-mode cellular unit. The mobile choices include service provider preference. If a CDMA system is selected, the mobile sets the CDMA channel parameters (CDMA_CH) to N_i, where N_i is either a primary or secondary CDMA channel number.
- **Pilot channel acquisition substate.** The mobile acquires the pilot channel of the CDMA system. The mobile tunes to the CDMA_CH, sets its code channel for the pilot channel, and searches for pilot channel. If the mobile acquires the pilot channel within T_{20m} seconds, the mobile enters the sync channel acquisition substate. If the mobile does not acquire the pilot channel within T_{20m} seconds, the mobile enters the system determination substate with an acquisition failure indication.
- **Synchronization (sync) channel acquisition substate.** The mobile acquires the synchronization channel and obtains system configuration and timing information for the

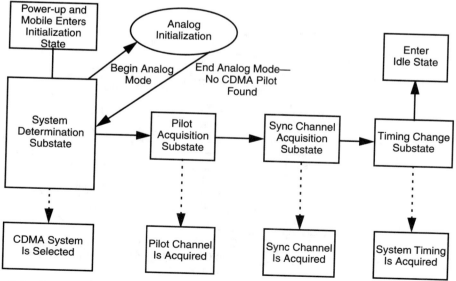

Figure 8-2 System Initialization State

CDMA system. Upon entering the sync channel acquisition substate, the mobile sets its code channel for the sync channel. If the mobile does not receive a valid sync channel message within T_{21m} seconds, the mobile enters the system determination substate with a protocol mismatch indication. If the mobile receives a valid sync channel message within T_{21m} seconds, the mobile stores system configuration and timing information.

- **Timing change substate.** The mobile synchronizes its long-code timing and system timing to those of the CDMA system after receiving and processing the synchronization message.

Fig. 8-3 shows the system states and the activities associated with them.

Synchronization occurs when the phase $(t - T)$ of the locally generated PN code is equal to the phase $(t - T_i - T_p)$ of the incoming code.

$$(t - T) = (t - T_i - T_p) \qquad (8.1)$$

$$\therefore T = T_i + T_p \qquad (8.2)$$

where T = phase of local PN code,
T_P = propagation delay, and
T_i = pilot offset for ith pilot.

At this stage of the system, we observe two points:

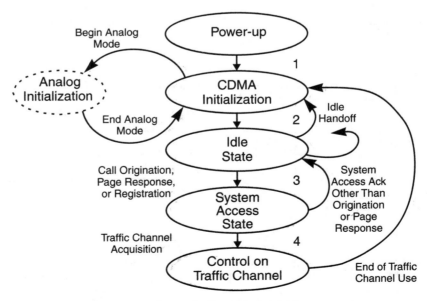

Figure 8-3 CDMA Call Processing States for Dual Mode Mobile

1. When the locally generated code phase matches with the incoming pilot, pilot acquisition occurs. Thus, PN $(t - T_i - T_p)$ is known.
2. Although total phase $(T_i + T_p)$ is known, the pilot offset T_i is not known. We get the pilot offset T_i from the Sync message of the sync channel.

Once the sync channel is acquired and the Sync message is received and processed, the mobile stores the following information from the Sync message:

- Protocol revision level (MIN_P_REV$_{stored}$ = MIN_P_REV$_{received}$)
- System identification (SID)
- Network identification (NID)
- Pilot PN offset (T_i)
- System time (T_s)
- Long-code state at system time (LC_STATE)
- Paging channel data rate (PRAT)
- Number of leap seconds that have occurred since the start of system time (LP_SEC)
- Offset of local time from system time (LTM_OFF)
- Daylight saving time indicator (DAY_LT)

In the timing change substate, the mobile uses the pilot offset, the system time, and the long-code state information, obtained from the Sync message, to synchronize its timing to the system time and to synchronize its long-code phase to that of the system.

In IS-95 the long code is generated with a 42-stage shift register. The mobile knows the generation polynomial. The problem is to get the correct code phase. The mobile obtains the long-code state from the Sync message. The long code is a 42-bit sequence that corresponds to the contents of a shift register at the system time, T_s. The mobile loads its shift registers with the 42-bit long-code state. The mobile waits and, at system time T_s, it starts shifting the contents of the shift registers at 1.2288 Mcps. At this point, long-code synchronization is achieved. The mobile may now tune to a paging channel in order to enter the idle state.

8.2.2 Idle State

In the idle state, the mobile monitors the paging channel. In this state, the mobile can

1. Receive messages and orders from the base station
2. Receive an incoming call
3. Initiate a registration process
4. Initiate a call
5. Initiate a message transmission

Fig. 8-4 summarizes the activities in the idle state. Upon entering the idle state, the mobile sets its Walsh code to the primary paging channel (W_1) and sets its paging channel rate to the rate obtained from the Sync message.

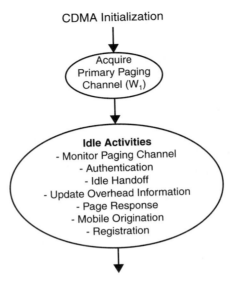

CDMA Initialization

Acquire
Primary Paging
Channel (W_1)

Idle Activities
- Monitor Paging Channel
- Authentication
- Idle Handoff
- Update Overhead Information
- Page Response
- Mobile Origination
- Registration

Figure 8-4   Idle State Activities in CDMA

The paging channel is subdivided into 80-ms slots called *paging channel slots*. In the non-slotted mode, paging and control data for the mobile can be received in any of the paging channel slots. Therefore, the mobile monitors all slots on a continuous basis. In IS-95, the paging channel protocol also allows scheduling the transmission of messages for a given mobile in certain assigned slots. A mobile station that monitors the paging channel only during certain assigned slots is referred to as operating in the *slotted mode*. During the slots in which the paging channel is not monitored, the mobile can stop or reduce its processing activities to save battery power. This is sometimes referred to as a discontinuous reception mode of operation.

In the slotted mode operation, the mobile monitors the paging channel for 1 or 2 slots per cycle. Slotted page messages contain a field called *more-pages*. When this field is set to 0, it indicates that the remainder of the slots will contain no more messages addressed to the mobile. This allows the mobile to stop monitoring the paging channel as soon as possible. If a Slotted Page message with the more-pages field set to 0 is not received in the assigned slot, the mobile continues to monitor the paging channel for one additional slot. For each of its assigned slots, the mobile begins monitoring the paging channel in time to receive the first bit of the assigned slot. The mobile then continues to monitor the paging channel until one of the following conditions is satisfied:

1. The mobile receives a Slotted Page message with the more-pages field set to 0.
2. The mobile monitors the assigned slot and the slot following the assigned slot and receives at least one valid message.

The mobile can specify its preferred slot cycle using the SLOT_CYCLE_INDEX field in the Registration message, Origination message, or Page Response message. When the mobile station is in the control on traffic channel state, it can also specify the preferred slot cycle using the SLOT_CYCLE_INDEX field of the terminal information record of the Status message. The length of the slot cycle, T, in units of 1.28 seconds, is given as

$$T = 2^i \tag{8.3}$$

where i = selected slot cycle index.

There are $16 \times T$ slots in a slot cycle for a particular mobile using some value of i, and four 20-ms full frames in 80-ms slot. SLOT_NUM is the paging slot number. To determine the assigned slots, the mobile uses a *Hash algorithm* to select a slot number in the range 0 to 2048. The minimum and maximum cycles are 16 slots (1.28 seconds) and 2048 slots (163.84 seconds), respectively. The value of SLOT_NUM is given as

$$\text{SLOT\_NUM} = \left(\frac{t}{4}\right)\text{mod}\,2048 \tag{8.4}$$

where t = system time in frames.

For each mobile station, the starting time of its slot is offset from the beginning of each of its slot cycles by a fixed, randomly selected number of slots, called PGSLOT. As an example, let the selected slot cycle index $i = 0$, $T = 2^0 = 1$, so $16 \times T = 1.28$ seconds for a slot cycle, and the computed value of PGSOLT equal to 6. The mobile begins monitoring the paging channel at the start of the slot in which SLOT_NUM equals 6 (see Fig. 8-5). The next slot in which the mobile must begin monitoring the paging channel is 16 slots later, i.e., the slot in which SLOT_NUM is 22, since the slot cycle length is 16 for $T = 1$.

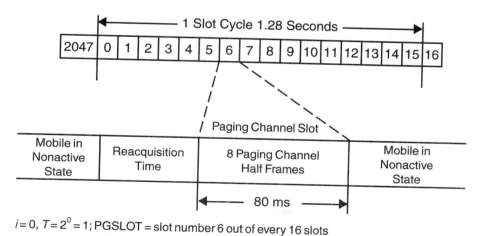

$i = 0$, $T = 2^0 = 1$; PGSLOT = slot number 6 out of every 16 slots

Figure 8-5 Slot Cycle with $i = 0$

8.2.2.1 Acknowledgment Procedure

We consider a message sent to the mobile by the base station on the paging channel. The message typically contains an address field, acknowledgment required sequence number, message sequence number, and other fields including the data field (see Fig. 8-6). Whether or not the mobile sends an acknowledgment to the base station depends on the value of the ACK_REQ field. Acknowledgment of messages received on the paging channel are sent to the base station on an access channel.

If the ACK_REQ field of the message from the base station is 1, the mobile transmits an acknowledgment with the following parameters (see Fig. 8-7):

- VALID_ACK field: 1
- ADDR_TYPE field: *I* (address type of message being acknowledged)
- ACK_SEQ field: *I* (message sequence number of the message being acknowledged)

When the Page messages or Slotted Page messages addressed to a mobile do not have an ACK_REQ field, the mobile transmits a Page Response message including an acknowledgment in response to each record of a page message or slotted message.

When a message does not include an acknowledgment, the mobile sets the VALID_ACK field to 0. The ADDR_TYPE and ACK_SEQ fields are then set to the ADDR_TYPE and MSG_SEQ values of the last message that requires an acknowledgment.

- VALID_ACK field: 0
- ADDR_TYPE field: *I* (address type of last message requiring acknowledgment)
- ACK_SEQ field: *I* (message sequence number of last message requiring acknowledgment)

When a message does not include an acknowledgment, the mobile sets the VALID_ACK field to 0. The ADDR_TYPE and ACK_SEQ fields then set to 000 and 111, respectively if there is no previous message that requires an acknowledgment.

8.2.2.2 Idle Handoff

An idle handoff, or change of paging channel, occurs when a mobile has moved from the coverage area of one base station to the coverage area of another base station during the idle

Other Field	ADDR_TYPE	ACK_REQ	MSG_SEQ

Figure 8-6 A Typical Message from Base Station to Mobile

Other Field	ADDR_TYPE	VALID_ACK	ACK_SEQ

Figure 8-7 A Typical Message with Acknowledgment from Mobile

state. The mobile determines that a handoff should occur when it detects a new pilot that is sufficiently stronger than the current pilot.

Pilot channels are identified by short PN offsets. They are grouped into sets describing their status with respect to pilot searching procedures. In the idle state, three pilot sets are maintained: active, neighbor, and remaining.

Using a strategy similar to a sliding correlator, it is possible to acquire a pilot if its short PN code's correct phase is known. For each pilot set, a search window is specified. This allows the mobile to search for the direct path as well as multipath components of the pilot signal. The search window is centered either on the earliest arriving multipath or the short PN offset.

If the mobile determines that a neighbor set or remaining set pilot is sufficiently stronger than the active set pilot, idle handoff is performed. While performing an idle handoff, the mobile operates in nonslotted mode until at least one valid message is received from the new paging channel. On receiving a valid message from the new paging channel, the mobile may resume slotted mode operation. After performing an idle handoff, the mobile discards all unprocessed messages received on the old paging channel.

The paging channel is used to transmit control information to the mobiles that have not been assigned to traffic channels. Two types of control messages are sent:

- Overhead messages that are broadcast messages for all mobile stations.
- Directed messages addressed to a particular mobile or a specific group of mobiles.

There are four overhead messages that are continuously broadcast on the paging channel.

1. System Parameter message
2. Neighbor List message
3. CDMA Channel List message
4. Access Parameters message

The first three messages are called *configuration messages*. A configuration message sequence number (CONFIG_MSG_SEQ) is associated with a set of configuration messages sent on the paging channel. When the contents of one or more configuration messages change, the configuration message sequence number is incremented. The mobile stores the sequence number contained in each configuration message received.

Access Parameters messages are sequenced by the access parameter sequence number (ACC_MSG_SEQ). The mobile stores the most recently received Access Parameters message sequence number in the ACC_MSG_SEQ field.

Configuration and access parameters from one paging channel are not used while monitoring a different paging channel. If the stored parameters are current, the mobile processes the messages on the paging channel. When a System Parameter message (SYS_PAR_MSG_SEQr) is received, its associated configuration message sequence number (CONFIG_MSG_SEQr) is compared to the stored value of the System Parameter message (SYS_PAR_MSG_SEQs). As long as there is a match, the received System Parameter message is ignored. If the result is a mismatch, the mobile stores the configuration message sequence number, the SID, NID, and the base station identification (BASE_ID).

When an Access Parameters message is received, the ACC_MSG_SEQr is compared to the stored value of the Access Parameters message sequence value (ACC_MSG_SEQs). If there is a match, the received Access Parameters message is ignored. However, if the result is a mismatch, the mobile stores the following parameters:

- Access Parameters message sequence number (ACC_MSG_SEQ)
- Number of access channel (ACC_CHAN)
- Nominal transmit power (NOM_PWR)
- Initial power offset for access (INIT_PWR)
- Power increment or power step (PWR_STEP)

8.2.2.3 Neighbor List Message

When a Neighbor List message (NGHBR_LST_MSG_SEQr) is received, its associated configuration message sequence number (CONFIG_MSG_SEQ) is compared with the stored value of the Neighbor List message (NGHBR_LST_MSG_SEQs). If there is a match, the received Neighbor List message is ignored. If the result is a mismatch, the mobile stores the following parameters:

- Configuration message sequence number (CONFIG_MSG_SEQ)
- Short PN offsets of neighbor list members
- Short PN sequence offset increment (PILOT_INC)

The mobile updates the idle handoff neighbor set so that it contains only the pilot offsets listed in the Neighbor List message.

8.2.2.4 CDMA Channel List Message

When a CDMA Channel List message (CHAN_LST_MSG_SEQr) is received, its associated configuration message sequence number (CONFIG_MSG_SEQ) is compared with the stored value of the CDMA Channel List message (CHAN_LST_MSG_SEQs). If there is a match, the received CDMA Channel List message is ignored. If the result is a mismatch, the mobile stores the following parameters:

- Configuration message sequence number (CONFIG_MSG_SEQ)
- CDMA channel list message sequence (CHAN_LST_MSG_SEQ)

8.2.3 System Access State

The system access state (see Fig. 8-8) includes the following substates:

- **Update overhead information substate.** In this substate, the mobile monitors the paging channel until it has received a current set of configuration messages
- **Mobile station origination attempt substate.** The mobile station sends an Origination message to the base station.

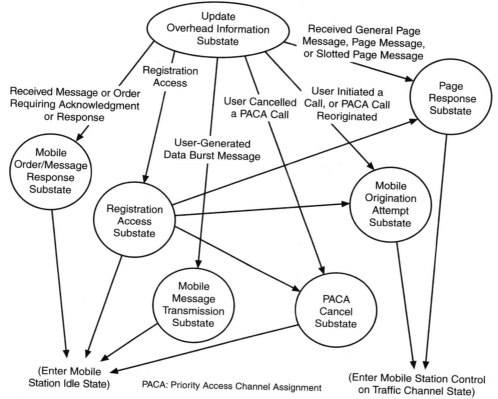

Figure 8-8 CDMA Mobile System Access State

- **Page response substate.** The mobile sends a Page Response message to the base station.
- **Registration access substate.** The mobile station sends a Registration message to the base station.
- **Mobile station order/message response substate.** The mobile sends a response to a message received from the base station.
- **Mobile station message transmission substate.** The mobile sends a Data Burst message to the base station.
- **PACA cancel substate.** The mobile sends a Priority Access Channel Assignment (PACA) Cancel message. If the base station responds with an authentication request, the mobile responds in this substate.

8.2.4 Mobile Control on Traffic Channel State

Call Origination: The mobile station control on the traffic channel state for call origination consists of the following substates (see Figs. 8-9 and 8-10):

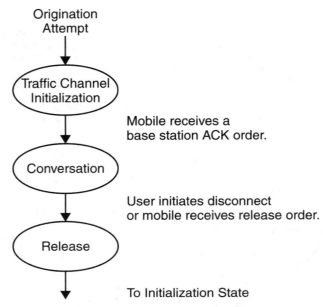

Figure 8-9   Mobile Control on Traffic Channel (Call Origination)

- *Traffic channel initialization substate.* In this substate, the mobile verifies that it can receive the forward traffic channel and begins to transmit on the reverse traffic channel.
- *Conversation substate.* The mobile station exchanges primary traffic packets with the base station.
- *Release substate.* The mobile station disconnects the call.

Call Termination: The mobile station control on traffic channel for call termination consists of the following substates (see Figs. 8-11 and 8-12):

- *Traffic channel initialization substate* (same as in "Call Origination")
- *Waiting for order substate.* In this substate, the mobile waits for an alert with an information message. The information may be data such as calling party number and/or voice.
- *Waiting mobile station answer substate.* The mobile waits for the user to answer the call.
- *Conversation substate.* The mobile station's primary service option application exchanges primary traffic packets with the base station.
- *Release substate.* The mobile disconnects the call.

Flow diagrams for CDMA far-end-initiated and mobile-initiated call release are shown in Figs. 8-13 and 8-14, respectively.

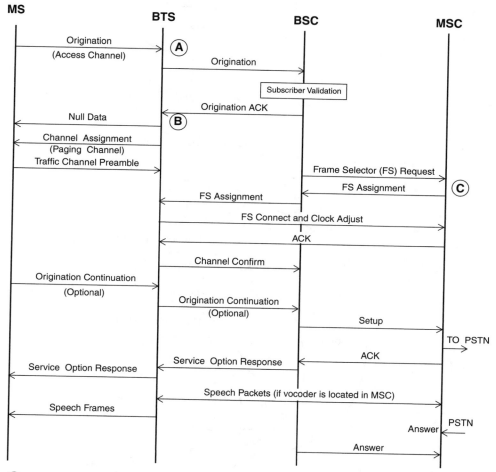

(A) Channel element assigned to call
(B) Walsh code assigned to call
(C) Speech handler assigned to call

Figure 8-10 Flow Diagram for CDMA Call Origination

8.3 CDMA Registration

The registration process is used by the mobile to notify the base station of its location, status, identification, slot cycle, and other characteristics. The base station can efficiently page the mobile station when establishing a mobile-terminated call. For operation in the slotted mode, the mobile supplies the SLOT_CYCLE_INDEX parameter so that the base station can determine

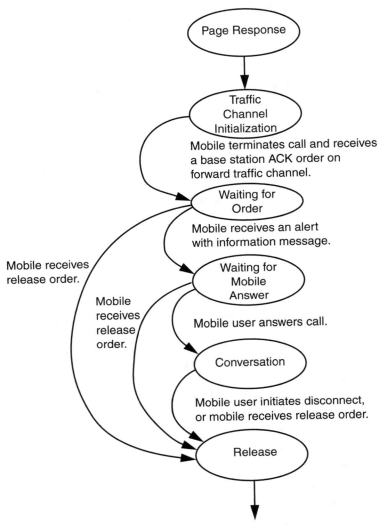

Figure 8-11 Mobile Control on Traffic Channel (Call Termination)

which slots the mobile is monitoring. The mobile supplies the station class mark and protocol revision number so that the base station knows the capabilities of the mobile station.

IS-95 supports nine different forms of registration:

1. **Power-up registration**. The mobile registers when it powers up, switches from using an alternative serving system, or switches from using the analog system with dual-mode cellular mobile.

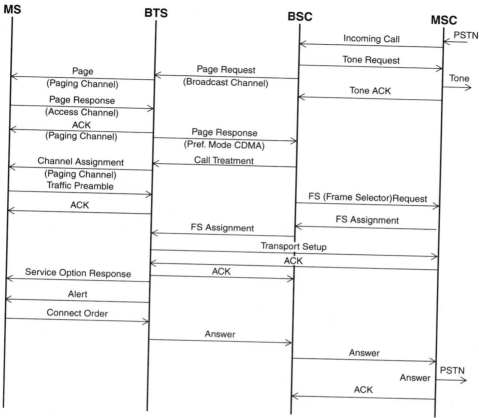

Figure 8-12   Flow Diagram for CDMA Call Termination

2. **Power-down registration.** The mobile registers when it powers down, informing the system that it is no longer active.

3. **Timer-based registration.** The mobile registers at regular intervals. Its use also allows the system to automatically deregister mobile stations that did not perform a successful power-down registration.

4. **Distance-based registration.** A mobile station registers when the distance between the current base station and the base station in which it last registered exceeds a threshold. The mobile determines that it has moved a certain distance by computing a distance based on the difference in latitude and longitude between the current base station and the base station where the mobile last registered. If this distance exceeds the threshold value, the mobile station registers.

5. **Zone-based registration.** A mobile registers when it enters a new zone. Zones are groups of base stations within a given system and network. A base station's zone

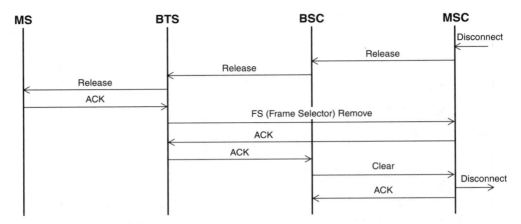

Figure 8-13   Flow Diagram for CDMA Call Release (Far-End Initiated)

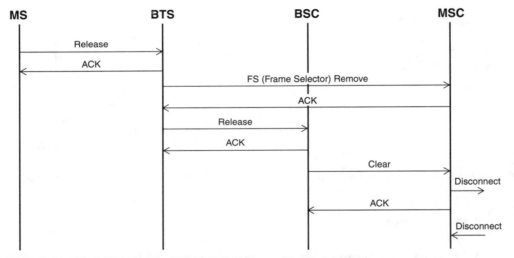

Figure 8-14   Flow Diagram for CDMA Call Release (Mobile Initiated)

assignment is identified by the REG_ZONE field of the system parameters message. Zone-based registration causes a mobile to register whenever it moves into a new zone that is not on its internally stored list of visited registration zones. A zone is added to the list whenever a registration (including implicit registration) occurs and is deleted upon expiration of a timer. After a system access, timers are enabled for every zone except one.

6. **Parameter-change registration.** This occurs when a mobile station modifies any of the following stored parameters:

- the preferred slot cycle index
- the station class mark
- the call termination enabled indicator

7. **Ordered registration.** The mobile registers when the base station requests that it do so.
8. **Implicit registration.** When a mobile successfully sends an Origination message or Page Response message, the base station can infer the mobile station's location. This is considered an implicit registration.
9. **Traffic channel registration.** Whenever the base station has registration information for a mobile that has been assigned to a traffic channel, the base station can notify the mobile that it is registered.

8.4 Authentication

To authenticate a mobile station, secret data known as the A-key is used. The A-key is known only to the Authentication Center (AC) of the mobile's home system and to the mobile station. It is the most secure piece of secret data. The A-key and a special random number (RANDSSD) can be used by the AC and mobile to generate Shared Secret Data (SSD). The AC may send SSD to the serving system, but it never sends it over the air link.

RAND is a 32-bit random number issued periodically by the base station in the system overhead data in two 16-bit segments: RAND_A and RAND_B. The mobile stores and uses the most recent version of RAND in the authentication process. The last RAND received by the mobile station is confirmed from the mobile with an 8-bit number RANDC, a part of RAND, since the current system RAND and the one used by the mobile station could differ when the base station receives mobile station results.

The Electronic Serial Number (ESN) is a 32-bit binary number that uniquely identifies the mobile to any system. It is factory set and not readily alterable in the field. Modification of the ESN requires a special facility not normally available to subscribers.

The Mobile Identification Number (MIN) is derived from mobile station's 10-digit directory telephone number. The first 3 digits map into the 10 most significant bits, the second 3 digits map into the next 10 bits, while the last 4 digits map into the remaining 14 bits.

The SSD is a 128-bit pattern stored in the semipermanent memory of the mobile and is known by the base station. SSD is a concatenation of two 64-bit subsets: SSD_A and SSD_B. SSD_A is used to support the authentication procedure, and SSD_B is used to support voice privacy and message confidentiality.

SSD is maintained during power off. It is generated using a 56-bit random number (RANDSSD created by the home AC), the mobile's A-key, and the ESN. The A-key is a 64-bit secret pattern assigned and stored in mobile station's permanent security and identification memory. The need to pass the A-key itself from system to system as the subscriber roams is eliminated. SSD updates are carried out only in the mobile station and its associated home system's HLR/AC, not in the serving system. The AC manages the encrypting keys associated with an individual subscriber when such functions are provided within the network.

All mobiles are assigned an ESN when they are manufactured. They are also assigned a 15-digit International Mobile Subscriber Identity (IMSI). When the mobile is turned on, it must register with the system. When it registers, it sends its IMSI and other data to the network. The VLR in the visited system then queries the home system's HLR for the security data and service profile information. The VLR then assigns a Temporary Mobile Subscriber Identity (TMSI) to the mobile station. The mobile station uses the TMSI for further accesses to that system. The TMSI provides anonymity of communications since only the mobile and the network know the identity of the mobile with a given TMSI. When a mobile roams into the new system, some air interfaces use TMSI to query the old VLR and then assign a new TMSI; other air interfaces request that the mobile station send its IMSI, and then they assign a new TMSI.

The network transmits a random number RAND that is received by all mobile stations. When a mobile station accesses that system, it calculates AUTHR, an encrypted version of RAND using SSD_A. It then transmits to the network the desired message with its authentication.

For additional details on authentication refer to chapter 11.

The network performs the same calculation and confirms the identity of the mobile. All communications between the mobile and network are encrypted to prevent someone's decoding of the data and using the data to clone other mobile stations. Furthermore, each time a mobile places or receives a call, a call history count is incremented (CHCNT). The counter is also used for clone detection since clones will not have a call history identical to the legitimate mobile.

Procedures have been designed to allow a system to challenge an individual mobile with a unique challenge and to update the SSD.

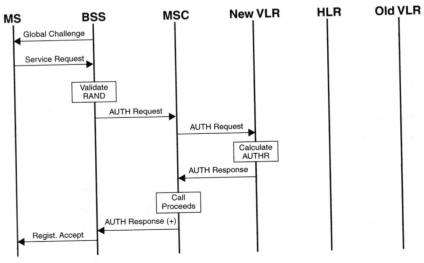

Figure 8-15   Call Flows for a Global Challenge

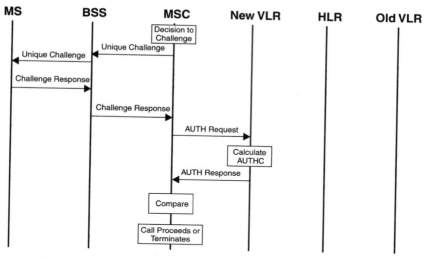

Figure 8-16 Call Flows for Unique Challenge

All mobile stations accessing the network must respond to the global challenge as part of their access. The global challenge response is an integral part of the network access (call origination, page response, registration, and so on). The call flow for global challenge is given in Fig. 8-15.

The unique challenge can be sent to a mobile station at any time. It is typically initiated by the MSC in response to the some event (registration failure and after a successful handoff are the most typical cases). This is used to challenge the mobile as to its identity. Fig. 8-16 shows call flows for the unique challenge.

8.5 Summary

This chapter discussed the CDMA call processing states and provided details of different messages that are used to exchange information in these states. We presented the registration and authentication procedures of CDMA and concluded the chapter by discussing mobile substates and flow diagrams for CDMA call origination, call termination, and call release.

8.6 References

1. Garg, V. K., Smolik, K. F., and Wilkes, J. E., *Applications of CDMA in Wireless Communications,* Prentice Hall PTR, Upper Saddle River, NJ, 1997.

2. Garg, V. K., and Wilkes, J. E., *Wireless and Personal Communications Systems*, Prentice Hall PTR, Upper Saddle River, NJ, 1996.

3. TIA IS-95 B, "Mobile Station and Radio Interface Specifications."

Figure 8-16 ...

8.5 Summary

8.6 References

Signaling Applications in IS-95 CDMA

9.1 Introduction

In chapter 5, we discussed the CDMA system architecture used for cellular and PCS in North America as embodied in the design standardized by TIA and ATIS in IS-95A for the cellular system and J-STD-008 for PCS. Chapter 8 covered IS-95 CDMA call processing states that a mobile station goes through in getting to a traffic channel. These states include the system initialization state, system idle state, system access state, and traffic channel state. It also covered CDMA registration and authentication procedures and included the messages that are used to exchange information in different call processing states. We then provided call flows for CDMA call origination, call termination, call release, and authentication.

In this chapter, we first describe the layering concept that has been used to develop the protocols for IS-95 CDMA and then focus on three functional entities—Mobile Station (MS), Base Station System (BS), and Mobile Switching Center (MSC)—discussing the standardized interfaces between these entities. The focus is mainly on the A-interface and TIA IS-634-defined MSC-BS messages, message sequencing, and mandatory timers at the BS and the MSC. The chapter concludes with call flow diagrams for typical supplementary services, handoff scenarios, and Over-the-Air Service Provisioning (OTASP).

9.2 Layered Structure

TIA IS-95 CDMA has a layered structure that is designed to provide voice, packet data (up to 64 kbps), simple circuit data (example async fax), and simultaneous voice and packet data services (see Fig. 9-1).

At the basic level, IS-95 provides protocols and services that correspond to the bottom two layers of the International Organization for Standardization (ISO)/open system interconnection (OSI) reference model, i.e., physical and link layer. The physical layer performs coding, interleaving, modulation, and spreading functions for the physical channels.

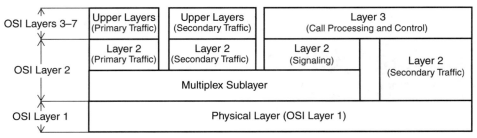

OSI = open system interconnections

Figure 9-1   TIA IS-95 Layered Structure

The link layer provides protocol support and control mechanisms for data transport services and maps the data transport needs of the upper layers into the specific capabilities and characteristics of the physical layer. It maps logical and signaling channels into code channels specifically supported by the coding and modulation functions of the physical layer.

The link layer (layer 2) is subdivided into the Link Access Control (LAC) and Medium Access Control (MAC) sublayers (see Fig. 9-2). Applications and upper-layer protocols corresponding to OSI layers 3 through 7 utilize the services provided by LAC. The LAC sublayer performs the functions essential to set up, maintain, and release a logical link connection.

The MAC sublayer provides a control function that manages resources supplied by the physical layer (e.g., physical code channels for communication of information over the air interface) and coordinates the usage of those resources desired by various LAC service entities. This coordination function (which operates under direct control of the BS MAC function) resolves contention issues between LAC service entities within a single mobile station, as well as between competing mobile stations. The MAC sublayer is also responsible for delivering the Quality of Service (QoS) level requested by a LAC service (e.g., by reserving air interface resources or by resolving priorities between competing LAC service entities).

In IS-95 CDMA the MAC uses Radio Link Protocol (RLP), which provides a highly efficient streaming service that makes a best effort to deliver data between peer entities. The RLP provides both a transparent and nontransparent mode of operation. In the nontransparent mode RLP uses Automatic Repeat Request (ARQ) protocols to retransmit data segments that were not delivered properly by the physical layer. In this mode RLP introduces some delay. In the transparent mode, RLP does maintain byte synchronization between the sender and receiver and notifies the receiver of the missing parts of the data stream. Transparent RLP does not introduce any transmission delay and is useful in implementing voice services over RLP.

Upper-layer entities provide support for multiple concurrent active sessions with any combination of service type, such as

- Voice service including voice telephony, PSTN access, mobile-to-mobile voice services, and Internet telephony

- End-user data-bearing services including packet data, circuit data services, and short message service (SMS)
- Signaling services that control all aspects of operation of the mobile station

Packet data services conform to industry standard connection-oriented and connectionless packet data including IP-based protocol (e.g., Transmission Control Protocol [TCP] and User Data Protocol [UDP]) and ISO/OSI Connectionless Interworking Protocol (CLIP).

Circuit data services emulate international-standard-defined connection-oriented services such as asynchronous dial-up access, fax, etc.

Fig. 9-2 shows IS-95 (2G CDMA system) and cdma2000 (3G CDMA) layered structure with protocols.

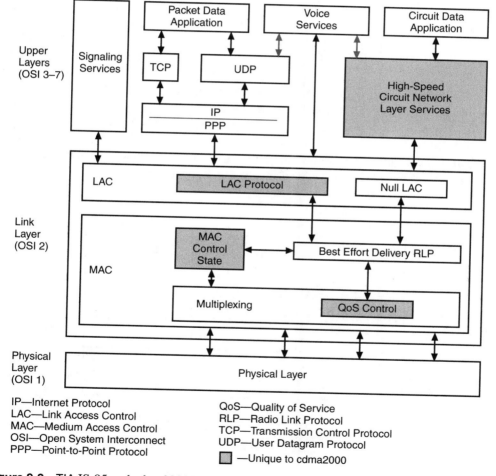

IP—Internet Protocol
LAC—Link Access Control
MAC—Medium Access Control
OSI—Open System Interconnect
PPP—Point-to-Point Protocol

QoS—Quality of Service
RLP—Radio Link Protocol
TCP—Transmission Control Protocol
UDP—User Datagram Protocol
▨ —Unique to cdma2000

Figure 9-2 TIA IS-95 and cdma2000 Layered Structure with Protocols

9.3 A-Interface

North American standards, until recently, have not addressed the standardization of the BS-MSC interface (the A-Interface in the network reference model). However, wireless service providers are experiencing explosive growth in North America and are consequently finding it necessary to purchase equipment from multiple equipment manufacturers. Thus, the wireless industry has pressed for standards specifying the A-Interface. At this time, however, the BTS-BSC interface, i.e., the A_{bis} interface, is not being addressed. TIA IS-634 defines MSC-BS messages, message sequencing, and mandatory timers at the base station and the mobile switching center. Call processing, radio resource management, mobility management, and transmission facilities management are separate functions that are supported by the applications layer (see Fig. 9-3).

The MSC-BS interface (A-Interface) utilizes point-to-point signaling between the BS and the MSC. The normal routing situation is that there are one or more signaling links available between the BS and the MSC, and these constitute a link set. They run in a load-sharing mode, and the change-over and change-back procedures are supported between these signaling links. Load sharing is performed on the BS with more than one signaling link by means of the *Signaling Link Selection* (SLS) field.

The underlying transport mechanism for the applications layer is ISDN with the physical layer specified by American National Standard Institute (ANSI) T1.101, the Message Transfer Part (MTP) specified by ANSI T1.111, and the Signaling Connection Control Part (SCCP) specified by ANSI T1.112. The physical interface supports one or more 1.544-Mbps digital transmission facilities, each providing twenty-four 56-kbps or 64-kbps channels. Each channel can be used for traffic or for signaling. The MTP and the SCCP support only signaling messages, while

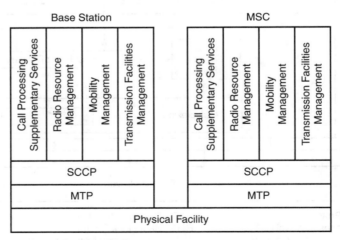

MTP: message transfer part
SCCP: signaling connection control part
MSC: mobile switching center

Figure 9-3 TIA IS-634 Functions

the physical layer supports both signaling messages and traffic messages. Traffic messages carry voice transmission. TIA IS-634 allows the transcoder (vocoder) to reside either at the base station or "very near" to the mobile switching center. In the first case, an entire DS0* (64-kbps) connection is required for each call, while the second case does not necessitate an entire DS0 connection.

At the applications layer, the call processing and mobility management functions are connected between the mobile station and the mobile switching center, while the radio resource management and the transmission facilities management functions are connected between the base station and the mobile switching center. Accordingly, the Base Station Application Part (BSAP), which is the applications-layer signaling protocol, is divided into two subapplications parts. The first is called the BS Management Application Part (BSMAP). BSMAP messages are sent between the base station and the mobile switching center. The second is the Direct Transfer Application Part (DTAP) in which messages are sent between the mobile station and the mobile switching center. The base station acts as a transparent conduit for DTAP messages. The base station merely maps the messages going to/from the mobile switching center into the appropriate air interface signaling protocol, e.g., TIA IS-95A. This approach simplifies the role of the base station for call processing and mobility management.

The DTAP is used to transfer call control and mobility management signaling messages to and from the MS. The BSMAP supports other procedures between the MSC and the BS related to the MS, or to a cell within the BS, or to the whole BS.

The base station associates the DTAP messages with a particular mobile station and call using a transaction identification. BSAP messages are transferred over an SCCP connection. The DTAP and BSMAP layer 3 messages between the base station and the mobile switching center are contained in the user data field of the SCCP frames. The data field is supported in Connection Request (CR), Connection Confirm (CC), Released (RLSD), and Data (DT) SCCP frames for mobile stations having one or more active transactions. The layer 3 user data field is partitioned into three components (see Fig. 9-4):

1. BSAP message header
2. Distribution data unit (this component includes the length indicator and the Data Link Connection Identifier [DLCI]—applies only to DTAP messages)
3. Layer 3 message

The BSAP message header consists of the message discrimination and the data link connection identifier, which is applicable only for DTAP messages. The *D*-bit (bit 0) of the message discrimination octet is set to 1 for a DTAP message and is set to 0 for a BSMAP message. The distribution data unit consists of the length indicator octet, which gives the number of octets following the length indicator.

DTAP messages are applicable only to mobility management and call processing (including supplementary services) functions, while BSAP messages are associated with radio

---

* A DS0 is a 64-kbps Pulse Code Modulation (PCM) transport facility and is a single 64-kbps time slot on a T1 carrier.

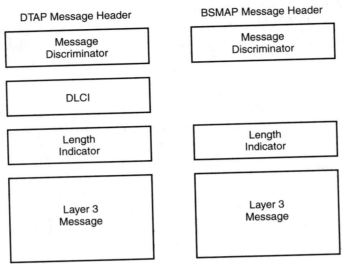

DLCI: data link connection identifier
BSMAP: base station management application part
DTAP: direct transfer application part

Figure 9-4   Layer 3 Data Field

resource management and to call processing (to a lesser degree). Each DTAP message contains the protocol discriminator octet, which identifies the associated procedure, i.e., call control, mobility management, radio resource management, and facilities management. All DTAP and BSMAP messages are identified by the message-type octet.

The remaining part of this section provides greater detail for each function supported by the BSAP.

9.3.1 Supported Architectural Configurations

TIA IS-634 makes a number of assumptions regarding the underlying CDMA architecture. The basic architecture is shown in Fig. 9-5.

The main entities are the mobile switching center, the transcoder (XC), the base station, the base transceiver system, and the mobile station. The MS is not shown in Fig. 9-5, but the MS communicates with the BTS over the air interface. TIA IS-634 assumes that the base station is really the BSC. One or more BTSs are connected to the BSC.

The transcoder supports both voice coding (vocoder) and diversity reception. Diversity reception allows the transcoder to pick the best frame when multiple connections are established during a soft handoff. Diversity reception distinguishes CDMA technology from other digital technologies. To be more specific, the XC is responsible for the following:

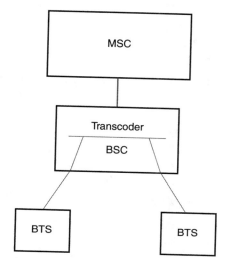

Figure 9-5   Basic CDMA Architecture

- Distribute speech/data on the forward traffic channel to all BTSs associated with a call. During a soft handoff, multiple BTSs are simultaneously assigned to the call. The XC selects the best speech/data frame from all the BTSs associated with the call on the reverse traffic channel. This implies that signal quality characteristics of the speech/data frame are provided to the transcoder.·
- Decode QCELP* format and change to PCM format for voice frames sent on the reverse traffic channel. If the call is a data call, this task is bypassed.
- Decode PCM format and change to QCELP format for voice frames sent on the forward traffic channel. If the call is a data call, this task is bypassed.
- Rate adapt voice frames to fully utilize the transmission bandwidth of the assigned terrestrial circuits. This task is bypassed for data calls.
- Rate adapt compressed voice PCM format into a circuit-switched subrate channel on a DS0 facility. One common compression approach is Adaptive Differential Pulse Code Modulation (ADPCM). Compression utilizes the fact that voice activity is less than 100% of the total duration. Typically, the actual voice activity is approximately 50%.
- Provide a control capability of inserting blank-and-burst or dim-and-burst signaling into the voice transmission on the forward traffic channel.

The transcoder is considered as a logical part of the BS, although the transcoder can be physically located at the BS, or at the MSC, or somewhere between the BS and the MSC. The

---

* Qualcomm Code-Excited Linear Prediction (QCELP) is the CDMA speech processing algorithm that is specified in TIA IS-96A. The algorithm is based upon code-excited linear prediction. See chapter 3 for more information.

terrestrial facility connects the transcoder to the MSC. The terrestrial facility may be full-rate (56 kbps or 64 kbps), subrate, PSTN, bypass, or PSTN/bypass. If only the BS is associated with a call, the terrestrial facility is connected to the PSTN. If a call is configured for a soft handoff between two base stations, another terrestrial facility is required to connect the transcoder at the target BS with the transcoder at the source BS. This terrestrial facility may be a full-rate, subrate, bypass, or bypass/PSTN facility. The connection between the transcoder and the BS is not addressed in TIA IS-634.

The transcoder and the BS may be physically co-located or may be externally connected by a full-rate or subrate facility if the transcoder is located near or at the MSC. The BSC may support multiple BTSs. Thus, if a call is in a soft handoff using only BTSs connected to a given BSC, no messaging between the MSC and the BS is necessary.

BTSs are uniquely identified by the Cell Global Identification (CGI). CGI is composed of four components:·

- Mobile Country Code (MCC)
- Mobile Network Code (MNC)
- Location Area Code (LAC)
- Cell Identity (CI)

TIA IS-634 supports addressing modes so that a BTS can be identified by the CGI, by the CI, by a combination of the LAC, MCC, and MNC, or by the associated BS.

9.3.2 Call Processing and Supplementary Services

TIA IS-634 supports call setup (mobile origination and mobile termination) as well as supplementary services (e.g., call waiting) and call release. However, the support of handoffs during a call is considered to be part of the radio resource management function.

Most of the messages that are associated with call processing and supplementary services are DTAP messages. For these messages, the role of the base station is minimized since the base station *passes* the messages to the mobile station.

The initial BS-MSC message in the call setup procedure includes the mobile identity. The mobile identity can be the Mobile Identification Number (MIN), the mobile station Electronic Serial Number (ESN), or the International Mobile Subscriber Identifier (IMSI). The identity type is selected by either the mobile station or the wireless network. For a mobile origination, the initial BS-MSC message is the Connection Management (CM) service request; and for a mobile termination, the initial BS-MSC message is the paging request. However, the initial message from the base station to the mobile switching center at call setup is an encapsulated DTAP message within a BSMAP message. The mobile switching center sends an assignment request message, which contains the terrestrial channel. Also, the mobile switching center may select the radio channel or provide channel parameters and permit the base station to choose the radio channel at the appropriate BTS.

Tables 9-1 and 9-2 list the messages defined by TIA IS-634 for call processing and supplementary services, respectively.

Table 9-1 Call Processing Messages

Message Name	Direction	Message Type
CM Service Request	BS → MSC	DTAP
Paging Request	MSC → BS	BSMAP
Paging Response	BS → MSC	DTAP
Setup	BS ↔ MSC	DTAP
Emergency Setup	BS → MSC	DTAP
Alerting	BS ↔ MSC	DTAP
Call Confirmed	BS → MSC	DTAP
Call Proceeding	MSC → BS	DTAP
Connect	MSC → BS	DTAP
Connect Acknowledge	BS ↔ MSC	DTAP
Progress	MSC → BS	DTAP
Release	MSC ↔ BS	DTAP
Release Complete	MSC ↔ BS	DTAP
Assignment Request	MSC ↔ BS	BSMAP
Assignment Complete	BS → MSC	BSMAP
Assignment Failure	BS → MSC	BSMAP
Privacy Mode Command	MSC → BS	BSMAP
Privacy Mode Complete	BS → MSC	BSMAP
Clear Request	BS → MSC	BSMAP
Clear Command	MSC → BS	BSMAP
Clear Complete	BS → MSC	BSMAP

Table 9-2 Supplementary Service Messages

Message Name	Direction	Message Type
Send Burst DTMF[a]	BS ↔ MSC	DTAP
Send Burst Acknowledge	BS ↔ MSC	DTAP
Start DTMF	BS → MSC	DTAP
Start DTMF Acknowledge	MSC → BS	DTAP
Stop DTMF	BS → MSC	DTAP
Stop DTMP Acknowledge	MSC → BS	DTAP
Flash with Information	BS ↔ MSC	DTAP

[a] Dual-tone multifrequency

9.3.3 Radio Resource Management

After call setup, the base station is responsible for maintaining a reliable radio link between the mobile station and the base station. This responsibility requires that the base station perform the following tasks:

- Radio channel supervision
- Radio channel management
- Initiation and execution of handoffs

The objective of each of these tasks is common for all radio technologies, although the actual implementation is dependent on the associated technology.

The support of soft handoffs is one capability that distinguishes CDMA from other multiple access technologies. Thus TIA IS-634 supports the procedures associated with soft handoffs. These procedures are:

- IS-95 add target procedure
- IS-95 drop target procedure
- IS-95 drop source procedure

The source BS is the BSC that controls the transcoder. If either a hard or soft handoff is to be configured with a target BTS that is connected to another BSC (target BS), then a handoff required message is sent to the mobile switching center. The mobile switching center then sends a handoff request to the target BS. At the same time, only one target BS can be addressed.

Table 9-3 summarizes messages associated with radio resource management.

9.3.4 Mobility Management

Mobility management is implemented using DTAP messages. The purpose of the mobility management function is to support registration and deregistration of a mobile. In addition, this function encompasses authentication and voice privacy. Authentication includes the authentication challenge and the SSD update. Supporting this function has little differential impact upon the BS-MSC architecture.

Messages associated with mobility management are listed in Table 9-4.

9.3.5 Transmission Facilities Management

The transmission facilities management function is responsible for the management of terrestrial circuits. Terrestrial circuits are transmission facilities that carry traffic (voice or data) and signaling information between the MSC and the BS. Furthermore, different facilities may carry traffic information from facilities carrying signaling information. Each facility may be blocked/unblocked and allocated/deallocated by the transmission facilities management function. For digital technologies, e.g., CDMA, this function can disable the transcoders at both the originating end and the terminating end for mobile-to-mobile calls. This action eliminates the need for vocoder tandeming, which degrades the voice quality of a call. However, TIA IS-634 does not explicitly address this capability for calls spanning multiple BSs.

Table 9-3 Handoff Messages

Message Name	Direction	Message Type
Strength Measurement Request	BS ↔ MSC	BSMAP
Strength Measurement Response	BS ↔ MSC	BSMAP
Strength Measurement Report	BS ↔ MSC	BSMAP
Handoff Required	BS → MSC	BSMAP
Handoff Request	MSC → BS	BSMAP
Handoff Request Acknowledge	BS → MSC	BSMAP
Handoff Failure	BS → MSC	BSMAP
Handoff Command	MSC → BS	BSMAP
Handoff Required Reject	MSC → BS	BSMAP
Handoff Commenced	BS → MSC	BSMAP
Handoff Complete	BS → MSC	BSMAP
Handoff Performed	BS → MSC	BSMAP
Soft Handoff Drop Target	BS → MSC → BS	BSMAP
Soft Handoff Drop Source	BS → MSC → BS	BSMAP

Table 9-4 Mobility Management Messages

Message Name	Direction	Message Type
Authentication Request	MSC → BS	DTAP
Authentication Reject	MSC → BS	DTAP
SSD Update Request	MSC → BS	DTAP
Base Station Challenge	BS → MSC	DTAP
Base Station Challenge Response	MSC → BS	DTAP
SSD Update Response	BS → MSC	DTAP
Location Updating Request	BS → MSC	DTAP
Location Updating Accept	MSC → BS	DTAP
Location Updating Reject	MSC → BS	DTAP
Parameter Update Request	MSC → BS	DTAP
Parameter Update Confirm	BS → MSC	DTAP

Table 9-5 summarizes the message types that are associated with transmission facilities management.

The MTP provides a mechanism that makes the transfer of signaling messages reliable. The SCCP is used to provide a referencing mechanism to identify a particular transaction relating to, for instance, a particular call. The SCCP can also enhance the message routing for operation and maintenance information.

At the BS, only messages with a correct Destination Point Code (DPC) are accepted. Other messages are discarded. At an MSC (with the capability of acting as a signal transfer point [STP]) each message received from a BS signaling link is passed through a screening function which checks that the DPC of the message is the same as the signaling point (SP) code of the exchange. If it is the same, the message is sent to the normal MTP message-handling functions; otherwise, the message is discarded. The SP code for signaling may be included in the national SP code scheme or in a separate signaling network.

The BS exchanges signaling messages only with its MSC, where a protocol conversion may be needed in some cases. Therefore, no SCCP translation function is required in the MS between the national and the local SCCP and the MTP within the MSC area.

Several functions of the SCCP (such as error detection, receipt confirmation, and flow control) are not used on the MSC-BS interface. The segmenting/reassembling function is used if the total message length exceeds the MTP's maximum allowed message length.

9.3.6 Use of the SCCP

The MTP and the SCCP are used to support signaling messages between the MSC and the BS. Fig. 9-6 shows the A-Interface signaling protocol stack.

Table 9-5 Transmission Management Messages

Message Name	Direction	Message Type
Overload	MSC ↔ BS	BSMAP
Block	BS → MSC	BSMAP
Block Acknowledge	MSC → BS	BSMAP
Unblock	BS → MSC	BSMAP
Unblock Acknowledge	MSC → BS	BSMAP
Reset	BS ↔ MSC	BSMAP
Reset Acknowledge	BS ↔ MSC	BSMAP
Reset Circuit	BS ↔ MSC	BSMAP
Reset Circuit Acknowledge	BS ↔ MSC	BSMAP
Transcoder Control Request	MSC ↔ BS	BSMAP
Transcoder Control Acknowledge	MSC ↔ BS	BSMAP

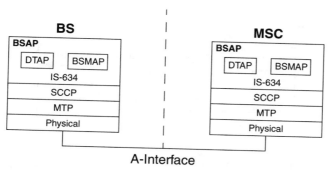

Figure 9-6 A-Interface Signaling Protocol Stack

The initial messages exchanged in call setup are used to establish an SCCP connection for subsequent signaling communications relating to the call. A new connection is established when individual information related to an MS transaction has to be exchanged between a BS and an MSC and no such transaction exists between the MSC and that BS. We need to distinguish the reasons for connection establishment:

1. A new transaction (e.g., location update, incoming or outgoing call) is initiated on the radio path.
2. Following an access request made by the MS on the random access channel, the connection establishment is then initiated by the MSC.

    The BS initiates a connection establishment when it receives the first layer 3 message from the MS. The message contains the mobile identity parameter (MIN, ESN, or IMSI). The BS then constructs the first MSC-BS Interface BSMAP message (Complete Layer 3 Information) which includes one of the appropriate DTAP messages (Location Update Request, CM Service Request, or Paging Response) depending on whether the MS is accessing the network for the purpose of registration, call origination, or call termination. The Complete Layer 3 Information message is sent to the MSC in the user data field of the SCCP Connection Request message. The Complete Layer 3 Information message includes cell identity and the layer 3 message that was received from the mobile.

    At the reception of the SCCP Connection Request message, the MSC may check, based on the received identity, whether another association already exists for the same MS. If that is the case, the connection establishment is refused. Otherwise, an SCCP Connection Confirm message is sent back to the BS. This message may optionally contain a BSMAP or DTAP message in the user data field (Fig. 9-7).
3. The MSC decides to perform an inter-BS handoff. The connection establishment is then initiated by the MSC.

    The connection establishment is undertaken by the MSC as soon as the MSC decides to perform an inter-BS handoff. An SCCP Connection Request message is sent

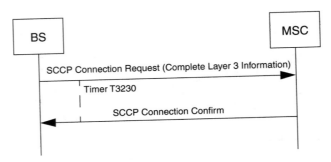

Figure 9-7   SCCP Connection Establishment

to the BS. The user data field of this message may contain the BSMAP Handoff Request Message. However, it is preferable to transfer the layer 3 messages in the user data field of the SCCP Connection Request in order to complete the establishment of the relationship between the radio channel requested and SCCP connection as soon as possible.

When it receives the SCCP Connection Request message, the BS performs the necessary checking and, in the successful case, reserves a radio channel for the requested handoff. An SCCP Connection Confirm message is also returned to the MSC and may contain the BSMAP Handoff Request Acknowledge message in the user field (Fig. 9-8).

This procedure is initiated by the MSC in normal conditions for all calls. A connection is released when a given signaling connection is no longer required. This may happen in normal cases:

- ◆ At the end of a transaction (call, location update)
- ◆ After completion of a successful external handoff—the connection with the old BS is released.

The MSC/BS sends an SCCP released (RLSD) message. The user data field of the message is optional and may contain a transparent layer 3 message (e.g., DTAP) or be empty.

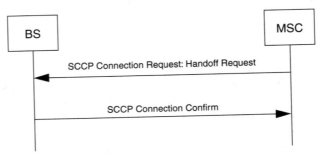

Figure 9-8   SCCP Connection Establishment during Handoff

When receiving this message, the BS/MSC releases all the radio resources allocated to the relevant MS, if there are still any left, and sends an SCCP release complete (RLC) back to the MSC/BS.

The normal release of SCCP connections is initiated by MSC. Under abnormal conditions, the SCCP connection may be released by the BS in order to clear resources. Whenever an SCCP connection is abnormally released, all resources associated with that connection are cleared. Abnormal release could result from, for example, resource failure, protocol error, or unexpected receipt of an SCCP released (RLSD) or SCCP RLC command.

The SCCP local reference number (source/destination) is a 3-byte element internally chosen by the MSC or BS to uniquely identify a signaling connection. In the direction MSC to BS, the source local reference is selected by the MSC and the destination local reference is chosen by the BS. In the direction BS to MSC, the source local reference is chosen by the BS and the destination local reference is chosen by the MSC. Note that it is the responsibility of the BS and MSC to insure that no two calls have identical SCCP local reference numbers.

9.4 Roaming

Roaming enables mobile stations to receive services outside of their home areas. When an MS is roaming, registration, call origination, and call delivery will take extra steps. Whenever the VLR tries to retrieve MS data, and data is not available, then the VLR sends a message to the appropriate HLR to retrieve the necessary data. The data consists of IMSI-to-MIN conversion, service profiles, SSD for authentication, and other data needed to process calls. The most logical time to retrieve this data is when the MS registers with the system.

Once the data on a roaming MS is stored in the VLR, then call processing for any originating services (basic or supplementary) is identical to that of home MS. However, there may be times when the MS originates a call before registration has been accomplished or when the VLR data is not available. At those times, an extra step will be added for the VLR to retrieve the data from the HLR. Thus any originating service has two optional steps where the VLR sends a message (using IS-41 signaling over SS7) to the HLR requesting data on the roaming MS, and the HLR returns a message with the proper call information.

Call delivery is not possible to an unregistered MS since the network does not know where the MS is located. Once the MS is registered with a system, then call delivery to the roaming MS is possible. This section will discuss call delivery to roaming MS in detail.

There are two types of call delivery to roaming MS—when the MS has a geographic-based directory number (indistinguishable from a wireline number), and when the MS has a nongeographic number.

We will describe the call flows for both operations.

When the MS has a geographic number, then the MSC is assigned a block of numbers that are within the local numbering plan for the area of the world where the MSC is located. Call routing to the MS is then done according to the procedures for that of a wireline telephone.[*] If an MS associated with an MSC is not in its home area, the MSC will query the HLR for the location of the MS. The MSC then invokes call forwarding to the MSC at the MS's location, and the connection is made to the second MSC where call-terminating services are delivered (refer to the procedures in section 8.2.4). This procedure is inefficient because it results in two sets of network connections—originating switch to home MSC, and home MSC to visited MSC.

Call delivery to a roaming MS is a cooperative effort among the home and visited MSCs, the VLR and HLR, and the radio system. The detailed call flow steps, for call delivery to a roaming MS with a geographic directory number (see Fig. 9-9 for the call flow diagram) are:

1. A user in the worldwide phone network (wired or wireless) dials the directory number of the MS.
2. The originating switch sends an SS7 initial address message (IAM) to the home MSC.
3. The home MSC queries the HLR for the location of the MS.
4. The HLR returns the location of the visited system.
5. The MSC invokes call forwarding to the MSC in the visited system, and the forwarding (home) MSC switch sends an SS7 IAM to the visited MSC.
6. Call processing continues to the terminating call flow (see Fig 8.12).

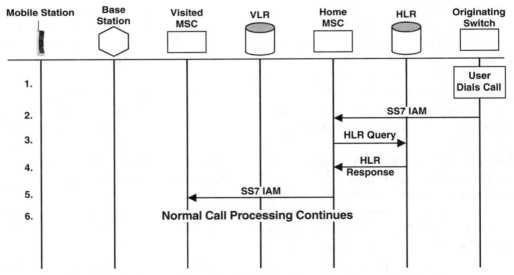

Figure 9-9   Call Termination to a Roaming Mobile Station with a Geographic Number

---

[*]For example, in Chicago, 312-944-XXXX is used by the local cellular provider for cellular phones in the downtown area. The wireline network routes calls to those numbers in a normal fashion, and calls terminate on the cellular switch.

When the MS has a nongeographic number, then calls can be directed from an originating switch directly to the visited switch. Call delivery to a nongeographic number requires the originating switch to recognize the number as a nongeographic number and do special call processing for routing. This special processing is known as Intelligent Network (IN) processing. If the originating switch does not support IN, then it will route the call to a switch that supports IN. With IN support, the originating switch will recognize the nongeographic number and send an SS7 message to the HLR with a request for the location of the MS. The HLR will return a temporary directory number (on the visited MSC) that can be used to route to the MS in the visited system. Calls then proceed according to normal termination. Call delivery to a roaming MS with a nongeographic number is, therefore, a cooperative effort between the visited MSC, the VLR and HLR, and the radio system. The detailed call flow steps for call delivery to a roaming MS with a nongeographic directory number (see Fig. 9-10 for call flow) are:

1. A user in the worldwide phone network (wired or wireless) dials the directory number of the MS.
2. The originating switch recognizes the number as a nongeographic number and sends an SS7 query message to the HLR at the home MSC.
3. The HLR returns the location of the visited system with a directory number to use for further call processing.
4. The originating switch sends an SS7 IAM to the visited MSC.
5. Call processing continues to the terminating call flow (see Fig 8.12).

9.4.1 Call Waiting

Call waiting notifies a wireless subscriber of an incoming call while the user's mobile station is in the busy state. The user can either answer or ignore the incoming call. Once the call is

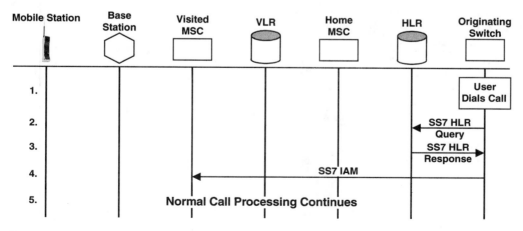

Figure 9-10 Call Termination to a Roaming Mobile Station with a Nongeographic Number

answered, the user can switch between the calls until one or more parties hang up. When either party hangs up, then the call reverts to a normal (non-call-waiting) call. If the MS user hangs up, then both calls are cleared according to normal call-clearing functions.

The detailed call flow steps for the delivery of call waiting to a mobile station (see Fig. 9-11) are:

1. User dials a call.
2. The originating switch sends an SS7 IAM to the MSC.
3. The MSC queries the VLR.
4. The VLR returns with a location of the MS that is within the serving system. If the MS is not inside the serving system, then the call is forwarded to the serving MSC.

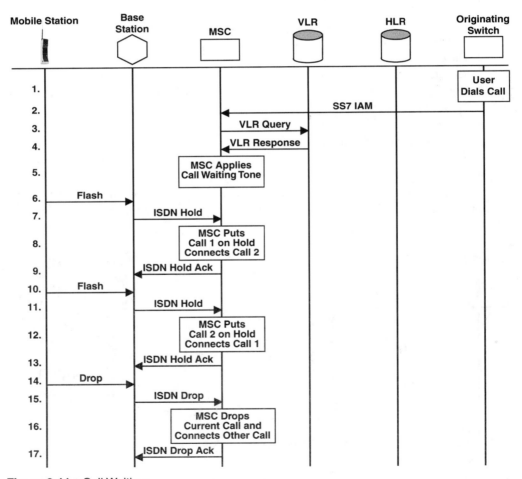

Figure 9-11   Call Waiting

5. The MSC determines that the MS is busy, subscribes to call waiting, and thus applies the call waiting tone.

6. The user presses the FLASH button (may be SEND on some MSs) to answer the call waiting indication, and the MS sends a Flash message to the base station.

7. The base station sends an ISDN Hold message to the MSC.

8. The MSC puts the first call on hold and connects the second call.

9. The MSC sends a Hold Acknowledge message to the base station.

10. The user presses the FLASH button (may be SEND on some MSs) to talk to caller 1, and the MS sends a Flash message to the base station.

11. The base station sends an ISDN Hold message to the MSC.

12. The MSC puts the second call on hold and connects the first call.

13. The MSC sends a Hold Acknowledge message to the base station.

14. The user wants to drop the current call (either 1 or 2) and pushes the DROP (or END) key, and the MS sends a Drop message to the base station.

15. The base station sends an ISDN Drop message to the MSC.

16. The MSC drops the current call and connects the other call (the one currently on hold).

17. The MSC sends an ISDN Drop Acknowledge message to the base station.

9.4.2 Handoffs

A wireless telephone (mobile station) moves around a geographic area. When the station is idle, it periodically reregisters with the system. When a call is active, then the combined mobile station, the base station, and the MSC manage the communications between the base station and mobile station so that good radio link performance is maintained. The process whereby a mobile station moves to a new traffic channel is called *handoff*. The original analog cellular system processed handoffs by commanding the mobile station to tune to a new frequency. For analog cellular, the handoff process caused a small break in time on the voice path and a noticeable "click" was heard by both parties in the telephone call. For data modems, the click often caused data errors or loss of data synchronization.

For CDMA systems, the characteristics of spread spectrum communications permit the system to simultaneously receive mobile transmissions on two or more base stations. In addition, the mobile station can simultaneously receive the transmissions of two or more base stations. With these capabilities, it is possible to process a handoff from one base station to another, or from one antenna face to another on the same base station, without any perceptible disturbance in voice or data communications.

During handoff, the signaling and voice information from multiple base stations must be combined (or bridged) in a common point with decisions made on the quality of the data. Similarly, voice and signaling information must be sent to multiple base stations and the mobile station must combine the results. The common point could be anywhere in the network, but it is typically at the mobile switching center. The call flows described here for handoff assume the MSC contains the bridging circuitry.

In CDMA, both the base station and the mobile station monitor the performance of the radio link and can request handoffs. Handoffs requested by a mobile station are called *mobile-assisted handoffs*; those requested by the base station are called *base-station assisted handoffs*. Either side can initiate the handoff process whenever the following triggers occur:

- **Base station traffic load not balanced.** The network can monitor loads at all base stations and trigger handoffs to balance loads between them to achieve higher traffic efficiency.
- **Distance limits exceeded.** Since all base stations and mobile stations are synchronized, both sides can determine base-to-mobile range. When the distance limit is exceeded, either side can request a handoff.
- **Pilot signal strength below threshold.** When the received signal strength of the pilot signal falls below a threshold, either side can initiate a handoff.
- **Power level exceeded.** When the base station commands a mobile station to increase its power and the maximum power level of the mobile station is exceeded, then either side can request a handoff.

The mobile station determines the parameters for the handoff request from the System Parameters message in the CDMA system. The message is transmitted on the system's paging channel.

As we have described, the handoff process is a cooperative effort between the old and new base stations, the mobile station, and the MSC. The following call flows are based on a frame relay A-Interface [9] between the base station and the MSC and are included as representative call flows. Actual call flows may be either standard or proprietary to an equipment vendor.

The detailed call flow steps for a CDMA soft handoff (beginning) (see Fig. 9-12) are:

1. The mobile station determines that another base station has sufficient pilot signal to be a target for handoff.
2. The mobile station sends a Pilot Strength Measurement message to the serving base station.
3. The serving base station sends an inter-BS Handoff Request message to the MSC.
4. The MSC accepts the handoff request and sends an inter-BS Handoff Request message to the target base station.
5. The target base station establishes communication with the mobile station by sending it a Null Traffic message.
6. The target base station sends a Join Request message to the MSC.
7. The MSC conferences the connections from the two base stations so the handoff can be processed without a break in the connection (i.e., soft handoff) and sends a Join Acknowledge message to the target base station.
8. The target base station sends an inter-BS Handoff Acknowledgment message to the MSC.
9. The MSC sends an inter-BS Handoff Acknowledgment message to the serving base station.

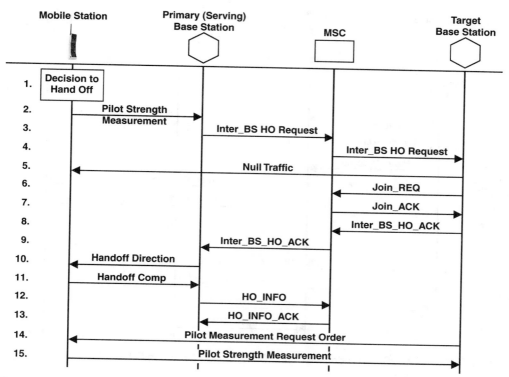

Figure 9-12 CDMA Soft Handoff—Beginning

10. The serving base station sends a Handoff Direction message to the mobile station.

11. The mobile station sends a Handoff Complete message to the serving base station.

12. The serving base station sends a Handoff Information message to the MSC.

13. The MSC confirms the message with a Handoff Information Acknowledgment message.

14. The target base station sends a Pilot Measurement Request Order message to the mobile station.

15. The mobile station sends a Pilot Strength Measurement message to the target base station.

The mobile unit is now communicating with two base stations (i.e., it is in soft handoff). Both base stations must communicate with the MSC, which then uses the highest quality signals from the two base stations and sends transmitted signals to both base stations.

After the mobile station is in soft handoff, one of the signals may fall below a predetermined threshold (based on information sent in overhead messages on the control channel) and the mobile stations will request that one base station be removed from the connection. The detailed call flow steps for a CDMA soft handoff with the serving base station dropping off (see Fig. 9-13) are:

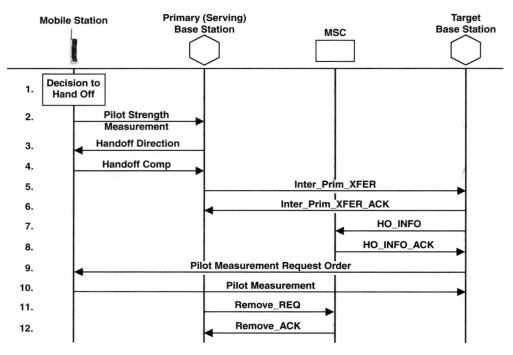

Figure 9-13 CDMA Soft Handoff—Serving Base Station Dropping Off

1. The mobile station determines that the serving base station has insufficient pilot signal to continue to be a base station in the soft handoff.
2. The mobile station sends a Pilot Strength message to the serving base station. The message requests that the base station drop off from the handoff.
3. The serving base station sends a Handoff Direction message to the mobile station that indicates which base station is to be dropped from the soft handoff (in this case, the serving base station).
4. The mobile station sends a Handoff Complete message to the serving base station.
5. The serving base station sends an Interface Primary Transfer message to the target base station with relevant call record information.
6. The target base station confirms the message with an Interface Primary Transfer Acknowledge message.
7. The target base station then sends a Handoff Information message to the MSC.
8. The MSC sends a Handoff Information Acknowledge message to the target base station.
9. The target base station sends a Pilot Measurement Request Order message to the mobile station.

10. The mobile station sends a Pilot Strength Measurement message to the target base station.

11. The old serving base station sends a Remove Request message to the MSC that requests that the base station be dropped from the connection.

12. The MSC confirms the message by sending a Remove Acknowledge message to the old serving base station.

The mobile station is now communicating with the target base station (new serving base station). If additional soft handoffs are needed, the handoff beginning procedure is repeated.

The procedures to drop a target base station from a soft handoff are similar to those that drop the serving base station. The detailed call flow steps for a CDMA soft handoff with the target base station dropping off (see Fig. 9-14) are:

1. The mobile station determines that the target base station has an insufficient pilot signal to continue to be a base station in the soft handoff.

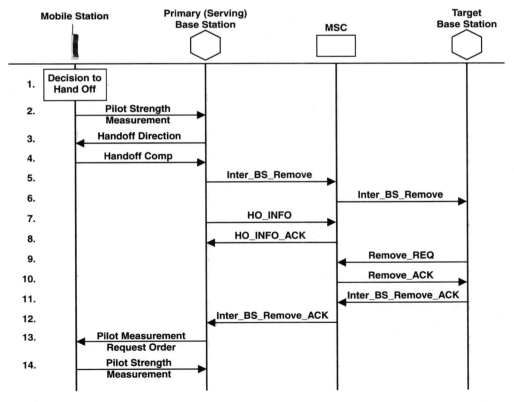

Figure 9-14 CDMA Soft Handoff—Target Base Station Dropping Off

2. The mobile station sends a Pilot Strength message to the serving base station. The message requests that the target base station drop off from the handoff.

3. The serving base station sends a Handoff Direction message to the mobile station that indicates which base station is to be dropped from the soft handoff (in this case, the target base station).

4. The mobile station sends a Handoff Complete message to the serving base station.

5. The serving base station sends an Inter-BS Remove message to the MSC.

6. The MSC sends an Inter-BS Remove message to the appropriate base station (in this case, the target base station).

7. The serving base station then sends a Handoff Information message to the MSC.

8. The MSC sends a Handoff Information Acknowledge message to the serving base station.

9. The target base station sends a Remove Request message to the MSC.

10. The MSC sends a Remove Acknowledge to the target base station.

11. After the target base station removes its resource from the call, it sends an Inter-BS Remove Acknowledge message to the MSC.

12. The MSC sends a Remove Acknowledge message to the serving base station.

13. The serving base station sends a Pilot Measurement Request Order message to the mobile station.

14. The mobile station sends a Pilot Strength Measurement message to the serving base station.

The mobile station is now communicating only with the serving base station. If additional soft handoffs are needed, the handoff beginning procedure is repeated.

9.4.3   Over-the-Air Service Provisioning (OTASP) [8]

Successful OTASP processing involves the following procedures and the exchange of some A-Interface messages (see Fig. 9-15):

- OTASP Call Setup
- OTASP Data Exchange, which includes exchange of the following messages:

  - ADDS Deliver
  - ADDS Deliver ACK

  Some existing procedures may also be applied as subprocedures:

  - SSD Update procedure (see chapter 11)
  - Privacy Mode Request procedure (see chapter 11)

- OTASP Call Clearing

1. The MS sends an Origination message over the access channel to the BS to initiate the OTASP process (see chapter 8).

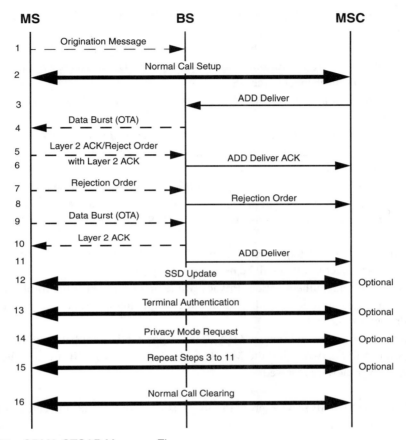

Figure 9-15 CDMA OTSAP Message Flow

2. The MSC and BS use a *normal call setup* procedure to establish the OTASP call.

3. Upon request from Over-the-Air Function (OTAF), the MSC encapsulates an OTASP data message within an ADD Deliver message and sends it to the BS.

4. The BS extracts the OTASP data message, places it in the CDMA Data Burst message, and transmits it over the traffic channel to the MS.

5. The MS may respond with a Layer 2 ACK or a Reject Order containing a Layer 2 ACK acknowledging the Data Burst message.

6. When the BS receives a Layer 2 ACK, or when a BS receives a Reject Order containing a Layer 2 ACK acknowledging the Data Burst message from the MS in response to an ADD Deliver message containing a Tag information element, it sends an ADD Deliver ACK message to the MSC with the corresponding Tag value.

7. The MS may return a Reject Order message.

8. The BS will send a Rejection message to the MSC to convey the information contained in the Reject Order message.

9. The OTASP application in the MS responds by sending an OTASP data message. The MS places the OTASP data message in the CDMA Data Burst message and transmits it over the traffic channel to the BS.

10. Upon reception of the CDMA Data Burst message, the BS responds with a Layer 2 ACK.

11. The BS extracts the OTASP data message and places it in the ADDS Deliver message to the MSC. Steps 12 through 15 are optional.

---

12. After the A-key has been derived from information transferred via the ADDS Deliver message, an SSD Update procedure over the traffic channel may also be used to exchange authentication information (RANDSSD, RANDBS, AUTHBS) (see chapter 11).

13. After an SSD Update procedure, terminal authentication needs to be performed to generate the cipher key that will be used for privacy (see chapter 11).

14. After terminal authentication, Privacy Mode procedures over the traffic channel may also be applied to specify the use of either Signaling Message Encryption (SME) or Privacy for the call.

15. Multiple forward and reverse OTASP messages can be sent between the OTASP end point in the network and the MS. The MSC and the BS will transfer the messages whenever they are received.

---

16. Once the OTASP service programming has been successful, the call can be cleared using a regular Call Clear procedure (see chapter 8).

9.5 Summary

First this chapter described the layering concept used to develop the protocols for the IS-95 CDMA and followed with a discussion of the signaling applications for a CDMA wireless telephony system. Since end-to-end call flows are not presented in any of the standards but are distributed across several standards, we described several basic and supplementary call flows.

An important component of wireless services is the ability to find and place calls to a roaming mobile station. Most mobile stations have a geographic number, but many will have nongeographic numbers in the future, so we described call flows for both. Geographic numbers are phone numbers that are located to a specific point on the worldwide phone system. Nongeographic numbers do not have a location associated with them, and the network maintains a database of the location of the phone. Additional routing steps are necessary to place a call to a mobile station with a nongeographic number.

While cellular and PCS systems (and CDMA, in particular) have a rich set of supplementary features, the most common feature is call waiting. The various standards describe additional procedures for all of the basic and supplementary services. We encourage you to consult the standards [1–7] for additional information.

Finally, since the CDMA system processes handoffs in a different way than analog cellular or TDMA cellular/PCS systems, we described the soft handoff process for CDMA and present call flows for soft handoff, beginning and ending. Also discussed was the Over-the-Air Service Provisioning (OTASP) procedure.

9.6 References

1. Committee T1, "Stage 2 Service Description for Circuit Mode Switched Bearer Services," Draft T1.704.

2. Committee T1—Telecommunications, "A Technical Report on Network Capabilities, Architectures, and Interfaces for Personal Communications," T1 Technical Report #34, May 1994.

3. EIA/TIA-553, "Cellular System Mobile Station–Land Station Compatibility Specification."

4. Garg, V. K., and Wilkes, J. E., *Wireless and Personal Communications Systems*, Prentice Hall PTR, Upper Saddle River, NJ, 1996.

5. Report of the Joint Experts Meeting on Privacy and Authentication for PCS, Phoenix, Arizona, November 8–12, 1993.

6. TIA Interim Standard, IS-41 C, "Cellular Radio Telecommunications Intersystem Operations."

7. TIA IS-95A, "Mobile Station–Base Station Compatibility Standard for Dual Mode Spread Spectrum Cellular System."

8. TIA IS-634, "MSC-BS Interface for 800 MHz," 1995.

9. TR-45 Contribution, "Frame Relay A-Interface."

Soft Handoff and Power Control in IS-95 CDMA

10.1 Introduction

Soft handoff is different from the traditional hard-handoff process. With hard handoff, a definite decision is made on whether to hand off or not. The handoff is initiated and executed without the user attempting to have simultaneous traffic channel communications with the two base stations. With soft handoff, a *conditional* decision is made on whether to hand off. Depending on the changes in pilot signal strength from the two or more base stations involved, a hard decision will eventually be made to communicate with only one. This normally happens after it is evident that the signal from one base station is considerably stronger than those from the others. In the interim period, the user has simultaneous traffic channel communication with all candidate base stations.

It is desirable to implement soft handoff in power-controlled CDMA systems because implementing hard handoff is potentially difficult in such systems. A system with power control attempts to dynamically adjust transmitter power while in operation. Power control is closely related to soft handoff. IS-95 uses both power control and soft handoff as an interference-reduction mechanism. Power control is the main tool used in IS-95 to combat the near-far problem. It is theoretically unnecessary to have power control if one can successfully implement a more intelligent receiver than that used in IS-95, which is the subject of the field of multiuser detection (MUD), a feature being proposed for the 3G CDMA systems. Power control is necessary in order for a CDMA system to achieve a reasonable level of performance in practice. The use of power control in the CDMA system necessitates the use of soft handoff when the original and new channels occupy the same frequency band. For power control to work properly, the mobile must attempt to be linked at all times to the base station from which it receives the strongest signal. If this does not happen, a positive power control feedback loop could inadvertently occur, causing system problems. Soft handoff can guarantee that the mobile is indeed linked at all times to the base station from which it receives the strongest signal, whereas hard handoff cannot guarantee this.

The performance of CDMA systems is very sensitive to differences in received signal powers from various users on the reverse link. Due to the nonorthogonality of the spreading PN codes used by different users, a strong interfering signal may mask out a weak desired signal, causing unreliable detection of the latter. This is called the *near-far problem.*

This chapter first covers handoff strategy used in IS-95 CDMA and then focuses on power control schemes for the reverse and forward link.

10.2 Types of Handoff

There are four types of handoff:

1. **Intersector or softer handoff.** The mobile communicates with two sectors of the same cell (see Fig. 10-1). A RAKE receiver at the base station combines the best versions of the voice frame from the diversity antennas of the two sectors into a single traffic frame.
2. **Intercell or soft handoff.** The mobile communicates with two or three sectors of different cells (see Fig. 10-2). The base station that has the direct control of call processing

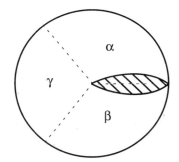

Figure 10-1 Softer Handoff

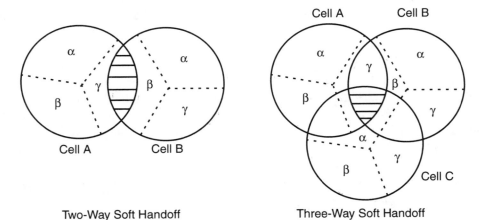

Two-Way Soft Handoff   Three-Way Soft Handoff

Figure 10-2 Soft Handoff

during handoff is referred to as the *primary base station*. The primary base station can initiate the forward control message. Other base stations that do not have control over call processing are called the *secondary base stations*. Soft handoff ends when either the primary or secondary base station is dropped. If the primary base station is dropped, the secondary base station becomes the new primary for this call. A three-way soft handoff may end by first dropping one of the base stations and becoming a two-way soft handoff.

The base stations involved coordinate handoff by exchanging information via SS7 links. A soft handoff uses considerably more network resources than the softer handoff.

3. **Soft-softer handoff.** The mobile communicates with two sectors of one cell and one sector of another cell (see Fig. 10-3). Network resources required for this type of handoff include the resources for a two-way soft handoff between cell A and B plus the resources for a softer handoff at cell B.

4. **Hard handoff.** Hard handoffs are characterized by the *break-before-make* strategy. The connection with the old traffic channel is broken before the connection with the new traffic channel is established. Scenarios for hard handoff include

   ◆ Handoff between base stations or sectors with different CDMA carriers
   ◆ Change from one pilot to another pilot without first being in soft handoff with the new pilot (disjoint active sets)
   ◆ Handoff from CDMA to analog, and analog to CDMA
   ◆ Change of frame offset assignment—CDMA traffic frames are 20 ms long. The start of frames in a particular traffic channel can be at 0 time in reference to a system or it can be offset by up to 20 ms (allowed in IS-95). This is known as the *frame offset*. CDMA traffic channels are assigned different frame offset to avoid congestion. The frame offset for a particular traffic channel is communicated to the mobile. Both forward and reverse links use this offset. A change in offset

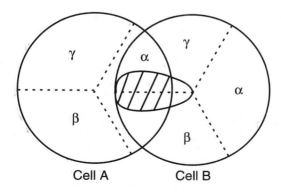

Figure 10-3 Soft-Softer Handoff

assignment will disrupt the link. During soft handoff the new base station must allocate the same frame offset to the mobile as assigned by the primary base station. If that particular frame offset is not available, a hard handoff may be required. Frame offset is a network resource and can be used up.

10.2.1 Soft Handoff (Forward Link)

In this case all traffic channels assigned to the mobile are associated with pilots in the active set and carry the same traffic information with the exception of power control subchannel. When the active set contains more than one pilot, the mobile provides diversity by combining its associated forward traffic channels.

10.2.2 Soft Handoff (Reverse Link)

During intercell handoff, the mobile sends the same information to both base stations. Each base station receives the signal from the mobile with appropriate propagation delay. Each base station then transmits the received signal to the vocoder/selector. In other words, two copies of the same frame are sent to the vocoder/selector. The vocoder/selector selects the better frame and discards the other.

10.2.3 Softer Handoff (Reverse Link)

During intersector handoff, the mobile sends the same information to both sectors. The channel card/element at the cell site receives the signals from both sectors. The channel card combines both inputs, and only one frame is sent to the vocoder/selector. It should be noted that extra channel cards are not required to support softer handoff as is the case for soft handoffs. The diversity gain from soft handoffs is more than the diversity gain from softer handoffs because signals from distinct cells are less correlated than signals from sectors of the same cell.

10.2.4 Benefit of Soft Handoff

A key benefit of soft handoff is the path diversity on the forward and reverse traffic channels. Diversity gain is obtained because less power is required on the forward and reverse links. This implies that total system interference is reduced. As a result, the average system capacity is improved. Also less transmit power from the mobile results in longer battery life and longer talk time.

In a soft handoff, if a mobile receives an *up* power control bit from one base station and a *down* control bit from the second base station, the mobile decreases its transmit power. The mobile obeys the *power down* command since a good communications link must have existed to warrant the command from the second base station.

10.3 Pilot Sets

The term *pilot* refers to a pilot channel identified by a pilot sequence offset and a frequency assignment. A pilot is associated with the forward traffic channels in the same forward CDMA link.

Each pilot is assigned a different offset of the same short PN code. The mobile search for pilots is facilitated by the fact that the offsets are the integer multiples of a known time delay (64 chips offset between adjacent pilots). All pilots in a pilot set have the same CDMA frequency assignment. The pilots identified by the mobile, as well as other pilots specified by the serving sectors (neighbors of the serving base stations/sectors), are continuously categorized by the mobile into four groups.

- **Active set.** It contains the pilots associated with the forward traffic channels (Walsh codes) assigned to the mobile. Because there are three fingers of the RAKE receiver in the mobile, the active set size is a maximum of three pilots. IS-95 allows up to six pilots in the active set, with two pilots sharing one RAKE finger.The base station informs the mobile about the contents of the active set by using the Channel Assignment message and/or the Handoff Direction message (HDM). An active pilot is a pilot whose paging or traffic channels are actually being monitored or used.
- **Candidate set.** This set contains the pilots that are not currently in the active set. However, these pilots have been received with sufficient signal strength to indicate that the associated forward traffic channels could be successfully demodulated. Maximum size of the candidate set is six pilots.
- **Neighbor set.** This set contains neighbor pilots that are not currently in the active or the candidate set and are likely candidates for handoff. Neighbors of a pilot are all the sectors/cells that are in its close vicinity. The initial neighbor list is sent to the mobile in the System Parameter message on the paging channel. The maximum size of the neighbor set is 20.
- **Remaining set.** This set contains all possible pilots in the current system, excluding pilots in the active, candidate, or neighbor sets.

While searching for a pilot, the mobile is not limited to the exact offset of the short PN code. The short PN offsets associated with various multipath components are located a few chips away from the direct path offset. In other words, the multipath components arrive a few chips later relative to the direct path component. The mobile uses the *search window* for each pilot of the active and candidate set, around the earliest arriving multipath component of the pilot. Search window sizes are defined in number of short PN chips. The mobile should center the search window for each pilot of the neighbor set and the remaining set around the pilot's PN offset using the mobile time reference.

10.4 Search Windows

The mobile uses the following three search windows to track the received pilot signals:

- SRCH_WIN_A: search window size for the active and candidate sets
- SRCH_WIN_N: search window size for the neighbor set
- SRCH_WIN_R: search window size for the remaining set

10.4.1 SRCH_WIN_A

SRCH_WIN_A is the search window that the mobile uses to track the active and candidate set pilots. This window is set according to the anticipated propagation environment—it should be large enough to capture all usable multipath signal components of a base station, and at the same time it should be as small as possible in order to maximize searcher performance.

EXAMPLE   10.1

Consider the propagation environment of a CDMA network, where the signal with a direct path travels 1 kilometer (km) to the mobile, whereas the multipath travels 5 km before reaching the mobile. What should be the size of SRCH_WIN_A?

$$\text{Direct path travels a distance of } \frac{1000}{244} = 4.1 \text{ chips}$$

$$\text{Multipath travels a distance of } \frac{5000}{244} = 20.5 \text{ chips}$$

The difference in distance traveled between the two paths = 20.5 – 4.1 = 16.4 chips
The window size $\geq 2 \times 16.4 = 32.8$ chips
Use window size = 33 chips

EXAMPLE   10.2

Consider cells A and B separated by a distance of 12 km. The mobile travels from cell A to cell B. The RF engineer wishes to contain the soft handoff area between points X and Y located at distance 6 and 10 km from cell A (see Fig. 10-4). What should be the search window size?

At point X the mobile is 6000/244 = 24.6 chips from cell A

At point X the mobile is 10,000/244 = 41.0 chips from cell B

Path difference = 41.0 – 24.6 = 16.4 chips

At point Y the mobile is 10,000/244 = 41.0 chips from cell A

At point Y the mobile is 6000/244 = 24.6 chips from cell B

Path difference = 41.0 – 26.4 = 16.4 chips

The SRCH_WIN_A > 2×16.4 > 32.8 chips

This way, as the mobile travels from cell A to cell B, the mobile can ensure that, beyond Y, the pilot from cell A drops out of the search window.

10.4.2 SRCH_WIN_N

SRCH_WIN_N is the search window that the mobile uses to monitor the neighbor set pilots. The size of this window is typically larger than that of SRCH_WIN_A. The window needs to be large enough not only to capture all usable multipath of the serving base station's signal, but also to capture the potential multipath of neighbors' signals. In this case, we need to take into account multipath and path differences between the serving base station and neighbor-

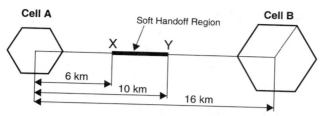

Figure 10-4 SRCH_WIN_A for Soft Handoff between X and Y

ing base stations. The maximum size of this search window is limited by the distance between two neighboring base stations. Let's consider two neighboring base stations located at a distance of 6 km. The mobile is located right next to base station 1, and, therefore, the propagation delay from base station 1 to the mobile is negligible. The distance between base station 2 and mobile is 6 km. The distance in chips is 6000/244 = 24.6 chips. The search window shows that the pilot from cell 2 arrives 24.6 chips later at the mobile. Thus, in order for a mobile (located within cells 1 and 2) to search pilots of potential neighbors, SRCH_WIN_N needs to be set according to the physical distances between the current base station and its neighboring base station. The actual size may not be this large, since this is an upper bound for SRCH_WIN_N.

10.4.3 SRCH_WIN_R

SRCH_WIN_R is the search window that the mobile uses to track the remaining set pilots. A typical requirement for the size of this window is that it is at least as large as SRCH_WIN_N.

10.5 Handoff Parameters

There are four handoff parameters. T_ADD, T_COMP, and T_DROP relate to the measurement of pilot E_c/I_t and T_TDROP is a timer. Whenever the strength of a pilot in the active set falls below a value of T_DROP, a timer is started by the mobile. If the pilot strength goes back above T_DROP, the timer is reset; otherwise the timer expires when a time T_TDROP has elapsed since the pilot strength has fallen below T_DROP. Mobile maintains a handoff drop timer for each pilot in the active set and in the candidate set.

10.5.1 Pilot Detection Threshold (T_ADD)

Any pilot that is strong but is not in the HDM is a source of interference. This pilot must be immediately moved to the active set for handoff to avoid voice degradation or a possible dropped call. T_ADD affects the percentage of mobiles in handoff. It should be low enough to quickly add useful pilots and high enough to avoid false alarms due to noise.

10.5.2 Comparison Threshold (T_COMP)

It has effect on handoff percentage similar to T_ADD. It should be low for faster handoff and should be high to avoid false alarms.

10.5.3 Pilot Drop Threshold (T_DROP)

It affects the percentage of mobiles in handoff. It should be low enough to avoid dropping a good pilot that goes into a short fade. It should be high enough not to quickly remove useful pilots in the active or candidate set. The value of T_DROP should be carefully selected by considering the values of T_ADD and T_TDROP.

10.5.4 Drop Timer Threshold (T_TDROP)

It should be greater than the time required to establish handoff. T_TDROP should be small enough not to quickly remove useful pilots. A large value of T_TDROP may be used to force a mobile to continue in soft handoff in a weak coverage area.

Table 10-1 provides typical values of the handoff parameters.

10.6 Handoff Messages

Handoff messages in IS-95 are Pilot Strength Measurement message (PSMM), Handoff Direction message (HDM), Handoff Completion message (HCM), and Neighbor List Update message (NLUM).

The mobile detects pilot strength (E_c/I_t) and sends the PSMM to the base station. The base station allocates the forward traffic channel and sends the HDM to the mobile. On receiving the HDM, the mobile starts demodulation of the new traffic channel and sends HCM to the base station.

The PSMM contains the following information for each of the pilot signals received by the mobile:

- Estimated E_c/I_t
- Arrival time
- Handoff drop timer

The HDM contains the following information:

- HDM sequence number
- CDMA channel frequency assignment
- Active set (now has old and new pilots [PN offsets])
- Walsh code associated with each pilot in the active set

Table 10-1 Handoff Parameter Values

Parameter	Range	Suggested Value
T_ADD	−31.5 to 0 dB	−13 dB
T_COMP	0 to 7.5 dB	2.5 dB
T_DROP	−31.5 to 0 dB	−15 dB
T_TDROP	0 to 15 seconds	2 seconds

- Window size for the active and candidate sets
- Handoff parameters (T_ADD, T_DROP, T_COMP, T_TDROP)

The HCM contains the following information:

- A positive acknowledgment
- PN offset of each pilot in the active set

The NLUM is sent by the base station. It contains the latest composite neighbor list for the pilots in the active set.

The mobile continuously tracks the signal strength for all pilots in the system. The signal strength of each pilot is compared with the various thresholds such as the pilot detection threshold, the pilot drop threshold, the comparison threshold, and the drop timer threshold.

A pilot is moved from one set to another depending on its signal strength relative to the thresholds. Fig. 10-5 shows a sequence on the threshold.

1. Pilot strength exceeds T_ADD. Mobile sends a PSMM and transfers pilot to the candidate set.
2. Base station sends an HDM to the mobile with the pilot to be added in active set.
3. Mobile receives HDM and acquires the new traffic channel. Pilot goes into the active set and mobile sends HCM to the base station.
4. Pilot strength drops below T_DROP; mobile starts the handoff drop timer.

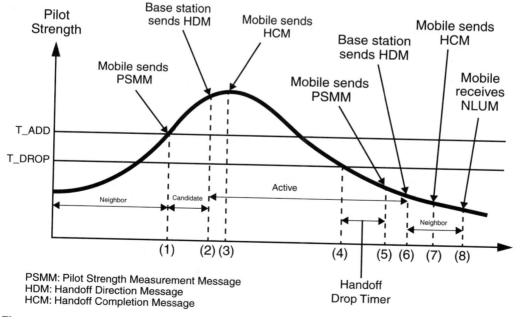

Figure 10-5 Handoff Threshold Example: Pilot Thresholds

5. Handoff drop timer expires. Mobile sends a PSMM to the base station.

6. Base station sends an HDM without related pilot to the mobile.

7. Mobile receives HDM. Pilot goes into the neighbor set and mobile sends HCM to the base station.

8. The mobile receives an NLUM which does not include the pilot. Pilot goes into the remaining set.

The mobile maintains a T_TDROP for each pilot in the active set and candidate set. The mobile starts the timer whenever the strength of the corresponding pilot becomes less than a preset threshold. The mobile resets and disables the timer if the strength of the corresponding pilot exceeds the threshold.

When a member of the neighbor or remaining set exceeds T_ADD, the mobile moves the pilot to candidate set (Fig. 10-6) and sends a PSMM to the base station. As the signal strength of candidate pilot P_c gradually increases, it rises above the active set pilot, P_a. A PSMM is sent to the base station only if

$$P_c - P_a > \text{T\_COMP x 0.5 dB}$$

where P_a and P_c are the strength of pilots in active and candidate sets.

10.7 Handoff Procedures

10.7.1 Mobile-Assisted Soft-Handoff (MASHO) Procedures

The mobile monitors the Forward Pilot Channel (FPICH) level received from neighboring base stations and reports to the network those FPICHs that cross a given set of thresholds. Two types of thresholds are used: the first to report FPICHs with sufficient power to be used for coher-

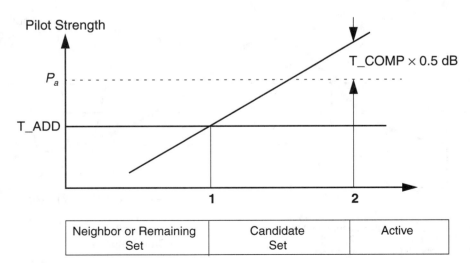

Figure 10-6   Pilot Movement from Neighbor or Remaining Set to Active Set

ent demodulation, and second to report those FPICHs whose power has dropped to a level where it is not beneficial to use them for coherent demodulation. The margin between the two thresholds provides a hysteresis to avoid a ping-pong effect due to variations in FPICH power. Based on this information, the network instructs the mobile to add or remove FPICHs from its active set.

The same user information, modulated by the appropriate base station code, is sent from multiple base stations. Coherent combining of different signals from different sectorized antennas, from different base stations, or from the same antennas but on different multiple path components is performed in the mobile using RAKE receivers. A mobile will typically place at least one RAKE receiver finger on the signal from each base station in the active set. If the signal from the base station is temporarily weak, then the mobile can assign the finger to a stronger base station.

The signal transmitted by a mobile is processed by base stations with which the mobile is in soft handoff. The received signal from different sectors of a base station is combined in the base station on a symbol-by-symbol basis. The received signal from different base stations can be selected in the infrastructure (on a frame-by-frame basis). Soft handoff results in increased coverage range and capacity on the reverse link.

10.7.2 Dynamic Soft-Handoff Thresholds

While soft handoff improves overall system performance, it may in some situations negatively impact system capacity and network resources. On the forward link, excessive handoff reduces system capacity whereas, on the reverse link, it costs more network resources (backhaul connections).

Adjusting the handoff parameters at the base stations will not necessarily solve the problem. Some locations in the cell receive only weak FPICHs (requiring lower handoff thresholds), and other locations receive a few strong and dominant FPICHs (requiring higher handoff thresholds). The principle of dynamic threshold for adding FPICHs is as follows:

- The mobile detects FPICHs that cross a given static threshold, T_1. The metric for the FPICH in this case is the ratio of FPICH energy per chip to total received power (E_c/I_t).
- On crossing the static threshold, the FPICH is moved to a candidate set. It is then searched more often and tested against a second dynamic threshold, T_2.
- Comparison with T_2 determines if the FPICH is worth adding to the active set. T_2 is a function of the total energy of FPICHs demodulated coherently (in the active set).
- The condition of an FPICH for crossing T_2 is expressed as

$$10\log(P_{cj}) \geq \text{Max}\left\{ \text{SOFT-SLOPE} \cdot 10\log\left(\sum_{i=1}^{N_A} P_{ai} \right) + \text{ADD-INTERCEPT}, T_1 \right\} \quad (10.1)$$

where P_{cj} = strength of the jth FPICH in the coordinate set,
P_{ai} = strength of the ith FPICH in the active set,
N_A = number of FPICHs in the active set, and
SOFT-SLOPE and ADD-INTERCEPT = adjustable system parameters.

When FPICHs in the active set are weak, adding an additional FPICH (even weak) will improve performance. However, when there is one or more dominant FPICHs, adding an additional weaker FPICH above T_1 will not improve performance, but will use more network resources. The dynamic soft-handoff thresholds reduce and optimize the network resource utilization.

- After detecting an FPICH above T_2, the mobile reports it back to the network. The network then sets up the handoff resources and orders the mobile to coherently demodulate this additional FPICH. Pilot 2 is added to active set.
- When the FPICH (pilot 1) strength decreases below a dynamic threshold T_3, the handoff connection is removed. The FPICH is moved back to the candidate set. The threshold T_3 is a function of the total energy of FPICHs in the active set. FPICHs not contributing sufficiently to total FPICH energy are dropped. If it decreases below a static threshold T_4, an FPICH is removed from the candidate set.
- An FPICH dropping below a threshold (e.g., T_3 and T_4) is reported back to the network only after being below the threshold for a specific period of time. This timer allows for a fluctuating FPICH not to be prematurely reported.

Fig. 10-7 shows a time representation of soft handoff and associated events when the mobile station moves away from a serving base station (FPICH 1) toward a new base station (FPICH 2). The combination of static and dynamic thresholds (vs. static thresholds only) results in reduced soft-handoff regions (see Fig. 10-7). The major benefit of this is to limit soft handoff to areas and times when it is most beneficial.

1. When pilot 2 exceeds T_1, mobile moves it to the candidate set.
2. When pilot 2 exceeds T_2 (dynamic), mobile reports it back to the network.
3. Mobile receives an order to add pilot 2 to the active set.
4. Pilot 1 drops below T_3 (relative pilot 2).
5. Handoff timer expires on pilot 1. Mobile reports pilot strength to the network.
6. Mobile receives an order to remove pilot 1.
7. Handoff timer expires after pilot 1 drops below T_4.

10.8 Setup and End of Soft Handoff

10.8.1 Setup

One of the major benefits of a CDMA system is the ability of a mobile to communicate with more than one base station at one time during a call. This functionality allows the CDMA network to perform soft handoff. In soft handoff a controlling primary base station coordinates with other base stations as they are added or deleted for the call. This allows the base stations (up to three, total) to receive/transmit voice packets with a single mobile for a single call.

Each base station transmits the received mobile voice packets to the BSC/MSC. The BSC/MSC selects the best voice frame from one of the three base stations. This provides the PSTN party with the best-quality voice.

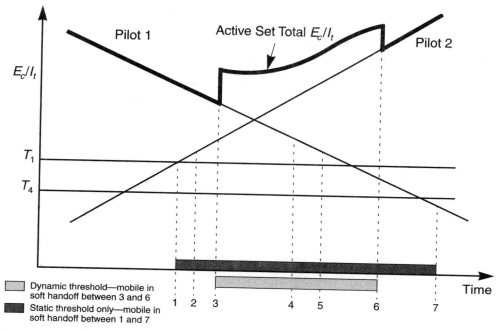

Figure 10-7 Dynamic Thresholds Handoff Procedure

Fig. 10-8 shows a mobile communicating with two base stations for one call. This is called a *two-way soft handoff*. Steps of soft handoff are

- The mobile detects a pilot signal from a new cell and informs primary base station A.
- A communications path from base station B to the original frame selector is established.
- The frame selector selects frames from both streams.
- The mobile detects that base station A's pilot is failing and requests that this path be dropped.
- The path from original base station A to the frame selector is dropped.

Base station B gives base station A its assigned Walsh code. Base station A gives the mobile the Walsh code of B as part of the HDM. Now the mobile can listen to base station B.

Base station A gives the user's long-code mask to base station B. Now B can listen to the mobile. Both base stations A and B receive forward link power control information back from the mobile and act accordingly. The mobile receives independent puncture bits from both A and B. If directions conflict, the mobile decreases power; otherwise the mobile obeys directions.

10.8.2 End of Soft Handoff

Fig. 10-9 shows the process used by a mobile communicating with two base stations A and B to end handoff when the signal from base station A is not strong enough. When the mobile

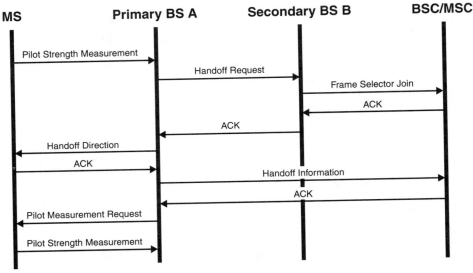

Figure 10-8   Soft Handoff Setup

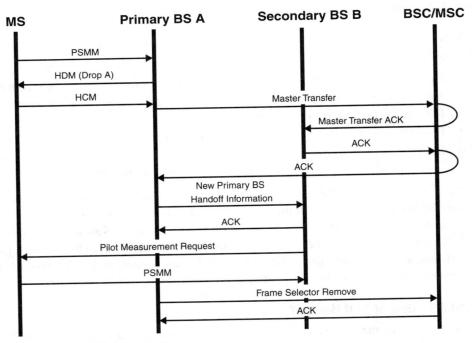

Figure 10-9   End of Soft Handoff

entered into soft handoff with base stations A and B, the primary base station was A. However, when the mobile drops A and starts communicating with base station B alone, B becomes the new primary base station.

10.9 Maintenance of Pilot Sets

10.9.1 Active Set Maintenance

The active set is initialized to contain only one pilot (e.g., the pilot associated with the assigned forward traffic channel). This occurs when the mobile is first assigned a forward traffic channel. As the mobile processes HDMs, it updates the active set with the pilots listed in the HDMs.

A pilot P_c from the candidate is added to the active set when P_c exceeds a member of the active set by T_COMP. A pilot P_a from the active set is removed when P_a has dropped below T_DROP and the drop timer (T_TDROP) has expired (see Fig. 10-10).

10.9.2 Candidate Set Maintenance

The candidate set is initialized to contain no pilot. This happens when the mobile is first assigned a forward traffic channel. A pilot P_n from the neighbor set is added to the candidate set when its strength exceeds T_ADD. Also, a pilot P_r from the remaining set is moved to the candidate set when its strength exceeds T_ADD. A pilot P_c is deleted from the candidate set when the handoff drop timer corresponding to P_c has expired. Also, when the candidate set size has been exceeded, the pilot P_c, whose handoff drop timer is close to expiring, is deleted from the candidate set (see Fig. 10-11).

10.9.3 Neighbor Set Maintenance

The neighbor set is initialized to contain the pilots specified in the most recently received Neighbor List message. This happens when the mobile is first assigned a forward traffic channel. The mobile maintains a counter—AGE—for each pilot in the neighbor set. If a pilot moves from the active set or candidate set to neighbor set, its counter is initialized to 0. However, if a pilot

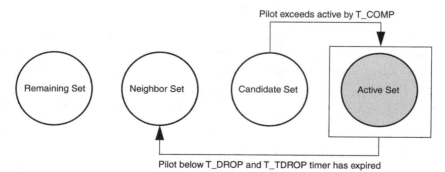

Figure 10-10 Active Set Maintenance

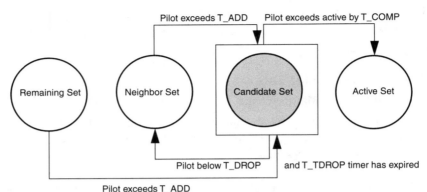

Figure 10-11   Candidate Set Maintenance

moves from the remaining set to the neighbor set, its counter is set to the maximum age value (see Fig. 10-12). The mobile adds a pilot in the neighbor set under the following conditions:

- A pilot in the active set is not contained in the HDM, and the corresponding handoff drop timer has expired.
- The handoff drop timer of a pilot in the candidate set has expired.
- A new pilot to the candidate set causes the candidate set size limit to be exceeded.
- The pilot is contained in the Neighbor List message and is not already a pilot of the candidate set or neighbor set.

The mobile deletes a pilot in the neighbor set under the following conditions:

- The HDM contains a pilot from the current neighbor set.
- The strength of a pilot in the neighbor set exceeds T_ADD.
- A new pilot to the neighbor set causes the size limit of the neighbor set to be exceeded.
- A neighbor set pilot's AGE exceeds the maximum value of the AGE counter.

10.10 The Need for Power Control

CDMA is an interference-limited system—since all mobiles transmit at the same frequency, internal interference generated within the system plays a critical role in determining system capacity and voice quality. The transmit power from each mobile must be controlled to limit interference. However, the power level should be adequate for satisfactory voice quality.

As the mobile moves around, the RF environment changes continuously due to fast and slow fading, shadowing, external interference, and other factors. The objective of power control is to limit transmitted power on the forward and reverse links while maintaining link quality under all conditions. Due to noncoherent detection at the base station, interference on the reverse link is more critical than it would be on the forward link. Reverse link power control is therefore essential for a CDMA system and is enforced by the IS-95 standard.

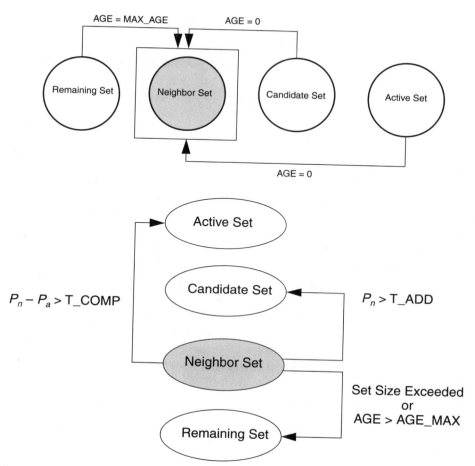

Figure 10-12 Neighborhood Set Maintenance

Power control is also needed in CDMA systems to resolve the near-far problem. To minimize the near-far problem, the goal in a CDMA system is to assure that all mobiles achieve the same received power levels at the base station. The target value for the received power level must be the minimum level possible that allows the link to meet user-defined performance objectives (BER, FER, capacity, dropped-call rate, and coverage). In order to implement such a strategy, the mobiles closer to the base station must transmit less power than those far away.

Voice quality is related to frame-error rate (FER) on both the forward and reverse link. The FERs are largely correlated to E_b/I_t. The FER also depends on vehicle speed, local propagation conditions, and distribution of other cochannel mobiles. Since the FER is a direct measure of signal quality, the voice quality performance in a CDMA system is measured in terms of FERs rather than E_b/I_t. Thus, to assure good signal quality, it is not sufficient to maintain a target E_b/I_t;

it is also necessary to respond to specific FERs as they occur. The recommended performance bounds are

- A typical recommended range for FER—0.2% to 3% (optimum power level is achieved when FER ≤ 1%)
- A maximum length of burst error—3 to 4 frames (optimum value of burst error ≈ 2 frames)

10.11 Reverse Link Power Control

The reverse link power control affects the access and reverse traffic channels. It is used for establishing the link while originating a call and reacting to large path-loss fluctuations. The reverse link power control includes the open-loop power control (also known as autonomous power control) and the closed-loop power control. The closed-loop power control involves the inner-loop power control and the outer-loop power control.

10.11.1 Reverse Link Open-Loop Power Control

The open-loop power control is based on the principle that a mobile closer to the base station needs to transmit less power as compared to a mobile that is farther away from the base station or is in fade. The mobile adjusts its transmit power based on total power received in the 1.23-MHz band (i.e., power in pilot, paging, sync, and traffic channels). This includes power received from all base stations on the forward link channels. If the received power is high, the mobile reduces its transmit power. On the other hand, if the power received is low, the mobile increases its transmit power.

In open-loop power control the base station is not involved. The mobile determines the initial power transmitted on the access channel and traffic channel through open-loop power control. A large dynamic range of 80 dB is allowed to provide an ability to guard against deep fades.

The mobile acquires the CDMA system by receiving and processing the pilot, sync, and paging channels. The paging channel provides the Access Parameters message which contains the parameters to be used by the mobile when transmitting to the base station on an access channel. The access parameters are

- The access channel number
- The nominal power offset (NOM_PWR)
- The initial power offset step size
- The incremental power step size
- The number of access probes per access probe sequence
- The time-out window between access probes
- The randomization time between access probe sequences

Based on the information received on the pilot, sync, and paging channels, the mobile attempts to access the system via one of several available access channels. During the access state, the mobile has not yet been assigned a forward link traffic channel (which contains the power control bits). Since the reverse link closed-loop power control is not active, the mobile initiates, on its own, any power adjustment required for a suitable operation.

The prime goal in CDMA systems is to transmit just enough power to meet the required performance objectives. If more power is transmitted than necessary, the mobile becomes a jammer to other mobiles. Therefore, the mobile tries to get the base station attention first by transmitting at very low power. The key rule is that the mobile transmits in inverse proportion to what it receives.

When receiving a strong pilot from the base station, the mobile transmits a weak signal back to the base station. A strong signal at the mobile implies a small propagation loss on the forward link. Assuming the same path loss on the reverse link, only a low transmit power is required from the mobile in order to compensate for the path loss.

When receiving a weak pilot from the base station, the mobile transmits back a strong signal. A weak received signal at the mobile indicates a high propagation loss on the forward link. Conversely, a high transmit power level is required from the mobile.

The mobile transmits the first access probe at a mean power level defined by

$$T_x = -R_x - K + (NOM\text{-}PWR - 16 \times NOM\text{-}PWR\text{-}EXT) + INIT\text{-}PWR \text{ (dBm)} \qquad (10.2)$$

where
T_x	= mean output transmit power (dBm),
R_x	= mean input receive power (dBm),
NOM-PWR	= nominal power (dB),
NOM-PWR-EXT	= nominal power for extended handoff (dB),
INIT-PWR	= initial adjustment (dB),
K	= 73 for cellular (Band Class 0), and
K	= 76 for PCS (Band Class 1).

If *INIT-PWR* were 0, then *NOM-PWR* – 16 × *NOM-PWR-EXT* would be the correction that should provide the correct received power at the base station. *NOM-PWR* – 16 × *NOM-PWR-EXT* allows the open-loop estimation process to be adjusted for different operating environment.

The values for *NOM-PWR*, *NOM-PWR-EXT*, *INIT-PWR*, and the step size of a single access probe correction *PWR-STEP* are system parameters specified in the Access Parameters message. These are obtained by the mobile station prior to transmitting. If, as the result of an Extended Handoff Direction message or a General Handoff Direction message, the *NOM-PWR* and *NOM-PWR-EXT* values change, the mobile uses the *NOM-PWR* and *NOM-PWR-EXT* values from the Extended Handoff Direction message or a General Handoff Direction message.

The total range of the *NOM-PWR* – 16 × *NOM-PWR-EXT* correction is –24 to 7 dB. While operating in Band Class 0, *NOM-PWR-EXT* is set to 0, making the total range of correction from –8 to 7 dB. The range of the *INIT-PWR* parameter is –16 to 15 dB, with a nominal value of 0 dB. The range of the *PWR-STEP* parameter is 0 to 7 dB. The accuracy of the adjustment to the mean output power due to *NOM-PWR*, *NOM-PWR-EXT*, *INIT-PWR*, or a single access probe correction of *PWR-STEP* should be ±0.5dB or ±20%, whichever is greater.

The major flaw with this criterion is that reverse link propagation statistics are estimated based on forward link propagation statistics. But, since the two links are not correlated, a significant error may result from this procedure. However, these errors will be corrected once the closed-loop power control mechanism becomes active as the mobile seizes a forward traffic channel and begins to process power control bits.

After the Acknowledgment time window (T_a) has expired, the mobile waits for an additional random time (RT) and increases its transmit power by a step size. The mobile tries again. The process is repeated until the mobile gets a response from the base station. However, there is a maximum number of probes per probe sequence and a maximum number of probe sequences per access attempt.

The entire process to send one message and receive an acknowledgment for the message is called an *access attempt*. Each transmission in the access attempt is referred to as an *access probe*. The mobile transmits the same message in each access probe in an access attempt. Each access probe contains an access channel preamble and an access channel capsule (see Fig. 10-13). Within an access attempt, access probes are grouped into access probe sequences. Each access probe sequence consists of up to 16 access probes, all transmitted on the same access channel.

There are two reasons that could prevent the mobile from getting an acknowledgment after the transmission of a probe.

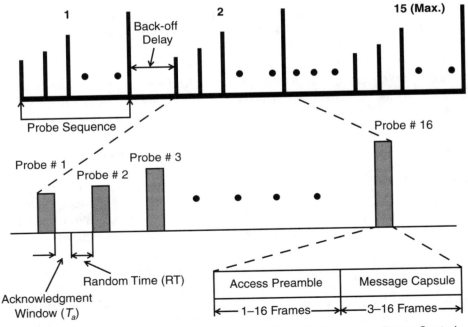

Figure 10-13   Access Attempt, Probe Sequence, and Probe in Open-Loop Power Control

1. The transmit power level might be insufficient. In this case, the incremental step power strategy helps to resolve the problem.
2. There might be a collision due to the random contention of the access channel by several mobiles. In this case, the random waiting time minimizes the probability of future collisions.

The process is shown by the access probe ladder in Fig. 10-14.

The transmit power is defined by

$$T_x = -R_x - K + (NOM\text{-}PWR - 16 \times NOM\text{-}PWR\text{-}EXT)$$
$$+ \text{Sum of Access Probe Corrections} \qquad (10.3)$$

where the access probe correction is the sum of all the appropriate incremental power steps prior to receiving an acknowledgment at the mobile.

For every access probe sequence, a back-off delay is generated pseudorandomly. Timing between access probes of an access probe sequence is also generated pseudorandomly. After transmitting each access probe, the mobile waits for T_a. If an acknowledgment is received, the access attempt ends. If no acknowledgment is received, the next access probe is transmitted after an additional random time (see Fig. 10-13).

If the mobile does not receive an acknowledgment within an access attempt, the attempt is considered as a failure and the mobile tries to access the system at another time. If the mobile receives an acknowledgment from the base station, it proceeds with the registration and traffic channel assignment procedures. The initial transmission on the reverse traffic channel shall be at a mean output power defined by Eq. (10.3).

The mobile station supports a total combined range of initial offset parameters, closed *NOM-PWR*, and access probe corrections of at least ±32dB for mobile stations operating in Band Class 0 and ±40dB for mobile stations operating in Band Class 1.

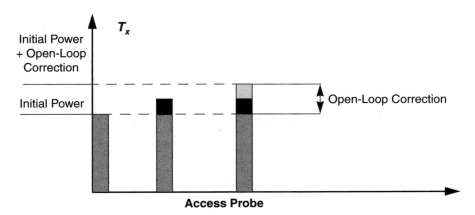

Figure 10-14 Access Probe Ladder

The sources of error in the open-loop power control are

- Assumption of reciprocity on the forward and reverse links
- Use of total received power including power from other base stations
- Slow response time ~ 30 ms to counter fast fading due to multipath

10.11.2 Reverse Link Closed-Loop Power Control

Fading sources in multipath require a much faster power control than the open-loop power control. The additional power adjustments required to compensate for fading losses are handled by the reverse link closed-loop power control mechanism, which has a response time of 1.25 ms for 1-dB steps and a dynamic range of 48 dB (covered in 3 frames). The quicker response time gives the closed-loop power control mechanism the ability to override the open-loop power control mechanism in practical applications. Together, two independent power control mechanisms cover a dynamic range of at least 80 dB. The closed-loop power control provides correction to the open-loop power control. Once on the traffic channel, the mobile and base stations engage in closed-loop power control.

The reverse link closed-loop power control mechanism consists of two parts—inner-loop power control and outer-loop power control. The inner-loop power control keeps the mobile as close to its target $(E_b/I_t)_{\text{setpoint}}$ as possible, whereas the outer-loop power control adjusts the base station target $(E_b/I_t)_{\text{setpoint}}$ for a given mobile.

To understand the operation of the closed-loop power control mechanism, let's review the structure of the forward traffic channel and its operation. The areas of focus are the output of the interleaver and the input to the MUX. A power control subchannel continuously transmits on the forward traffic channel. This subchannel runs at 800 power control bits per second. Therefore, a power control bit (0 or 1) is transmitted every 1.25 ms. A 0 bit indicates to the mobile that it should increase its mean output power level, whereas 1 indicates to the mobile to decrease its mean output power level.

A 20-ms frame is organized into 16 time intervals of equal duration (see Fig. 10-15). These time intervals, each of 1.25 ms, are called Power Control Groups (PCGs). Thus, a frame has 16 PCGs. Prior to transmission, the reverse traffic channel interleaver output data stream is gated with a time filter. The time filter allows transmission of some symbols and deletion of

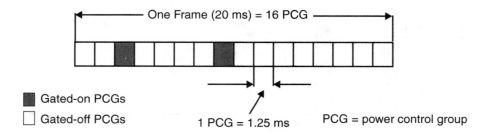

Figure 10-15 Power Control Groups

others. The duty cycle of the transmission gate varies with the transmit data rate, i.e., variable rate vocoder output, which, in turn, depends on the voice activity. Table 10-2 indicates the number of PCGs that are sent at different frame rates.

The assignment of the gated-on and gated-off groups is determined by the Data Burst Randomizer (DBR). At the base station, the reverse link receiver estimates the received signal strength by measuring E_b/I_t during each power group (1.25 ms).

- If the signal strength exceeds a target value, a power-down power control bit 1 is sent.
- Otherwise a power-up control bit 0 is transmitted to the mobile via the power control subchannel on the forward link.

Similar to the reverse link transmission, the forward link transmissions are organized in 20-ms frames. Each frame is subdivided into 16 PCGs. The transmission of a power control bit occurs on the forward traffic channel in the second PCG following the corresponding reverse link PCG in which the signal strength was estimated. For example, if the signal strength is estimated on PCG #2 of a reverse link frame, then the corresponding power control bit must be sent on PCG #4 of the forward link frame (see Fig. 10-16). Once the mobile receives and processes the forward link channel, it extracts the power control bits from the forward traffic channel. The power control bits then allow the mobile to fine-tune its transmit power on the reverse link.

Based on the power control bit received from the base station, the mobile either increases or decreases transmit power on the reverse traffic channel as needed to approach the target value of $(E_b/I_t)_{nom}$ or set point that controls the long-term FER. Each power bit produces a 1-dB change in mobile power, i.e., it attempts to bring the measured E_b/I_t value 1 dB closer to its target value. Note that it might not succeed because I_t is also always changing. Therefore, further adjustments

Table 10-2 Power Control Groups vs. Frame Rate

Frame Rate	Rate (kbps)	No. of PCGs Sent
Full	9.6	16
1/2	4.8	8
1/4	2.4	4
1/8	1.2	2

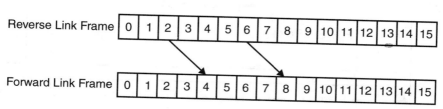

Figure 10-16 PCG Location in Reverse and Forward Link Frames

may be required to achieve the desired E_b/I_t. The base station, through the mobile, can directly change only E_b, not I_t, and the objective is the ratio of E_b to I_t, not any particular value for E_b or I_t.

The base station measures E_b/I_t 16 times in each 20-ms frame. If the measured E_b/I_t is greater than the current target value of E_b/I_t, the base station informs the mobile to decrease its power by 1 dB. Otherwise, the base station orders the mobile to increase its power by 1 dB (see Fig. 10-17).

The relationship between E_b/I_t and the corresponding FER is nonlinear and varies with vehicle speed and RF environment. Performance deteriorates with increasing vehicle speed. The best performance corresponds to a stationary vehicle where additive white Gaussian noise dominates. Thus, a single value of E_b/I_t is not satisfactory for all conditions. The use of a single, fixed value for E_b/I_t could reduce channel capacity by 30% or more by transmitting excessive, unneeded power.

The value of the variable a is kept very small (see Fig. 10-18), so it may take 35 frames to reduce the E_b/I_t set point by 1 dB. Typically, the value of $100a$ is set at about 3 dB. The set point value is reduced by a for each consecutive frame until a frame error occurs. The set point is then increased by a relatively large amount and the process is repeated. The set point can range from 3 dB to 10 dB. A value of $E_b/I_t \geq 5$ dB corresponds to good voice quality.

Since FER is a direct measure of link quality, the system is controlled using the measured FERs rather than E_b/I_t. FER is the key parameter in controlling and assuring a satisfactory voice quality. It is not sufficient to maintain a target E_b/I_t, but it is necessary to control FERs as they occur. The objective of the Reverse Outer-Loop Power Control (ROLPC) is to balance the desired FER on the reverse link and system capacity. System capacity can be controlled with the

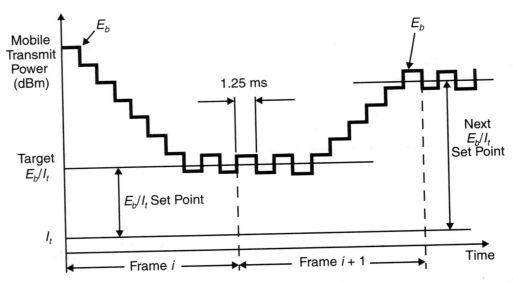

Figure 10-17 Target E_b/I_t

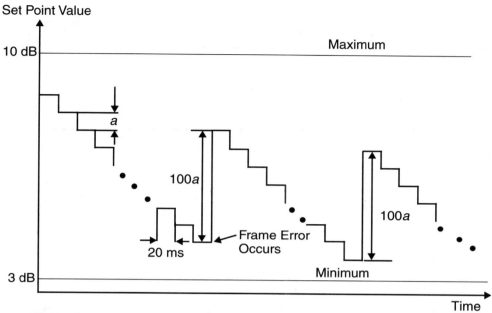

Figure 10-18 Set Point Value vs. Time

ROLPC parameters by increasing the acceptable FER. Changing FER can be accomplished by setting the ratio of *down_frr* to *up_frr*. The *down_frr* is calculated by the system by using the desired reverse FER (*rfer*) and *up_frr* as

$$down\_frr = (rfer \times up\_frr)/2 \tag{10.4}$$

Based on simulations, the following values for *up_frr* are suggested:

If $(0.2\% \leq rfer \leq 0.4\%)$, $up\_frr = 6000$

If $(0.6\% \leq rfer \leq 1.0\%)$, $up\_frr = 5000$

If $(1.2\% \leq rfer \leq 2.0\%)$, $up\_frr = 3000$

If $(2.2\% \leq rfer \leq 3.0\%)$, $up\_frr = 1000$

Tables 10-3 and 10-4 lists the range and default values of different parameters for RS1 and RS2.

The inner-loop power control is also responsible for detecting the mobile that fails to respond to power control and that may be causing interference to other mobiles. The base station counts the number of consecutive power decrease commands, and, if the count exceeds the specified threshold value, the base station will send a Lock until Power Cycle message to the mobile. This message disables the mobile until the user turns the power off and on. Fig. 10-19 gives the flow chart for the reverse link closed-loop power control.

Table 10-3 ROLPC Parameters for RS1

Parameter	Range	Suggested Value	Description of Parameter
rfer 1	0.2–3.0%	1%	target reverse link FER (*rfer*)
$(E_b/I_t)_{\text{nom 1}}$ (dB)	3.5–8.0%	6.5%	initial $(E_b/I_t)_{\text{set point}}$
$(E_b/I_t)_{\text{max 1}}$ (dB)	5.5—9.5%	8.5%	maximum $(E_b/I_t)_{\text{set point}}$
$(E_b/I_t)_{\text{min 1}}$ (dB)	3.0–5.8%	3.5%	minimum $(E_b/I_t)_{\text{set point}}$

Table 10-4 ROLPC Parameters for RS2

Parameter	Range	Suggested Value	Description of Parameter
rfer 2	0.2–6.0%	1%	target reverse link FER
$(E_b/I_t)_{\text{nom 2}}$ (dB)	3.8–8.3%	6.8%	initial $(E_b/I_t)_{\text{set point}}$
$(E_b/I_t)_{\text{max 2}}$ (dB)	5.8–9.8%	8.8%	maximum $(E_b/I_t)_{\text{set point}}$
$(E_b/I_t)_{\text{min 2}}$ (dB)	3.0–5.8%	3.8%	minimum $(E_b/I_t)_{\text{set point}}$

The mobile power output with both open-loop and closed-loop power control is given as

$$T_x = -R_x - K + (NOM\text{-}PWR - 16 \times NOM\text{-}PWR\text{-}EXT) + INIT\text{-}PWR$$
$$+ \text{ Sum of Access Probe Corrections}$$
$$+ \text{ Sum of all Closed-Loop Power Control Corrections} \qquad (10.5)$$

10.12 Forward Link Power Control

Forward link power control (FLPC) aims at reducing interference on the forward link. The FLPC not only limits the in-cell interference, but it is especially effective in reducing other cell/sector interference.

The forward link power control attempts to set each traffic channel transmit power to the minimum required to maintain the desired FER at the mobile. The mobile continuously measures forward traffic channel FER. It reports this measurement to the base station on a periodic basis. After receiving the measurement report, the base station takes the appropriate action to increase or decrease power on the measured logical channel. The base station also restricts the power dynamic range so that the transmitter power never exceeds a maximum value that would cause excessive interference or so that it never falls below the minimum value required for adequate voice quality.

Since FERs are measured (not E_b/I_t as in the closed inner-loop strategy), this process is a direct reflection of voice quality. However, it is a much slower process. Because orthogonal Walsh codes are employed for the forward link instead of long PN codes, cochannel interference is not an urgent issue. Therefore, slow measurements do not add much degradation to system performance. Fig. 10-20 is a flow chart for the FLPC process.

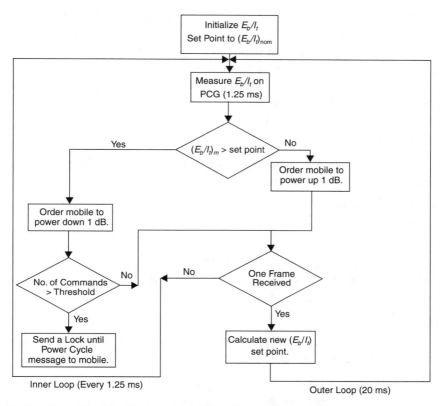

Figure 10-19 Flow Chart for Reverse Link Closed-Loop Power Control

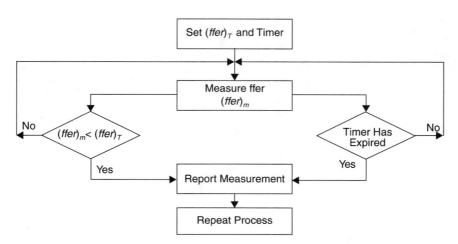

Figure 10-20 Flow Chart for Forward Link Power Control

Forward link power control is expressed in terms of parameters N, D, U, and V (see Fig. 10-21), which may be adjusted to various values for the operation of an actual system.

For RS1, the Power Measurement Report message (PMRM) contains the number of error frames received and the total number of frames received during the interval covered by the report (frame counters then are reset for the next report interval). The FER is equal to the number of error frames divided by the total number of frames received in the reporting interval. The following are the steps for forward link power control for RS1 (see Fig. 10-21).

Action by Mobile

- Mobile keeps track of the number of error frames in a period of length *pwr_rep_frame*.
- If error frames > a specified number, the mobile sends a PMRM containing:

  - ◆ Total number of frames in *pwr_rep_frame*
  - ◆ Number of error frames in *pwr_rep_frame*
  - ◆ FER

- If error frames < a specified number, a PMRM is not sent.
- After sending a PMRM, the mobile waits for a period—*pwr_rep_delay*—before starting a new period.

Action by Base Station

- On receiving the PMRM, the base station compares the reported FER as follows and adjusts traffic channel power.

  - ◆ FER < *fer_small* → reduce power by D
  - ◆ *fer_small* < FER < *fer_big* → increase power by U
  - ◆ FER > *fer_big* → increase power by V

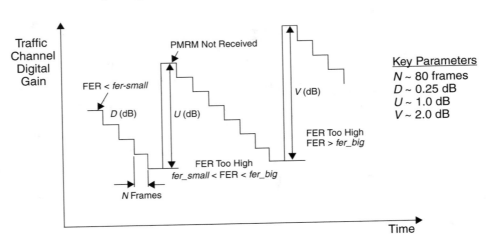

Figure 10-21 Forward Link Power Control for RS1

- If no PMRM is received

  - ◆ Base station starts a timer *fpc_step*.
  - ◆ When timer expires, power level is reduced by *D*.
  - ◆ The timer resets after it expires or after receipt of a PMRM.

- Digital gain is never set below *min_gain* or above *max_gain*.
- If *flpc_enable* = 0, digital gain is set to *nom_gain*.

For RS2, 1 bit per reverse link frame (the E or erasure bit) is dedicated to inform the base station whether or not the last forward link frame was received without error at the mobile. This allows more rapid and precise control of forward link power than the scheme used for RS1. The following are the steps for forward link power control for RS2 (see Fig. 10-22).

Forward Link Power Control with RS2

- Uses erasure indicator bit instead of PMRM
- Much faster than RS1 implementation

  - ◆ Forward link power control could change every 2 frames; thus, its response is very fast.

- Process

  - ◆ In each frame, the mobile sends an erasure indicator bit showing whether the previous forward frame had an erasure bit or not.
  - ◆ If an erasure is indicated by the mobile, the base station increases traffic channel digital gain by *dn_adj*.

Tables 10-5 and 10-6 list the values of the parameters for forward link power control for RS1 and RS2, respectively.

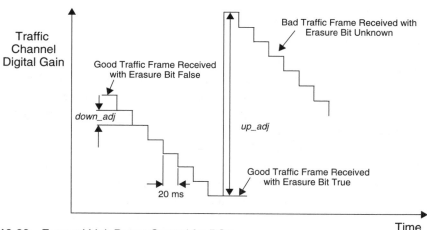

Figure 10-22 Forward Link Power Control for RS2

Table 10-5 Forward Link Power Control Parameters for RS1

Parameters	Range	Suggested Value	Description
FER	0.2–3%	1%	target forward FER
fer_small	0.2–5%	2%	lower forward link FER threshold minimum PMRM FER required to increase gain by U
fer_big	2–10%	6%	upper forward link FER threshold minimum PMRM FER required to increase gain by V
min_gain	34–50	40	minimum traffic channel digital gain
max_gain	50–108	80	maximum traffic channel digital gain
nom_gain	34–108	57	nominal traffic channel digital gain
fpc_step	20–5000 ms	1600 ms	forward power control timer value which determines when gain is decreased by D

Table 10-6 Forward Link Power Control Parameters for RS2

Parameter	Range	Suggested Value	Description
FER	0.2–6%	1%	target forward FER
up_adj	1–50	15	gain increase when forward erasure is observed
dn_adj	—	N/A	gain decrease when no forward erasure is observed
min_gain	30–50	30	minimum traffic channel digital gain
max_gain	50–127	127	maximum traffic channel digital gain
nom_gain	40–108	80	nominal traffic channel digital gain

where $dn\_adj = (up\_adj \times \text{FER})/100$

10.13 Summary

This chapter covered soft handoff and power control in IS-95 CDMA. Soft handoff provides path diversity on the forward and reverse links. Diversity gains are achieved because less power is required on the forward and reverse links. This results in the reduction of total system interference and an increase in system capacity.

Since the RF environment changes continuously due to fast and slow fading, shadowing, external interference, and other factors, the aim of power control is to adjust the transmitted power on the forward and reverse link while maintaining link quality under all operating conditions. Power control in the CDMA system is required to resolve the near-far problem. To minimize the near-far problem, the goal in a CDMA system is to assure that all mobile stations achieve the same received power levels at the base station.

The reverse link power control includes the open-loop power control and the closed-loop power control. The open-loop power control is too slow to counter fast fading due to multipath. The closed-loop power control provides correction to the open-loop power control. It begins after acquiring the traffic channel and is directed by the base station. The closed-loop power control occurs every 1.25 ms and is much faster and more effective than the open-loop power control. With the closed-loop power control, power can change ±16 dB per 20-ms frame.

10.14 References

1. Garg, V. K., Smolik, K. F., and Wilkes, J. E., *Application of CDMA in Wireless/Personal Communications,* Prentice Hall PTR, Upper Saddle River, NJ, 1997.

2. TIA/EIA IS-95A, "Mobile Station–Base Station Compatibility Standard for Dual-Mode Wideband Spread Spectrum Cellular Systems," May 1995.

3. TIA/EIASP-3693, "Mobile Station–Base Station Compatibility Standard for Dual-Mode Wideband Spread Spectrum Cellular Systems," November 18, 1997.

4. Wang, S. W., and Wang, I., "Effects of Soft Handoff, Frequency Reuse and Non-Ideal Antenna Sectorization on CDMA System Capacity," *Proc. IEEE VTC*, Secaucus, NJ, May 1993, pp. 850–54.

5. Wong, Daniel, and Lim, T. J., "Soft Handoff in CDMA Mobile Systems," *IEEE Personal Communications,* 4(6), December 1997.

Security and Identification in IS-95 CDMA

11.1 Introduction

This chapter looks at various parameters that are used to identify a mobile station, including International Mobile Station Identity (IMSI), Mobile Directory Number (MDN), Electronic Serial Number (ESN), and station class mark. Then the focus shifts to authentication procedures—the authentication of MS registration, of MS originations, of MS terminations, of MS data bursts, and of Temporary Mobile Station Identity (TMSI) assignment—and unique challenge response procedures. This chapter also covers the procedure to update Shared Secret Data (SSD) as well as the parameter update and voice privacy procedures.

11.2 Mobile Identification Parameters

11.2.1 Mobile Station Identification Number

Mobile stations operating in the CDMA mode are identified by International Mobile Station Identity (IMSI). The IMSI contains up to 15 numerical characters (0–9). The first three digits of the IMSI are the Mobile Country Code (MCC), and the remaining digits are the National Mobile Station Identity (NMSI). The NMSI consists of the Mobile Network Code (MNC) and Mobile Station Identification Number (MSIN). The IMSI structure is shown in Fig. 11-1.

An IMSI that is 15 digits in length is called a Class 0 IMSI (the NMSI is 12 digits in length); an IMSI that is less than 15 digits in length is called a Class 1 IMSI (the NMSI is less than 12 digits in length).

IMSI_M is an IMSI that contains a Mobile Identification Number (MIN) in the lower 10 digits of the NMSI. An IMSI_M can be a Class 0 or Class 1 IMSI.

IMSI_T is an IMSI that is associated with the MIN assigned to the mobile. An IMSI_T can be a Class 0 or Class1 IMSI. When operating in the CDMA mode, MS sets its operational IMSI value, IMSI_O, to either IMSI_M or IMSI_T depending on the capabilities of the BS.

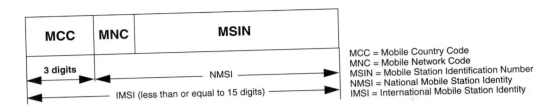

Figure 11-1 IMSI Structure

An IMSI_S is a 10-digit (34-bit) number derived from an IMSI. When an IMSI has 10 or more digits, IMSI_S is equal to the last 10 digits. When an IMSI has fewer than 10 digits, the least significant digits of IMSI_S are equal to the IMSI and 0s are added to the most significant side to obtain a total of 10 digits. A 10-digit IMSI_S consists of 3- and 7-digit parts, called IMSI_S2 and IMSI_S1, respectively, as shown in Fig. 11-2.

The IMSI_S derived from IMSI_M is designated IMSI_M S. The IMSI_S derived from IMSI_T is designated IMSI_T S. The IMSI_S derived from IMSI_O is designated IMSI_O S. When an IMSI has 12 or more digits, IMSI_11 12 is equal to the 11th and 12th digits of the IMSI. When an IMSI has fewer than 12 digits, digits with value equal to 0 are added to the most significant side to obtain a total of 12 digits and the IMSI_11 12 is equal to the 11th and 12th digits of the resulting number. For encoding various types of IMSI, refer to the IS-95B standards [5].

11.2.2 Mobile Directory Number

A Mobile Directory Number (MDN) is a dialable number associated with the mobile station through a service subscription. An MDN is not necessarily the same as the mobile station's identification on the air interface, i.e., MIN, IMSI_M, or IMSI_T. An MDN consists of up to 15 digits. The MS should have memory to store at least one MDN.

11.2.3 Electronic Serial Number (ESN)

The ESN is a 32-bit binary number that uniquely identifies the mobile station to any wireless system.

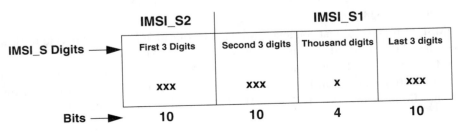

Figure 11-2 IMSI_S Binary Mapping

11.2.4   Station Class Mark

Class-of-station information referred to as the station class mark is stored in an MS. The digital representation of this class mark is given for Band Class 0 and Band Class 1 in Tables 11-1 and 11-2.

11.2.5   Registration Memory

The MS should have memory to store one element in the zone-based registration list. This stored element includes both REG_ZONE and the corresponding (SID, NID) pair. The data should be retained for at least 48 hours under power-off conditions. If, after 48 hours, the data integrity cannot be guaranteed, then the entry should be deleted upon power-on.

The MS should also have memory to store one element in the system/network registration list SID_NID_LIST. The data should be retained for at least 48 hours under power-off conditions. If, after 48 hours, the data integrity cannot be guaranteed, then the entry in SID_NID_LIST should be deleted upon power-on.

Table 11-1   Station Class Mark for Band Class 0

Function	Bits	Setting	
Extended SCM indicator	7	always 0	0xxxxxxx
Dual mode	6	CDMA only	x0xxxxxx
		dual mode	x1xxxxxx
Slotted class	5	nonslotted	xx0xxxxx
		slotted	xx1xxxxx
IS-54 power class	4	always 0	xxx0xxxx
25-MHz bandwidth	3	always 1	xxxx1xxx
Transmission	2	continuous	xxxxx0xx
		discontinuous	xxxxx1xx
Power class	1–0	Class I	xxxxxx00
		Class II	xxxxxx01
		Class III	xxxxxx10
		Class IV	xxxxxx11

Table 11-2   Station Class Mark for Band Class 1

Function	Bits	Setting	
Extended SCM indicator	7	always 1	1xxxxxxx
Reserved	6	always 0	x0xxxxxx
Slotted class	5	nonslotted	xx0xxxxx
		slotted	xx1xxxxx
Reserved	$4 - 0_s$	all 0s	xxx00000

The MS should have memory to store the distance-based registration variables BASE_LAT_REG, BASE_LONG_REG, and REG_DIST_REG. The data should be retained for at least 48 hours under power-off conditions. If, after 48 hours, the data integrity cannot be guaranteed, then REG_DIST_REG should be set to 0 upon power-on.

11.2.6   Access Overload Class

The 4-bit access overload class (ACCOLC) indicator is used to identify which overload class controls access attempts by the MS; it is also used to identify redirected overload classes in global service redirection. For mobile stations that are classified as overload classes ACCOLC 0 through ACCOLC 9, the MS's 4-bit ACCOLC indicator is derived from the last digit of the associated decimal representation of the IMSI_M by a decimal-to-binary conversion (see Table 11-3). When a mobile station's IMSI_M is updated, the mobile station must recalculate the ACCOLC as indicated above. Mobile stations designated for test use are assigned to ACCOLC 10; mobile stations designated for emergency use are assigned to ACCOLC 11. ACCOLC 12 through ACCOLC 15 are reserved. The 4-bit ACCOLC indicators for overload Classes 10 through 15 are specified in Table 11-4.

11.2.7   Home System and Network Identification

In addition to the HOME_SID parameter that the mobile stores for 800-MHz analog operation, the mobile station stores at least one home (SID, NID) pair. The mobile station stores the 1-bit parameters MOB_TERM, MOB_TERM_FOR_SID, and MOB_TERM_FOR_NID.

11.2.8   Local Control Option

When a mobile station supports the local control option, a means is provided within the mobile station to enable or disable the local control option.

Table 11-3   ACCOLC Indicator Mapping for Overload Classes 0 through 9

Last Digit of the Decimal Representation of the IMSI	ACCOLC
0	0000
1	0001
2	0010
3	0011
4	0100
5	0101
6	0110
7	0111
8	1000
9	1001

Table 11-4 ACCOLC Indicator Mapping for Overload Classes 10 through 15

Overload Class	ACCOLC
10	1010
11	1011
12	1100
13	1101
14	1110
15	1111

11.2.9 Preferred Operation Selection

When the mobile supports operation in Band Class 0, a means is provided within the mobile to identify the preferred system as either system A or system B. In addition, the mobile station may provide a means to allow operation with only system A or system B.

When the mobile station supports operation in Band Class 0, means may be provided within the mobile station to identify the preferred operation type—either CDMA or analog mode. In addition, the mobile station may provide a means to allow operation with only the analog or CDMA mode.

11.2.10 Discontinuous Reception

The mobile station provides memory to store the preferred slot cycle index: SLOT_CYCLE_INDEX.

11.3 Authentication Procedures

In the authentication process, information is exchanged between an MS and a BS to confirm the identity of the MS. A successful authentication process occurs only when it is demonstrated that the MS and BS possess identical sets of Shared Secret Data (SSD).

The MS uses the operational IMSI (IMSI_O) for authentication purposes, and the BS uses the IMSI associated with the last MS registration. Table 11-5 summarizes the input parameters of the AUTH_SIGNATURE procedure for each of its uses in IS-95 standards.

SSD is a 128-bit quantity that is stored in semipermanent memory in the MS and is readily available to the BS. The SSD is not passed across the air interface between the MS and the network nor across the MSC-BS interface. As shown in Fig. 11-3, SSD is divided into two subsets—SSD_A and SSD_B. SSD_A is used to support the authentication procedures and SSD_B is used to support voice privacy and signaling message encryption. SSD is generated using the procedure discussed in section 11.4.

- **Rand Challenge Memory (RAND).** When operating in 800-MHz analog mode, RAND is a 32-bit value held in the MS. When operating in CDMA mode, it is equal to

Table 11-5   AUTH_SIGNATURE Input Parameters

Procedure	RAND_ CHALLENGE	ESN	AUTH_DATA	SSD_AUTH	SAVE_ REGISTERS
Registration	RAND	ESN	IMSI_S1	SSD_A	False
Unique challenge	RANDU & 8 LSBs of IMSI_S2	ESN	IMSI_S1	SSD_A	False
Originations	RAND	ESN	Digits	SSD_A	True
Terminations	RAND	ESN	IMSI_S1	SSD_A	True
MS data bursts	RAND	ESN	Digits	SSD_A	False
BS challenge	RANDBS	ESN	IMSI_S1	SSD_A_NEW	False
TMSI assignment	RAND	ESN	IMSI_S1	SSD_A	False

SSD_A	SSD_B
64 bits	64 bits

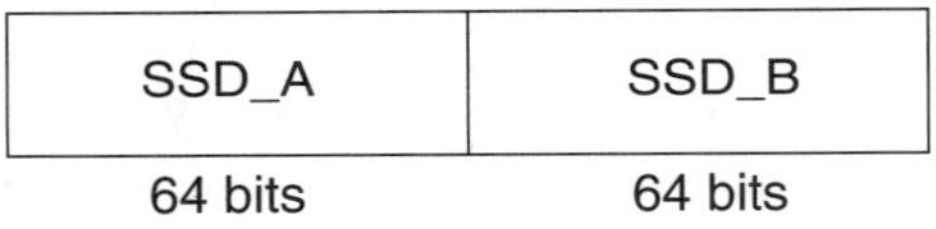

Figure 11-3   Partitioning of SSD

the RAND value received in the last Access Parameters message of the CDMA paging channel. RAND is used along with SSD_A and other parameters, as appropriate, to authenticate MS originations, terminations, and registrations.

- **Call History Parameter (COUNT).** COUNT is a modulo-64 count held in the MS and is updated by the MS when a Parameter Update order is received on the CDMA forward traffic channel.
- **A-Key.** A-key is 64 bits long and is assigned to the mobile station (Fig. 11-4). It is stored in the mobile station's permanent security identification memory and is known only to the mobile station and its associated HLR/AC.
- **Temporary Mobile Station Identity (TMSI).** TMSI is a temporary, locally assigned number used to address the MS. The MS obtains a TMSI when assigned by the BS. As a number, the TMSI does not have any association with the MS, IMSI, ESN, or directory numbers, all of which are permanent identifications.

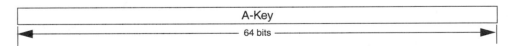

A-Key
◄──────────────── 64 bits ────────────────►

Figure 11-4   A-key

11.3.1 Authentication of MS Registrations

The following authentication procedures are performed when AUTH is set to 01 (standard authentication mode), and the MS attempts to register by sending a Registration message on the access channel.

The MS sets the input parameters of the AUTH_SIGNATURE procedure (see Fig. 11-5) and sets the SAVE_REGISTERS input parameter to FALSE. The MS then executes the AUTH_SIGNATURE procedure. The 18-bit output AUTH_SIGNATURE is used to fill the AUTHR field of the Registration message. The RANDC (the 8 most significant bits of the RAND) and COUNT fields of the message are filled with the current values stored in the MS.

The BS compares the received value of RANDC to the most significant 8 bits of its internally stored value of RAND. The BS may also compare the received value of COUNT with its internally stored value associated with the received IMSI/ESN. The BS computes the value of AUTHR in the same manner as the MS, but it uses its internally stored value of SSD_A. The BS compares its computed value of AUTHR to the value received from the MS.

If any comparisons fail, the BS may deem the registration attempt unsuccessful, initiate the unique challenge response procedure, or commence the process of updating the SSD.

11.3.2 Unique Challenge Response Procedure

The unique challenge response procedure is initiated by the BS and can be carried out either on the paging and access channels or on the forward and reverse traffic channels. The BS generates the 24-bit RANDU and sends it to the MS in the Authentication Challenge message on either the paging channel or the forward traffic channel. Upon receipt of the Authentication Challenge message, the MS sets the input parameters of the AUTH_SIGNATURE procedure (Fig. 11-6). The 24 most significant bits of the RAND_CHALLENGE input parameter are used with RANDU, and the 8 least significant bits of RAND_CHALLENGE are used with the 8 least significant bits of IMSI_S2. The MS sets the SAVE_REGISTERS input parameter to FALSE.

The MS then executes the AUTH_SIGNATURE procedure. The 18-bit output AUTH_SIGNATURE is used to fill the AUTHU field of the Authentication Challenge Response message, which is sent to the BS. The BS computes the value of AUTHU in the same way as the MS, but uses its internally stored value of SSD_A. The BS compares its computed value of

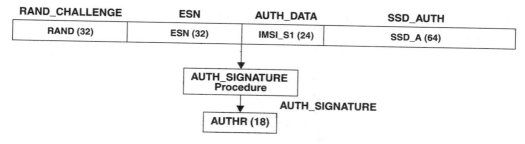

Figure 11-5 Calculation of AUTHR for Authentication of MS Registration

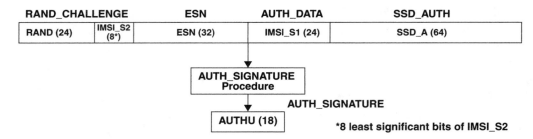

Figure 11-6   Calculation of AUTHU for Unique Challenge Response Procedure

AUTHU to the value received from the MS. If the comparison fails, the BS may deny further access attempts by the MS, drop the call in process, or initiate the process of updating SSD.

11.3.3   Authentication of MS Originations

When AUTH is set to 01 and the MS attempts to originate a call by sending an Origination message on the access channel, the following authentication procedure is performed. The MS sets the input parameters of the AUTH_SIGNATURE procedure, as shown in Fig. 11-7. The AUTH_DATA input parameter consists of the last 6 digits contained in the CHAR field of the Origination message. If fewer than 6 digits are included in the Origination message, the most significant bits of IMSI_S1 are used to replace the missing digits. IMSI_S1 is used initially to fill the AUTH_DATA input parameter, and then the last dialed digits entered by the subscriber are used to replace all or part of this initial value. If a full 6 digits are dialed, the first digit of the six that were dialed is used as the most significant 4 bits of AUTH_DATA, the second digit is the next less-significant 4 bits of AUTH_DATA, and so forth. If fewer than 6 digits are dialed, then the least significant 4 bits of AUTH_DATA are the last dialed digit, the second-to-the-last dialed digit becomes the next more-significant 4 bits of AUTH_DATA, and so on up to the first of the dialed digits. The MS sets the SAVE_REGISTERS input parameter to TRUE.

The MS then executes the AUTH_SIGNATURE procedure. The 18-bit output AUTH_SIGNATURE is used to fill the AUTHR field of the Origination message. The RANDC

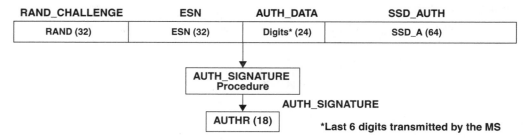

Figure 11-7   Calculation of AUTHR for Authentication of MS Originations

(8 most significant bits of RAND) and COUNT fields of the message are filled with the current value stored in the MS. The BS compares the received value of the RANDC to the most significant 8 bits of the internally stored value of the RAND. The BS may also compare the received value of COUNT with its internally stored value associated with the received IMSI/ESN. The BS computes the value of AUTHR in the same manner as the MS, but uses its internally stored value of SSD_A. The BS compares its computed value of AUTHR to the value received from the MS. If the comparisons executed at the BS are successful, the BS may initiate the appropriate channel assignment procedures. After channel assignment, the BS may issue a Parameter Update order on the forward traffic channel, updating the value of COUNT in the MS. If any of the comparisons fail, the BS may deny service, initiate the unique challenge procedure, or commence the process of updating the SSD.

11.3.4 Authentication of MS Terminations

When AUTH is set to 01 and the MS responds to a page by sending a Page Response message on the access channel, the following authentication procedures are performed.

The MS sets the input parameters of the AUTH_SIGNATURE procedure (refer to Fig. 11-8) and the SAVE_REGISTERS input parameter to TRUE. The MS then executes the AUTH_SIGNATURE procedure. The 18-bit output AUTH_SIGNATURE is used to fill the AUTHR field of the Page Response message. The RANDC (8 most significant bits of RAND) and COUNT fields of the message are filled with the current values stored in the MS. The BS compares the received value of RANDC to the 8 most significant bits of its internally stored value of RAND. The BS may also compare the received value of COUNT with its internally stored value associated with the received IMSI/ESN.

The BS computes the value of AUTHR in the same manner as the MS, but uses its internally stored value of SSD_A. The BS compares its computed value of AUTHR to the value received from the MS. If the comparisons executed at the BS are successful, the BS initiates the appropriate channel assignment procedures. After channel assignment, the BS may issue a Parameter Update order on the forward traffic channel updating the value of COUNT in the MS. If any of the comparisons fail, the BS may deny service, initiate the unique challenge response procedure, or commence the process of updating the SSD.

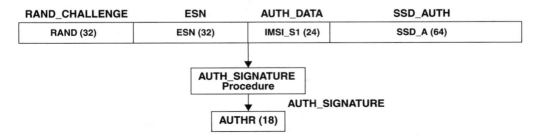

Figure 11-8 Calculation of AUTHR for Authentication of MS Terminations

11.3.5 Authentication of MS Data Bursts

When AUTH is set to 01 and the MS attempts to send a Data Burst message on the access channel, the following authentication procedures are performed.

The MS sets the input parameters of AUTH_SIGNATURE procedure (refer to Fig. 11-9). The AUTH_DATA input is generated by first filling the AUTH_DATA parameter with the 24 bits of IMSI_S1 and then replacing part or all of the prefilled value with up to six 4-bit digits that are provided by the procedure (according to BURST_TYPE) requesting the Data Burst message.

The MS sets the SAVE_REGISTERS input parameter to FALSE. The MS then executes the AUTH_SIGNATURE procedure. The 18-bit output AUTH_SIGNATURE is used to fill the AUTHR field of the Data Burst message. The RANDC (8 most significant bits of RAND) and COUNT fields of the message are filled with the current values stored in the MS. The BS compares the received value of RANDC to the 8 most significant bits of its internally stored value of RAND. The BS may also compare the received value of COUNT with its internally stored value associated with the received IMSI/ESN.

The BS computes the value of AUTHR in the same way as the MS, but using its internally stored value of SSD_A and by generating the AUTH_DATA input in the same way as discussed for the MS. The BS compares its computed value of AUTHR to the value received from the MS. If comparisons executed at the BS are successful, the BS may process the message. If any of the comparisons fail, the BS may ignore the message, initiate the unique challenge response procedure or commence the process of updating the SSD.

11.3.6 Authentication of TMSI Assignment

The MS sets the input parameters of AUTH_SIGNATURE procedure (see Fig. 11-5). The MS sets the SAVE_REGISTERS input parameter to FALSE and executes the AUTH_SIGNATURE procedure. The 18-bit output AUTH_SIGNATURE is used to fill the AUTHR field of the TMSI Assignment Completion message. The RANDC (8 most significant bits of RAND) and COUNT fields of the message are filled with the current values stored in the MS. The BS compares the received value of RANDC to the 8 most significant bits of its internally stored value of RAND.

The BS may also compare the received value of COUNT with its internally stored value associated with the received IMSI/ESN. The BS computes the value of AUTHR in the same way

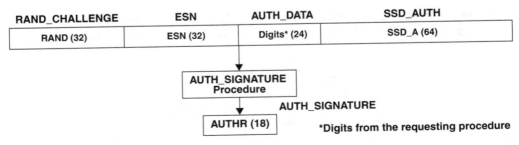

Figure 11-9 Calculation of AUTHR for Authentication of MS Data Burst

as the MS, but uses its internally stored value of SSD_A. The BS compares its computed value of AUTHR to the value received from the MS.

If any of the comparisons fail, the BS may deem the TMSI assignment unsuccessful, initiate the unique challenge response procedure, or commence the process of SSD update.

11.4 Shared Secret Data

Air interface procedures have been defined to update the SSD at the MS. A new SSD is generated at the HLR/AC, which initiates the SSD update procedure. The SSD update procedure involves exchange of the following MSC-BS messages:

- SSD Update Request message
- Base Station Challenge Order message
- Base Station Challenge Response message
- SSD Update Response message

The call flow for the SSD update procedure, shown in Fig. 11-10, is as follows:

1. The MSC sends an SSD Update Request message to the BS to indicate that the SSD at the MS needs updating. The update information is in the form of a random number (RANDSSD).
2. The BS sends an SSD Update Order message to the MS on either the paging channel or the forward traffic channel. The RANDSSD field of the message contains the same value used for the HLR/AC computation of SSD.
3. Upon receipt of the SSD Update Order message from the BS, the MS uses the RANDSSD as input to the SSD_Generation procedure (algorithm) to generate the new SSD.

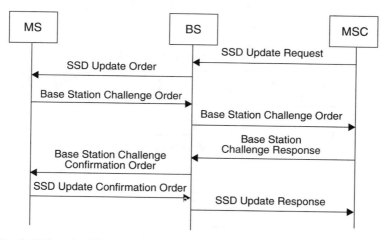

Figure 11-10 Call Flow for SSD Update—Successful Case

4. The MS sets SSD_A_NEW and SSD_B_NEW to the outputs of the SSD_Generation procedure.

5. The MS selects a 32-bit random number (RANDBS) and sends it to the BS in a Base Station Challenge Order message on the access channel or reverse traffic channel.

6. The BS forwards the Base Station Challenge Order message to the MSC to verify that the new SSD calculated at the MS is the same as the number in the MSC.

7. On reception of the Base Station Challenge Order message, the MSC uses the new SSD as input to the algorithm to generate the authentication response signature (AUTHBS) (Fig. 11-11). The MSC then sends the authentication response signature (AUTHBS) to the BS in the Base Station Challenge Response message.

8. Upon receipt of the Base Station Challenge Response message from the MSC, the BS transmits this information in a Base Station Challenge Confirmation Order message to the MS on the paging channel or the forward traffic channel.

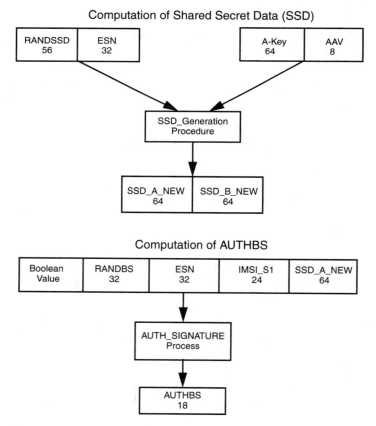

Figure 11-11   Computation of SSD and AUTHBS

9. Upon receipt of the Base Station Challenge Confirmation Order message, the MS compares the received value of AUTHBS to its internally computed value.

10. If the MS receives a Base Station Challenge Confirmation Order message when an SSD update is not in progress, the MS responds with an SSD Update Rejection order.

11. If the comparison is successful, the MS executes the SSD_update procedure to set SSD_A and SSD_B to SSD_A_NEW and SSD_B_NEW, respectively. The MS then sends an SSD Update Confirmation Order message to the BS, indicating a successful completion of the SSD update.

12. If the comparison is not successful, the MS discards SSD_A_NEW and SSD_B_NEW and then sends an SSD Update Rejection Order message to the BS indicating unsuccessful completion of the SSD update.

13. Upon receipt of the SSD Update Confirmation Order message, the BS sets SSD_A and SSD_B to the value received from the HLR/AC.

14. If the MS fails to receive the Base Station Challenge Confirmation Order message within T_{64m} seconds of the reception of the acknowledgment to the Base Station Challenge Order message, the MS discards SSD_A_NEW and SSD_B_NEW. The mobile then terminates the SSD update process.

Message flow for an SSD update is shown in Fig. 11-12.

11.5 Parameter Update

This procedure is performed when the MSC needs to instruct the MS to update the call history count (COUNT). This is done at the earliest convenient time after a traffic channel is allocated for either call origination or termination.

The MSC sends a Parameter Update Request message to the BS and starts timer T3220. Upon receipt of this message, the BS instructs the MS to update its count by sending a Parameter Update Order. When the MS receives this order, it increments its call history count and immediately sends the Parameter Update Confirm message to the MSC. Upon receipt of this message, the MSC increments its count and stops timer T3220.

11.6 Voice Privacy

If provided by the air interface, there can be voice privacy between the BS and MS. In order to provide this protection, the relevant device used for the voice privacy needs to be loaded with the appropriate key. This is supplied by the MSC over the MSC-BS interface. Voice privacy cannot be invoked unless broadcast authentication is activated in the system.

The privacy mode procedure should be completed either before handoff is initiated or after a handoff operation is complete.

The following call flow indicates voice privacy invoked during an established call (see Fig. 11-13).

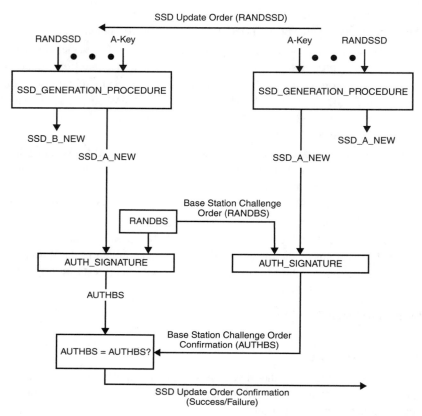

Figure 11-12   SSD Update Message Flow

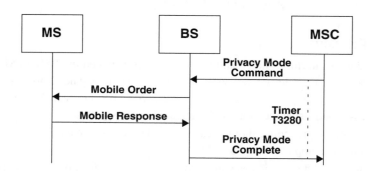

Figure 11-13   Privacy Mode Procedure

1. At any point during the call following the receipt of the Assignment Complete message the MSC may send the Privacy Mode Command message to the BS to specify that privacy is to be provided for traffic information. The MSC then starts timer T3280.

2. After the radio traffic channel has been acquired, voice privacy can be established when the BS transmits a voice privacy request order to the MS.

3. The MS performs the required privacy mode procedures and acknowledges the BS with a voice privacy response order.

4. The BS returns the Privacy Mode Complete message to the MSC to indicate successful receipt of the Privacy Mode Command message. The MSC stops timer T3280 upon receipt of the Privacy Mode Complete message.

The Privacy Mode Command Optional BSMAP message may be sent by the MSC to the BS after receipt of the Assignment Complete message. Its typical application is to specify the use of encryption/privacy parameters for the call. It may be sent in the following cases:

- To preload the BS with encryption/privacy parameters during call setup
- To enable or disable the use of encryption/privacy during conversation

The preloading of the BS with parameters allows the BS to immediately initiate privacy upon request by the mobile user or immediately following assignment to a traffic channel.

Where the MSC needs to provide encryption/privacy information to the BS during call setup, it may place the information in the Privacy Mode Command message or place it in the Assignment Request message. Which message is chosen for including encryption information is a BS and MSC manufacturers' choice.

The privacy mode procedures may be invoked by the MSC during the conversation state to enable or disable the use of encryption/privacy. This may be initiated by the MSC or sent in response to a request for privacy by mobile user.

11.7 Summary

This chapter introduced various parameters used to identify a mobile station. Because fraud is a serious problem in the analog AMPS system, IS-95 CDMA uses cryptographic methods for combating fraud. At any time during call processing, the MSC can present a unique challenge to a mobile station to confirm its identity. We discussed the unique challenge response procedures used by a mobile station.

The chapter described various types of authentication procedures in which information is exchanged between an MS and a BS to confirm the identity of the MS. A successful authentication process occurs only when it is demonstrated that the MS and BS possess identical sets of Shared Secret Data (SSD).

Also discussed was the procedure to update the SSD, including a call flow diagram. The procedure involves exchange of the SSD update request, BS challenge, BS challenge response, and SSD update response messages between the MSC and the BS. The chapter concluded with a presentation of procedures used when the MSC instructs the mobile station to update the call history count (COUNT) and when voice privacy is requested between the BS and MS.

11.8 References

1. Garg, V. K., and Wilkes, J. E., *Wireless and Personal Communications Systems*, Prentice Hall PTR, Upper Saddle River, NJ, 1996.

2. Report of the Joint Experts Meeting on Privacy and Authentication for PCS, Phoenix, Arizona, November 8–12, 1993.

3. TIA Interim Standard, IS-41C, "Cellular Radio Telecommunications Intersystem Operations," 1997.

4. TIA IS-95A, "Mobile Station–Base Station Compatibility Standard for Dual Mode Spread Spectrum Cellular System," 1992.

5. TIA IS-95B, "Mobile Station–Base Station Compatibility Standard for 800 MHz Cellular Mobile Telecommunications System and 1.8 to 2.0 GHz CDMA PCS Systems," 1998.

RF Engineering and Network Planning

12.1 Introduction

This chapter presents basic guidelines for engineering a CDMA system. The topic is extremely complex and cannot be covered extensively in a single chapter. This chapter discusses several topics that are germaine to the engineering of a CDMA system: propagation models, link budgets, the transition from analog operation to CDMA operation, radio link capacity, facility engineering, border cells on a boundary between two service providers, and interfrequency handoff.

12.2 Radio Design for a Cellular/PCS Network

A designer needs to consider many factors early in the design of a cellular/PCS network for an urban area. For example, all prospective service providers must carefully evaluate the extent of radio coverage for indoor locations, the quality of service for different environments, efficient use of the spectrum, and the evolution of the network. Often, these factors are further complicated by the constraints imposed by the operating environments and regulatory issues. A system designer must carefully balance all the trade-offs to ensure that the network is robust, future proof, and of high service quality.

12.3 Radio Network Planning

The grade-of-service (GOS) performance measures include *area coverage probability* and *blocking*. The area coverage probability is related to the quality of network planning and the network capacity. Blocking is based on the available resources. Area coverage probability can also be defined by *outage*. Outage occurs when the network is not able to provide the specified quality of service. If the system is *coverage limited*, outage can be defined as the probability when path loss and shadowing exceed the difference between the maximum transmitted power and the required received signal level. Coverage and capacity objectives require a trade-off between the desired quality and overall network cost. A smaller signal outage probability means smaller cells and therefore higher overall network cost; smaller interference outage probability implies lower capacity and thus higher network cost. An outage probability of 5 to 10% corresponding to 90 to 95% coverage probability is often used. The coverage probability could be different for different services.

Many factors must be included in radio network planning. The network planning should address issues such as traffic distribution, macro- and micro-cell deployment, provision for indoor and high-bit-rate coverage, cell locations and cost of sites, and environmental concerns such as cell tower appearance.

12.4 Radio Link Design

For any wireless communications system, the first important step is to design the radio link. This is required to determine base station density in different environments as well as the corresponding radio coverage. For a wireless network to provide good-quality indoor and outdoor service in an urban environment, flexibility and resilience should be incorporated into the design. The transmit power of handsets will be the determining factor for a CDMA system with balanced up/downlink power.

Although the mobile antenna gain does not affect the balancing of the link budget, it is an important factor in the design of the power budget for handset coverage. From a user point of view, a cellular/PCS network should imply that there is little restriction on making or receiving calls within a building or in a moving vehicle using handsets. A system should be designed to allow the antenna of a handset to be placed in nonoptimal positions. In addition, the antenna may not even be extended when calls are being made or received. In normal system designs, it is assumed that the gain of a mobile antenna is 0 dBi.[*] However, allowing for the handset antennas to be placed in suboptimal positions, a more conservative gain of –3 dBi should be used. In reality, because of the positioning of an antenna in an arbitrary position or with the antenna retracted into the handset housing antenna gain could be as low as –6 to –8 dBi depending on specific handsets and their corresponding housing designs.

12.5 Estimation of Cell Count

The number of users and offered traffic load per user are used to determine the overall traffic load. Knowing the cell capacity and cell coverage, an estimate of the number of cells can be made.

The cell capacity is determined from simulations or analytical formulas. User information rate, traffic characteristics, QoS requirements (delay, BER/FER), and outage probability are the important factors in determining the cell capacity.

The link budget is used to determine maximum cell coverage. In addition to E_b/I_t, equipment-specific factors such as cable losses, antenna gains, and the receiver noise figure are needed for link budget calculations.

---

* dBi refers to the gain relative to an isotropic antenna.

Soft-handoff gain has a large impact on link budget. The soft-handoff gain depends on shadowing correlation and coverage probability. Soft handoff provides macroscopic diversity gain through increased diversity. The actual gain depends on the radio environment and number of RAKE receiver fingers. Since each radio environment has its own characteristics, for the detailed coverage prediction, some correction factors for the path-loss models are needed.

For the reverse link, the impact of load factor ρ in the link budget for interference margin I_m (dB) can be determined from

$$I_m = 10\log\left(\frac{1}{1-\rho}\right)$$

(12.1)

Since the interference margin increases with ρ, cell range would decrease with the increasing load factor. Asymmetric traffic load should be taken into consideration in link budget calculations. CDMA can trade the reverse link capacity for coverage. This is useful since usually the mobile transmission power limits the maximum cell range.

After the cell count has been obtained, detailed radio network planning can be initiated by taking into account the exact radio environment where each cell will be located. Due to cost of sites, zoning requirements, building restrictions, or other reasons, it may not be possible to achieve optimum cell sites in a real network. This may impact the initial coverage plan. For detailed network planning, a network planning tool should be used. A network planning tool has a digital map of the area to be planned. Building heights and antenna pattern are also modeled. The optimization process of radio network coverage generally includes

- Detailed description of radio environment
- Control channel power planning
- Soft-handoff parameters planning
- Interfrequency handoff planning
- Iterative network coverage analysis
- Network testing

12.6 Radio Coverage Planning

The most important design objective of a cellular/PCS network is to provide near-ubiquitous radio coverage. One of the most crucial considerations in the radio coverage planning process is the propagation model. The accuracy of the prediction by a particular model depends on its ability to account for the detailed terrain, vegetation, and buildings. This accuracy is of vital importance in determining the path loss and, hence, the cell sizes and the infrastructure requirement of a cellular/PCS network. An overestimation will lead to an inefficient use of the network resources, whereas an underestimation will result in poor radio coverage. Propagation models generally tend to oversimplify real-life propagation conditions and may be grossly inaccurate in complex metropolitan urban environments. Empirical propagation models only provide general guidelines and are too simplistic for accurate network design. Accurate field measurements must be made to provide information on radio coverage in an urban environment. Measured data can be used either directly in the planning process to access the feasibility of individual cell sites or to calibrate the coefficients of the empirical propagation model to achieve better characterization of a specific environment.

Radio propagation in an urban environment is subject to shadowing. To ensure that the signal level in 90% of the cell area is equal to or above the specified threshold, a shadow fading margin, which is dependent on the standard deviation of the signal level, must be included in the link budget. For a typical urban environment, a shadow fading margin of 8 to 9 dB should be used based on the assumption that the path loss follows an inverse 2–5 exponent law—the path loss is inversely proportional to the distance of separation raised to a power between 2 and 5. The value of the power is dependent upon propagation characteristics.

Another critical factor that affects radio coverage is the penetration loss for both buildings and vehicles. If radio coverage for the outer portion of a building is sufficient, then an assumed penetration loss of 10 to 15 dB should be adequate. However, if calls will be received and originated within the inner core of the building, a penetration loss of about 30 dB should be used. Similarly, for in-vehicle coverage, the penetration loss is equally important. A car could experience a penetration loss of 3 to 6 dB, whereas vans and buses have even larger variations. The penetration loss at the front of a van should be no more than that experienced in a car, but the loss at the back of a van could be as high as 10 to 12 dB, depending on the amount of window space. Thus, the designer should assume a high penetration loss to ensure good service quality. For an urban environment, because building penetration loss is the dominant factor in designing the system, in-vehicle penetration will generally be sufficient as a consequence.

12.7 Propagation Models

Designers use propagation models to determine how many cell sites they need to satisfy the coverage requirements for the network. Initial network design typically engineers for coverage. Later, growth engineers for capacity. Some systems may need to start with wide area coverage and high capacity and therefore may be at a later stage of the growth curve right from start-up.

The coverage requirement is coupled with traffic loading requirements, which rely on the propagation model chosen to determine the traffic distribution and the off-loading from an existing cell site to new cell sites as part of the capacity relief program. The propagation model helps to determine where the cell sites should be located to achieve an optimal position in the network. If the propagation model used is not effective in helping to place cell sites correctly, the probability of incorrectly deploying a cell site into the network is high.

The performance of the network is affected by the propagation model chosen because that model is used for interference predictions. As an example, if the propagation model is inaccurate by 4 dB, then E_b/I_t could be 11 dB or 3 dB (assuming that $E_b/I_t = 17$ dB is the design requirement). Based on traffic loading condition, designing for a high E_b/I_t level could negatively affect financial feasibility. On the other hand, designing for a low E_b/I_t would degrade the quality of service.

The propagation model is also used in other system performance aspects including hand-off optimization, power-level adjustments, and antenna placements. Although no propagation model can account for all perturbations experienced in the real world, it is essential to use one or several models for determining the path losses in the network. Each of the propagation models being used in the industry has pros and cons. It is through a better understanding of the limitations of each of the models that a good RF engineering design can be achieved in a network.

12.7.1 Modeling for the Outside Environment

12.7.1.1 Analytical Model

The propagation loss between the base station and the mobile station in the outside environment has been studied extensively. Propagation loss is generally expressed by the following expression [2,4,7]:

$$P(r) = N(r_0, \sigma) + 10\gamma\log\frac{r}{r_0} \text{ dB} \qquad (12.2)$$

where $P(r)$ = loss at distance r relative to the loss at a reference distance r_0,
  γ = path-loss exponent, and
  σ = standard deviation, typically 8 dB.

The second term on the right side of Eq. (12.2) represents a constant attenuation in the outside environment between the base station and the mobile station. Typically, γ approximately equals 4, although it may range between 2 (which equals the loss in free space) and 5. If γ is equal to 4, then the signal will be attenuated 40 dB if the distance increases ten times with respect to the reference distance. The first term in Eq. (12.2) represents the variation in the loss about the average path loss. This function is an approximate log-normal distribution with an average equal to the second term and a standard deviation of approximately 8 dB. It has been found that this value is applicable for a wide range of radio environments, including urban and rural areas.

12.7.1.2 Empirical Models

Several empirical models have been suggested and used to predict propagation path losses. The two most widely used models are the Hata-Okumura model and the Walfisch-Ikegami Model.

The Hata-Okumura Model [3]

Most of the propagation tools use a variation of the Hata model. Hata's model is an empirical relationship derived from the technical report made by Okumura [6] so that the results could be used in computational tools. Okumura's report consists of a series of charts that have been used in radio communication modeling. The following are the expressions used in the Hata model in order to determine the mean loss L_{50}:

Urban area:

$$L_{50} = 69.55 + 26.16\log f_c - 13.82\log h_b - a(h_m) + (44.9 - 6.55\log h_b)\log r \text{ dB} \qquad (12.3)$$

where f_c = frequency (MHz),

L_{50} = mean path loss (dB),

h_b = base station antenna height (m),

$a(h_m)$ = correction factor for mobile antenna height (dB), and

r = distance from base station (km).

The range of the parameters for which the Hata model is valid is

$$150 \leq f_c \leq 1{,}500 \text{ MHz}$$

$$30 \leq h_b \leq 200 \text{ m}$$

$$1 \leq h_m \leq 10 \text{ m}$$

$$1 \leq r \leq 20 \text{ km}$$

$a(h_m)$ is computed as

For a small or medium-sized city:

$$a(h_m) = (1.1 \log f_c - 0.7) h_m - (1.56 \log f_c - 0.8) \text{ dB} \qquad (12.4)$$

For a large city:

$$a(h_m) = 8.29 (\log 1.54 h_m)^2 - 1.1 \text{ dB}, f_c \leq 200 \text{ MHz} \qquad (12.5)$$

or

$$a(h_m) = 3.2 (\log 11.75 h_m)^2 - 4.97 \text{ dB}, f_c \geq 400 \text{ MHz} \qquad (12.6)$$

Suburban area:

$$L_{50} = L_{50}(\text{urban}) - 2 \left[\log \left(\frac{f_c}{28} \right)^2 - 5.4 \right] \text{dB} \qquad (12.7)$$

Open area:

$$L_{50} = L_{50}(\text{urban}) - 4.78 (\log f_c)^2 + 18.33 \log f_c - 40.94 \text{ dB} \qquad (12.8)$$

Hata's model does not account for any of the path-specific correction used in Okumura's model.

Okumura's model [6] tends to average some of the extreme situations and does not respond sufficiently quickly to rapid changes in the radio path profile. The distance-dependent behavior of Okumura's model is in agreement with the measured values. Okumura's measurements are valid only for the building types found in Tokyo. Experience with comparable measurements in the United States has shown that the typical U.S. suburban situation is often somewhere between Okumura's suburban and open areas. Okumura's suburban definition is more representative of a U.S. residential metropolitan area with large groups of row houses.

Okumura's model requires that considerable engineering judgment be used, particularly in the selection of the appropriate environmental factors. Data is needed in order to be able to predict the environmental factors from the physical properties of the buildings surrounding a mobile receiver. In addition to the appropriate environmental factors, path-specific corrections are required to convert Okumura's mean path-loss predictions to the predictions applicable to the specific path under study. Okumura's techniques for correction of irregular terrain and other path-specific features require engineering interpretations and are thus not readily adaptable for computer use.

The Walfisch-Ikegami (or COST 231) Model [14]

This model is used to estimate the path loss in an urban environment for cellular communication (see Fig. 12-1). It is a combination of the empirical and deterministic model for estimating the path loss in an urban environment over the frequency range of 800 MHz to 2000 MHz. This model is used primarily in Europe for the GSM system and in some propagation models in the United States. The model contains three elements: free-space loss, rooftop-to-street diffraction and scatter loss, and multiscreen loss. The expressions used in this model are

$$L_{50} = L_f + L_{rts} + L_{ms} \qquad (12.9)$$

or

$$L_{50} = L_f \text{ when } L_{rts} + L_{ms} \leq 0 \qquad (12.10)$$

where L_f = free-space loss,
L_{rst} = rooftop-to-street diffraction and scatter loss, and
L_{ms} = multiscreen loss.

Free-space loss is given as

$$L_f = 32.4 + 20\log r + 20\log f_c \text{ dB} \qquad (12.11)$$

The rooftop-to-street diffraction and scatter loss is given as

$$L_{rts} = -16.9 - 10\log W + 10\log f_c + 20\log \Delta h_m + L_0 \text{ dB} \qquad (12.12)$$

where W = street width (m), and
$\Delta h_m = h_r - h_m$ (m).

$$L_0 = -9.646 \text{ dB} \qquad 0 \leq \phi \leq 35 \text{ degrees}$$

$$L_0 = 2.5 + 0.075(\phi - 35) \text{ dB} \qquad 35 \leq \phi \leq 55 \text{ degrees}$$

$$L_0 = 4 - 0.114(\phi - 55) \text{ dB} \qquad 55 \leq \phi \leq 90 \text{ degrees}$$

where ϕ = incident angle relative to the street.

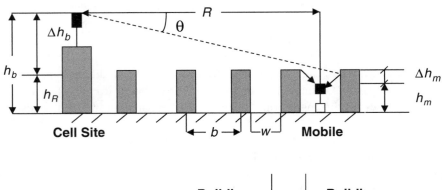

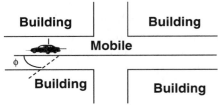

Figure 12-1   The Walfisch-Ikegami Propagation (COST 231) Model

The multiscreen loss is given as

$$L_{\mathrm{ms}} = L_{\mathrm{bsh}} + k_a + k_d \log r + k_f \log f_c - 9 \log b \tag{12.13}$$

where b = distance between buildings along the radio path (m),

$L_{\mathrm{bsh}} = -18 \log 11 + \Delta h_b, \quad h_b > h_r,$

$L_{\mathrm{bsh}} = 0, \quad h_b < h_r,$

$k_a \quad = 54, \quad h_b > h_r,$

$k_a \quad = 54 - 0.8 h_b, \quad r \geq 500 \text{ m}, h_b \leq h_r,$

$k_a \quad = 54 - 1.6 \Delta h_b r, \quad r < 500 \text{ m}, h_b \leq h_r,$

$k_d \quad = 18, \quad h_b < h_r,$

$k_d \quad = 18 - \dfrac{15 \Delta h_b}{\Delta h_m}, \quad h_b \geq h_r,$

$k_f \quad = 4 + 0.7 \left(\dfrac{f_c}{925} - 1 \right)$ for midsize city and suburban area with moderate tree density, and

$k_f \quad = 4 + 1.5 \left(\dfrac{f_c}{925} - 1 \right)$ for metropolitan center.

Note: Both L_{bsh} and k_a increase the path loss with lower base station antenna heights.

The range of parameters for which the Walfisch-Ikegami model is valid is

$800 \leq f_c \leq 2000$ (MHz)

$4 \leq h_b \leq 50$ (m)

$1 \leq h_m \leq 3$ (m)

$0.02 \leq r \leq 5$ (km)

The following default values can be used for the model:

$b = 20\text{--}50$ (m)

$W = b/2$

$\phi = 90$ degrees

Roof = 3 m for pitched roof and 0 m for flat roof

$h_r = 3$ (number of floors) + roof

Table 12-1 uses the following data to show a comparison of the path loss from the Hata and Walfisch-Ikegami models.

$f_c = 880$ MHz; $h_m = 1.5$ m; $h_b = 30$ m; roof = 0 m; $h_r = 30$ m; $\phi = 90$ degrees; $b = 30$ m; and $W = 15$ m.

The path losses predicted by Hata's model are about 13 to 16 dB lower (see Fig. 12-2) than those predicted by the Walfisch-Ikegami model. Hata's model ignores effects from street width, street diffraction, and scatter losses, which the Walfisch-Ikegami model includes.

- **Correction Factor for Attenuation Due to Trees.** Weissberger [15] has developed a modified exponential delay model that can be used where a radio path is blocked by dense, dry, in-leaf trees found in temperate climates. The additional path loss can be calculated from the following expression:

$$L_t = 1.33(f_c)^{0.284}(d_f)^{0.588} \text{ dB}, \qquad \text{for } 14 \leq d_f \leq 400 \text{ m} \tag{12.14}$$

$$= 0.45(f_c)^{0.284}d_f \text{ dB}, \qquad \text{for } 0 \leq d_f \leq 14 \text{ m} \tag{12.15}$$

where L_t = loss in dB,
f_c = frequency in GHz, and
d_f = tree height in meters.

Table 12-1 A Comparison of Path Loss from Hata and Walfisch-Ikegami Models

Distance (km)	Path Loss (dB)	
	Hata's Model	Walfisch-Ikegami Model
1	126.16	139.45
2	136.77	150.89
3	142.97	157.58
4	147.37	162.33
5	150.79	166.01

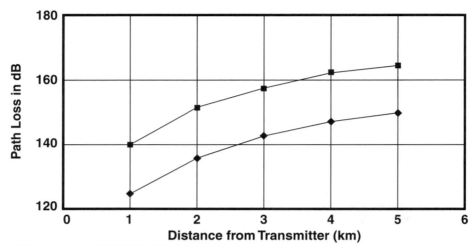

Figure 12-2   Comparison of Hata and Walfisch-Ikegami Models (Path Loss vs. Distance)

The difference in path loss for trees with and without leaves has been found to be about 3 to 5 dB. For a frequency of 900 MHz, the above equations are reduced to

$$L_t = 1.291(d_f)^{0.588} \text{ dB,} \qquad \text{for } 14 \le d_f \le 400 \text{ m} \qquad (12.16)$$

$$= 0.437 d_f \text{ dB,} \qquad \text{for } 0 \le d_f \le 14 \text{ m} \qquad (12.17)$$

12.7.2   Models for Indoor Environment

Experimental studies have indicated that a portable receiver moving in a building experiences Rayleigh fading for obstructed propagation paths and Ricean fading for line-of-sight (LOS) paths, regardless of the type of building. Rayleigh fading is short-term fading resulting from signals traveling separate paths (multipath) that partially cancel each other. An LOS path is clear of building obstructions; in other words, there are no reflections of the signal. Ricean fading results from the combination of a strong LOS path and a ground path plus numerous weak reflected paths. A TIA IS-95A mobile station, however, is capable of discerning signals that travel different paths since a RAKE receiver is incorporated. In fact, TIA IS-95A does not address equalization of delay spread on the radio link, and thus the mobile station's receiver does not have an equalizer.

Quantification of propagation between floors is important for an in-building wireless system of multifloor buildings that need to share frequencies within the building. Frequencies are reused on different floors to avoid cochannel interference. The type of building material, the aspect ratio of building sides, and types of windows have shown to impact the RF attenuation between floors. Measurements have indicated that loss between floors does not increase linearly in dB with increasing separation distance. The greatest floor attenuation in dB occurs when the receiver and transmitter are separated by a single floor. The overall path loss increases at a smaller rate as the number of floors increases. Typical values of attenuation between floors is 15 dB for one floor of separation and an additional 6 to 10 dB per floor separation up to four floors of separation. For five or more floors of separation, path loss will increase by only a few dB for each additional floor (see Table 12-2).

The signal strength received inside a building due to an external transmitter is important for wireless systems that share frequencies with neighboring buildings or with an outdoor system. Experimental studies have shown that the signal strength received inside a building increases with height [9]. At lower floors of a building, the urban cluster induces greater attenuation and reduces the level of penetration. At higher floors, an LOS path may exist, thus causing a stronger incident signal at the exterior wall of the building. RF penetration is found to be a function of frequency as well as height within a building. Penetration loss decreases with increasing frequency. Measurements made in front of a window showed 6 dB less penetration loss on coverage than those measurements made in parts of the building without windows. Experimental studies also showed that building penetration loss decreased at a rate of about 2 dB per floor from ground level up to the 10th floor and then began to increase around the 10th floor. The increase in penetration loss at the higher floors was attributed to shadowing effects of adjacent buildings.

Table 12-2 Mean Path-Loss Exponents and Standard Deviations

Type		γ	σ (dB)
All buildings	All locations	3.14	16.3
	Same floor	2.76	12.9
	Thru 1 floor	4.19	5.1
	Thru 2 floors	5.04	6.5
Office building 1	Entire building	3.54	12.8
	Same floor	3.27	11.2
	West wing 5th floor	2.68	8.1
	Central wing 5th floor	4.01	4.3
	West wing 4th floor	3.18	4.4
Grocery store		1.81	5.2
Retail store		2.18	8.7
Office building 2	Entire building	4.33	13.3
	Same floor	3.25	5.2

The mean path loss is a function of distance to the γth power [7].

$$L_{50}(r) = L(r_0) + 10 \times \gamma \log\left(\frac{r}{r_0}\right) \text{ dB} \qquad (12.18)$$

where $L_{50}(r)$ = mean path loss (dB),
$\quad L(r_0)$ = path loss from transmitter to reference distance r_0 (dB),
$\quad \gamma \quad$ = mean path-loss exponent,
$\quad r \quad$ = distance from the transmitter (m), and
$\quad r_0 \quad$ = reference distance from the transmitter (m).

We choose r_0 equal to 1 m and assume $L(r_0)$ due to free-space propagation from the transmitter to a 1-m reference distance. Next, we assume the antenna gain equals the system cable losses[*] and get a path loss, $L(r_0)$, of 31.7 dB at 914 MHz over a 1-m free-space path.

The path loss was found to be log-normally distributed about Eq. (12.18). The mean path-loss exponent γ and standard deviation σ are the parameters that depend on building type, building wing, and number of floors between the transmitter and receiver. The path loss at a transmitter-receiver (T-R) separation of r meters can be given as

$$L(r) = L_{50}(r) + X_\sigma \text{ dB} \qquad (12.19)$$

where $L(r)$ = path loss at a T-R separation distance r meters, and
$\quad X_\sigma$ = 0 mean log-normally distributed random variable with standard deviation σ dB.

Table 12-2 gives a summary of the mean path-loss exponents and standard deviation about the mean for different environments [9].

In a multifloor environment, Eq. (12.18) can be modified to emphasize that the mean path-loss exponent is a function of the number of floors between the transmitter and the receiver. The value of γ (multifloor) is given in Table 12-2.

$$L_{50}(r) = L(r_0) + 10 \times \gamma(\text{multifloor})\log\left(\frac{r}{r_0}\right) \qquad (12.20)$$

Another path-loss prediction model suggested in [7] uses a Floor Attenuation Factor (FAF). A constant floor attenuation factor (in dB), which is a function of the number of floors and building type, was included in the mean path loss predicted by a path-loss model that uses the same-floor path-loss exponent for the particular building type.

$$L_{50}(r) = L(r_0) + 10 \times \gamma(\text{same-floor})\log\left(\frac{r}{r_0}\right) + FAF \quad \text{dB} \qquad (12.21)$$

where r is in meters and $L(r_0)$ = 31.7 dB at 914 MHz.

---

[*] This is obviously not always true.

Table 12-3   Average Floor Attenuation Factor

Type		FAF (dB)	σ
Office building 1	Thru 1 floor	12.9	7.0
	Thru 2 floor	18.7	2.8
	Thru 3 floor	24.4	1.7
	Thru 4 floor	27.0	1.5
Office building 2	Thru 1 floor	16.2	2.9
	Thru 2 floor	27.5	5.4
	Thru 3 floor	31.6	7.2

Table 12-3 provides the floor attentuation factors and the standard deviation (in dB) of the difference between the measured and predicted path loss. Values for the floor attenuation factor in Table 12-3 are an average (in dB) of the difference between the path loss observed at multi-floor locations and the mean path loss predicted by the simple r^γ model (Eq. [12.18]), where γ is the same-floor exponent listed in Table 12-2 for the particular building structure and r is the shortest distance, measured in three dimensions, between the transmitter and receiver.

12.7.2.1   Soft-Partition and Concrete-Wall Attenuation Factor Model

The path-loss effects of soft partitions and concrete walls (in dB) between the transmitter and receiver for the same floor were modeled in [5] and have been given as

$$L_{50}(r) = 20\log\left(\frac{4\pi r}{\lambda}\right) + p \times AF(\text{soft-partition}) + q \times AF(\text{concrete-wall}) \qquad (12.22)$$

where p = number of soft partitions between the transmitter and receiver,
  q = number of concrete walls between the transmitter and receiver,
  λ = wavelength (m),
  AF = 1.39 dB for each soft partition, and
  AF = 2.38 for each concrete wall.

EXAMPLE   12.1

Use the two models (Eqs. [12.20] and [12.21]) to predict the mean path loss at a distance $r = 30$ m through three floors of an office building; assume the mean path-loss exponent for same-floor measurements in the building is $\gamma = 3.27$, the mean path-loss exponent for three-floor measurements is $\gamma = 5.22$, and the average FAF is 24.4 dB.

From Eq (12.20):

$$L_{50}(30) = 31.7 + 10 \times 5.22\log\left(\frac{30}{1}\right) = 108.8 \text{ dB}$$

From Eq. (12.21):

$$L_{50}(30) = 31.7 + 10 \times 3.27 \log\left(\frac{30}{1}\right) + 24.4 = 104.4\,\text{dB}$$

The results obtained by the two models are fairly close.

12.7.3 IMT-2000 Models

Since ITU IMT-2000 will be the worldwide standard, the proposed models to evaluate radio transmission technologies consider a broad range of environmental characteristics including large and small cities, suburbs, tropical, rural, and desert areas. IMT-2000 operating environments have been identified by an appropriate subset consisting of indoor office environment, outdoor-to-indoor and pedestrian environment, and vehicular environment. The key parameters of each propagation model are:

- delay spread, its structure, and its statistical variation (see section 12.8)
- geometrical path-loss rule and excess path loss
- shadow fading
- multipath fading characteristics (e.g., Doppler spectrum, Rician vs. Rayleigh) for the envelope of channels
- operating radio frequency

12.7.3.1 Indoor Office Environment Model

This environment is characterized by small cells and low transmit powers. The base stations and pedestrian users are located indoors. Root-mean-square (RMS) delay spread ranges from around 35 ns to 460 ns. The path-loss rule varies due to scatter and attenuation by walls, floors, and metallic structures such as partition and filing cabinets. These objects also produce shadowing effects. A log-normal shadowing with a standard deviation of 12 dB can be expected. The fading characteristic ranges from Rician to Rayleigh, with Doppler frequency offsets set by walking speeds. The path-loss model for this environment is

$$L_{50} = 37 + 30 \log r + 18.3 F^{[(F+2)/(F+1) - 0.46]} \tag{12.23}$$

where r = separation between transmitter-receiver (m), and
  F = number of floors in the path.

12.7.3.2 Outdoor-to-Indoor and Pedestrian Environment

This environment is characterized by small cells and low transmit power. Base stations with low antenna heights are located outdoors; pedestrian users are located on streets and inside buildings and residences. Coverage into buildings in high-power systems is included in the vehicular environment discussed in the next section. RMS delay spread varies from 100 to 1800 ns. A geometrical path-loss rule of r^{-4} is applicable. If the path is a line of sight on a canyon-like street, the path loss follows an r^{-2} rule, when there is Fresnel zone clearance. For the region with longer Fresnel zone clearance, a path-loss rule of r^{-4} is appropriate, but a range of up to r^{-6} may be encountered due to trees and other obstructions along the path. Log-normal shadow fading with a standard deviation of 10 dB for outdoors and 12 dB for indoors is reasonable. Average building penetration loss of 18 dB with a standard deviation of 10 dB is appropriate. Rayleigh and/or Rician fading rates are generally set by walking speeds, but faster fading due to reflections from moving vehicles may occur sometimes. The following path-loss model has been suggested for use in this environment:

$$L_{50} = 40 \log r + 30 \log f_c + 49 \text{ dB} \tag{12.24}$$

where f_c = carrier frequency (MHz).

This model is valid for a no-LOS case only and describes worst-case propagation with log-normal shadow fading with a standard deviation of 10 dB. The average building penetration loss is 18 dB with a standard of 10 dB.

12.7.3.3 Vehicular Environment

This environment consists of larger cells and higher transmit power. RMS delay spread from 0.4 ms to about 12 ms may occur on elevated roads in hilly or mountainous terrain. A geometrical path-loss rule of r^{-4} and log-normal shadow fading with a standard deviation of 10 dB are used in urban and suburban areas. Building penetration loss averages 18 dB with a standard deviation of 10 dB. In rural areas with flat terrain, the path loss is lower than that of urban and suburban areas. In mountainous terrain, if path blockages are avoided by placement of base stations, the path-loss rule is closer to r^{-2}. Rayleigh fading rates are set by vehicle speeds. Lower fading rates are appropriate for applications employing stationary terminals. The following model is used for this environment:

$$L_{50} = 40(1 - 4 \times 10^{-2} \Delta h_b) \log r - (18 \cdot \log \Delta h_b) + 21 \cdot \log f_c + 80 \text{ dB}$$

where r = base station and mobile separation (km),
 f_c = carrier frequency (MHz), and
 Δh_b = base station antenna height measured from average rooftop level (m).

12.8 Delay Spread

The radio signal follows different paths because of multipath reflection. Each path has a different path length, so the time of arrival for each path is different. The effect, which smears or spreads out the signal, is called *delay spread*. As an example, if an impulse is transmitted by the transmitter, by the time this impulse is received at the receiver, it is no longer an impulse but rather a pulse that is spread (refer to Fig. 12-3). In a digital system, the delay spread causes intersymbol interference, thereby limiting the maximum symbol rate of a digital multipath channel [8].

The mean delay spread τ_d is

$$\tau_d = \frac{\int_0^\infty t D(t) dt}{\int_0^\infty D(t) dt} \tag{12.25}$$

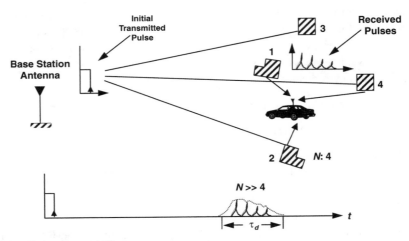

Figure 12-3   Delay Spread Phenomenon

where $D(t)$ is the delay probability density function, and

$$\int_0^\infty D(t)dt = 1.$$

Some representative delay functions are

- **Exponential**

$$D(t) = \frac{1}{\tau_d} e^{-\frac{t}{\tau_d}}$$

- **Uniform**

$$D(t) = \frac{\tau_d}{2}, \qquad 0 \le t \le 2\tau_d$$

$$D(t) = 0, \text{ elsewhere}$$

The measured data suggest that the mean delay spreads are different in different environments (refer to Table 12-4).

Table 12-4   Measured Data for Delay Spread

Type of Environment	Delay Spread τ_d (μs)
Open area	< 0.2
Suburban area	0.5
Urban area	3

A majority of the time, the RMS delay spreads are relatively small, but, occasionally, there are worst-case multipath characteristics that lead to much larger RMS delay spreads. Measurements in outdoor environments show that RMS delay spread can vary over an order of magnitude within the same environment. Delay spreads can have a major impact on system performance. To accurately evaluate the relative performance of radio transmission technologies, it is important to model the variability of delay spread as well as the worst-case locations where delay spread is relatively large. Three multipath channels are defined by IMT-2000 for each environment. Channel A represents the low-delay-spread case that occurs frequently; channel B corresponds to the medium-delay-spread case that also occurs frequently; channel C is the high-delay-spread case that occurs only rarely. Table 12-5 provides the average RMS delay spread values for each channel and for each environment.

12.8.1   Coherence Bandwidth

The coherence bandwidth B_c is a statistical measure of the range of frequencies over which the channel passes all spectral components with approximately equal gain and linear phase. The coherence bandwidth represents a frequency range for which either the amplitudes or phases of two received signals have a high degree of correlation. A signal's spectral components in that frequency range are affected by the channel in a similar manner—for example, exhibiting fading or no fading. As an approximation

$$B_c \approx 1/\tau_{dmax} \tag{12.26}$$

where τ_{dmax} = maximum delay spread.

The maximum delay spread, τ_{dmax}, is not necessarily the best indicator of how any given system will perform on a channel because different channels with the same value of τ_{dmax} may exhibit very different profiles of signal intensity over the delay span. A more useful measurement is often expressed in terms of the RMS delay spread, τ_{drms}. An exact relationship between coherence bandwidth and delay spread does not exist. Several approximations have been proposed.

The coherence bandwidth is defined as that bandwidth where the correlation function $|R_T(\Omega)| = 0.5$, for two fading signal envelopes at two frequencies f_1 and f_2, respectively, where

Table 12-5   Average RMS Delay Spread Values (IMT-2000)

Environment	Channel A		Channel B		Channel C	
	τ_{rms} (ns)	% Occurrence	τ_{rms} (ns)	% Occurrence	τ_{rms} (ns)	% Occurrence
Indoor office	35	50	100	45	460	5
Outdoor to indoor and pedestrian	100	40	750	55	800	5
Vehicular high antenna	400	40	4000	55	12,000	5

$$\Delta f = |f_1 - f_2| \tag{12.27}$$

Two frequencies that are farther apart than the coherence bandwidth, B_c, will fade independently. This concept is also useful for diversity reception.

The coherence bandwidth for two fading amplitudes of two received signals is

$$\Delta f > B_c = \frac{1}{2\pi\tau_{drms}} \tag{12.28}$$

If coherence bandwidth is defined as the frequency interval over which the channel's complex frequency transfer function has a correlation of at least 0.9, the coherence bandwidth is approximately given as

$$B_c \approx \frac{1}{50\tau_{drms}} \tag{12.29}$$

For mobile radio, the model generally accepted as a useful model for the urban environment has an array of radially uniformly spaced scatters, all with equal magnitude reflection coefficients but independent, randomly occurring reflection phase angles. This model is referred to as the dense-scatter channel model. Using such a model, coherence bandwidth has been defined for a bandwidth interval over which the channel's complex frequency transfer function has a correlation of at least 0.5 [10,11], or

$$B_c \approx \frac{0.276}{\tau_{drms}} \tag{12.30}$$

A more popular approximation of B_c corresponding to a bandwidth interval with a correlation of at least 0.5 is

$$B_c \approx \frac{1}{5\tau_{drms}} \tag{12.31}$$

A channel is a *frequency-selective* channel if $B_c < 1/T_s = B_w$, where the symbol rate $1/T_s$ is nominally taken to be equal to the signal bandwidth B_w. Frequency-selective fading distortion occurs when a signal's spectral components are not all affected equally by the channel.

Frequency-nonselective or *flat-fading* degradation occurs whenever $B_c > B_w$. Hence, all of the signal's spectral components are affected by the channel in a similar manner. Flat fading does not introduce channel-induced intersymbol interference (ISI) distortion, but performance degradation can still be expected due to loss in SNR whenever the signal is fading. In order to avoid channel-induced ISI distortion, the channel must be a flat-fading channel by ensuring that

$$B_c > B_w = \frac{1}{T_s} \tag{12.32}$$

Thus, the channel coherence bandwidth sets an upper limit on the transmission rate that can be used without incorporating an equalizer in the receiver.

The GSM symbol rate (or bit rate, since modulation is binary) is 271 ksps and the bandwidth is $B_w = 200$ kHz. If the typical RMS delay spread in an urban environment is $\tau_{drms} = 2\mu s$, then using Eq. (12.31) the coherence bandwidth $B_c \approx 100$ kHz. It is therefore apparent that since $B_c < B_w$, the GSM receiver must use some form of mitigation to overcome frequency-selective distortion. To accomplish this goal, the Viterbi equalizer [1] is typically used.

EXAMPLE 12.2

Assume vehicle speed equal to 60 mph (88 ft/sec), carrier frequency $f_c = 860$ MHz, and delay spread $\tau_{drms} = 2$ μsec. Calculate coherence time and coherence bandwidth. At a coded symbol rate of 19.2 kbps (IS-95), what kind of symbol distortion will be experienced? What type of fading will be experienced by the IS-95 channel?

$$v = 60 \text{ mph } (= 88 \text{ ft/sec})$$

$$\lambda = \frac{c}{f} = \frac{9.84 \times 10^8}{860 \times 10^6} = 1.1442 \text{ ft/sec}$$

$$f_m = \frac{v}{\lambda} = \frac{88}{1.1442} = 77 \text{ Hz}$$

$$T_c = \frac{1}{2\pi f_m} = \frac{1}{2\pi \times 77} = 0.0021 \text{ seconds}$$

$$T_s = \frac{10^6}{19,200} = 52 \text{ μsec}$$

The symbol interval is much smaller than the channel coherence time. Therefore, symbol distortion is minimal. In this case fading is slow.

$$B_c \approx \frac{1}{2\pi\tau_{drms}} = \frac{1}{2\pi \times 2 \times 10^{-6}} = 79.56 \text{ kHz}$$

This shows that IS-95 is a wideband system in this multipath situation and experiences selective fading over only 6.5% ($79.57/1228.8 = 0.0648$) of its bandwidth.

12.9 Doppler Spread

The width of the Doppler power spectrum is called *spectral broadening* or *Doppler spread*, is denoted f_d, and is sometimes called the *fading bandwidth* of the channel. The Doppler shift of each arriving path is generally different from that of another path. The effect on the received signal is seen as a Doppler spreading or spectral broadening of the transmitted signal frequency rather than a shift. Doppler spread and coherence time T_0 are reciprocally related as

$$T_0 = \frac{1}{f_d} \tag{12.33}$$

f_d is regraded as the typical fading rate of the channel. When T_0 is defined as the time duration over which the channel's response to a sinusoid has a correlation greater than 0.5, then

$$T_0 \approx \frac{9}{16\pi f_d} \tag{12.34}$$

A popular rule of thumb is to define T_0 as the geometric mean of Eqs. (12.33) and (12.34)

$$T_0 = \sqrt{\frac{9}{16\pi f_d^2}} = \frac{0.423}{f_d} \tag{12.35}$$

The time required to traverse a distance $\lambda/2$ (equal to fade interval) when traveling at a constant velocity v is

$$T_0' = \frac{\lambda/2}{v} = \frac{0.5}{(v/\lambda)} = \frac{0.5}{f_d} \approx T_0 \tag{12.36}$$

With frequency equal to 900 MHz and velocity equal to 120 km/h, the coherence time is about 5 ms and Doppler spread is approximately 100 Hz. For a voice-grade channel with a typical transmission rate of 10 ksps, the fading rate is considerably less than the symbol rate. Under such conditions, the channel would manifest slow-fading effects. A channel is referred to as *fast fading* if the symbol rate $1/T_s$ is less than the fading rate $1/T_0$ i.e., the fast fading is characterized by

$$B_w < f_d \tag{12.37a}$$

or

$$T_s > T_0 \tag{12.37b}$$

A channel is referred to as slow fading if the signaling rate is greater than the fading rate. Thus, to avoid signal distortion caused by fast fading, the channel must be made to exhibit slow fading by ensuring that the signaling rate exceeds the channel fading rate

$$B_w > f_d \tag{12.38a}$$

or

$$T_s < T_0 \tag{12.38b}$$

The channel fading rate f_d sets a lower limit on the signaling rate that can be used without suffering fast-fading distortion. A better way to state the requirement for mitigating the effects of fast fading would be that we desire $B_w \gg f_d$ (or $T_s \ll T_0$). If this condition is not satisfied, the random frequency (FM) due to varying Doppler shifts will limit system performance significantly. The Doppler effect yields an irreducible error rate that cannot be overcome by simply increasing E_b/I_t. This irreducible error rate is most pronounced for any modulation that involves switching the carrier phase. For voice-grade applications with error rates of 10^{-3} to 10^{-4}, a large value of Doppler shift is considered to be on the order of 0.01 B_w. Thus, to avoid fast-fading distortion and Doppler-induced irreducible error rate, the signaling rate must exceed the fading rate by a factor of 100 to 200. The exact factor depends on the signal modulation, receiver design, and required bit error rate.

12.10 Intersymbol Interference

In practical radio systems, the presence of a transmitter band-pass filter is essential to save spectrum as much as possible. However, such a band-limited channel could degrade the transmission performance due to ISI. Therefore, we should reduce the signal bandwidth as much as possible without producing any ISI.

In a time-dispersive medium, the transmission rate R_b for a digital transmission is limited by delay spread. If a low bit-error-rate performance is required, then

$$R_b < \frac{1}{2\tau_d}$$

(12.39)

In a real situation, R_b is determined based upon the required bit error rate.

12.11 Link Budget and Cell Coverage

The L_{50} path-loss estimates provide the average signal strength at a given distance from the transmitter. The signal varies from that average by large amounts, both lower and higher. Field data show that statistically the signal strength at a given distance from the transmitter has a log-normal distribution with standard deviation equal to 8 to 10 dB. These variations are called *shadow losses* and are caused by obstructions between the transmitter and the receiver. Field data also indicate that an obstruction affects the path loss for an average distance of 500 m in the suburban environment and 50 m in the urban environment. Fig. 12-4 shows a typical plot of the path loss vs. distance in meters with a mean path loss (MPL) and an MPL ±10.24 dB. The path loss has a normal distribution with a standard deviation of ±8 dB. A normal distribution curve has ±32% of the values outside ±8 dB. This means 16% of the values will be more than 8 dB above the mean. We must consider this variation when designing a system so that the received signal to interference ratio is greater than the minimum value required for an acceptable voice quality over most of the coverage area. For example, if the system objectives are to provide adequate voice quality over 90% of the coverage area, the system must be designed with at least a 10.24-dB margin.

To determine the maximum cell range, designers must calculate the maximum allowable path loss that will provide adequate signal strength at the cell boundary for acceptable voice quality over 90% of the coverage area. The allowable path loss is the difference between the transmitter's effective radiated power and minimum signal strength required at the receiver for acceptable voice quality. The components that determine the path loss are called the *link budget*.

Link budgets are used to calculate the coverage and performance for a base station and a mobile station. The components include propagation factors to calculate path loss and system parameters (transmitter power, receiver noise figure, antenna gains, receiver bandwidth, processing gain, and interference). Other losses such as power control errors, building penetrations, body/orientation losses, and interference from other sources are also included.

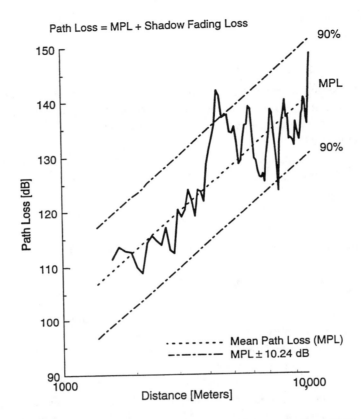

Note: 10% of the path loss will exceed the MPL by more than 10.24 dB.

Figure 12-4 Shadow Margin

For CDMA, the link budget is used to

- Decide an appropriate network loading
- Allocate appropriate power to various forward link channels

The following procedure is used for link budget analysis:

- Identify parameters affecting the forward and reverse links

    ◆ access technology-specific parameters
    ◆ product-specific parameters
    ◆ morphology-based parameters

- Determine the maximum allowable path loss to maintain communication on the forward and reverse links

- Balance the forward and reverse link

We illustrate the procedure for calculating link budget and determining the range of a base station.

EXAMPLE 12.3

Refer to Figure 12-5 and use the following parameters to calculate the maximum allowable path loss:

- Information rate = 9600 bps
- Mobile station's effective radiated power (P_m) = 200 mW (23 dBm)
- Base station antenna gain (G_b) = 14 dBi
- Base station receiver antenna cable loss (L_c) = 2.5 dB
- PCS minicell receiver noise figure (F_b) = 5 dB
- Required margin (E_b/N_t) = 6.8 dB (with diversity antenna at base station)
- Base station noise floor (N_0) = –174 dBm/Hz
- Log-normal shadowing margin = 8 dB
- Body/orientation loss = 2 dB
- Building penetration loss = 10 dB

Base station noise floor

$$N_T = N_0 + F_b = -174 + 5 = -169 \text{ dBm/Hz} \tag{12.40}$$

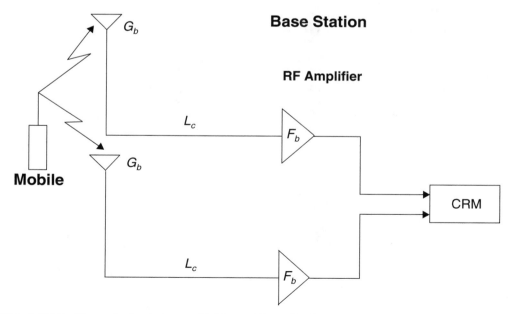

Figure 12-5 Transmission between Mobile and Base Station

Minimum bit energy required for specified E_b/N_0

$$(E_b)_{min} = N_T + (E_b/N_t)_{reqd} = -169 + 6.8 = -162.2 \text{ dBm/Hz} \tag{12.41}$$

Minimum signal strength required

$$S_{min} = (E_b)_{min} + 10\log R = -162.2 + 10\log 9600 = -122.4 \text{ dBm} \tag{12.42}$$

Mean path loss (L_{50}) with S_{min}

$$L_{50} = P_m - S_{min} + G_b - L_c = 23 + 122.4 + 14 - 2.5 = 156.9 \text{ dB} \tag{12.43}$$

To provide margin for shadowing

$$\text{Path loss} = L_{50} - 10.2 = 156.9 - 10.2 = 146.7 \text{ dB} \tag{12.44}$$

To provide margin for body/orientation loss and building penetration loss

$$\text{Allowable path loss} = 146.7 - 2.0 - 10 = 134.7 \text{ dB} \tag{12.45}$$

EXAMPLE   12.4

Using the allowable path-loss value from Example 12.3, determine the coverage for the mini-PCS cell. Assume the following data:

- PCS frequency (f_c) = 1800 MHz
- Street width (W) = 20 m
- Spacing between buildings (b) = 40 m
- Average roof height of building (h_r) = 40 m
- Mobile antenna height (h_m) = 2 m
- Base station antenna height (h_b) = 40 m
- Street orientation, ϕ = 90 degrees

We use the COST 231 model

$$\Delta h_m = h_b - h_m = 40 - 2 = 38 \text{ m}$$

$$\Delta h_b = h_b - h_r = 40 - 40 = 0 \text{ m}$$

$$L_0 = 4 - 0.114(\phi - 55) = 4 - 0.114(90 - 55) = 0$$

$$L_{bsh} = -18\log 11 + \Delta h_b = -18.75 \text{ dB}$$

$$k_a = 54, k_d = 18 - \frac{15\Delta h_b}{\Delta h_m} = 18, k_f = 4 + 1.5\left(\frac{f_c}{925} - 1\right) = 4 + 1.5\left(\frac{1800}{925} - 1\right) = 5.42$$

$$L_f = 32.4 + 20\log r + 20\log 1800 = 97.5 + 20\log r \text{ dB}$$

$$L_{rts} = -16.9 - 10\log 20 + 10\log 1800 + 20\log 38 + 0 = 34.25 \text{ dB}$$

$$L_{ms} = -18\log 11 + 0 + 54 + 18\log r + 5.42\log 1800 - 9\log 40 = 38.47 + 18\log r \text{ dB}$$

$$\therefore 134.7 = 97.5 + 20\log r + 34.25 + 38.47 + 18\log r$$

$$38\log r = -35.52$$

$$\log r = -0.935, \ r = 0.116 \, \text{km or } 116 \, \text{m}$$

12.12 Dual-Mode CDMA Mobiles

The nominal CDMA channel requires 41 contiguous analog channels with a 9-channel spacing as a guard between the edge of a CDMA channel and the adjacent analog channels. This implies that 59 contiguous analog channels should be removed from the service to introduce the first CDMA channel, 100 analog channels for the first and second CDMA channels, and so on. In an analog system with a reuse factor of 7, 3 analog channels per sector will be removed to add the first CDMA channel (i.e., [59/7]/3 ≈ 3) and an additional 2 analog channels per sector will be removed for each additional CDMA channel ([42/7]/3 = 2).

Consider an example where the base station is equipped with 78 channel elements. Four channel elements on each sector are dedicated to the pilot channel, sync channel, and two paging channels, leaving 22 channel elements per sector remaining for traffic channels. Assume that 22 channel elements are permanently assigned to each sector. The real system pools all 66 channel elements, so any channel element can be assigned to any sector; this method is referred to as *dynamic channel assignment*. Dynamic channel assignment improves the base station capacity as it reduces the probability of blocking on a sector since an idle channel element can be used in any sector. Blocking occurs when the interference from an additional cell will reduce the voice quality below the acceptable limit. Blocking is determined by the channel capacity and not by the amount of hardware used. Using the method of permanently assigning channel elements to a sector gives the lower bound on the total traffic carried by the base station. The traffic carried by the same channel elements will be higher in the real system.

Consider the case where a 3-sector analog base station is equipped with 57 channels that are assigned an N = 7 reuse pattern. We remove 3 analog channels to introduce 1 CDMA channel to serve 22 users. Assume no overflow traffic from CDMA to analog channels, assume 2% probability of blocking, and model the traffic using Erlang B-statistic. Also assume that each sector can serve 22 simultaneous calls with the same voice quality as an analog channel. Table 12-6 provides the calculated capacity of the analog base station and the combined CDMA and analog base station. It may be noted that the traffic capacity of the combined base station is double the capacity of the analog base station.

EXAMPLE 12.5

Estimate the cell capacity as a function of CDMA user penetration for an overlay system with 2% blocking for both analog and digital subscribers. Assume N = 7 as the reuse pattern for the analog channels and N = 1 for CDMA channels. The average traffic per subscriber is 0.02 Erlangs. See Table 12-7.

Table 12-6 Comparison of BS Capacity with Analog and Combined CDMA and Analog Channels

Analog Voice Channels	CDMA Traffic Channels	Analog Traffic per Sector (Erlangs)	Total Analog Traffic (Erlangs)	CDMA Traffic per Sector (Erlangs)	Total CDMA Traffic (Erlangs)	Total BS Traffic (Erlangs)
19	0	12.34	37.0	—	—	37.0
16	22	9.83	29.5	14.9	44.7	74.2

Table 12-7 Capacity of a cell vs. CDMA User Penetration

Analog Channels per Cell, N_a	CDMA Channels per Cell, N_c	CDMA Users per Cell, CE	% CDMA Traffic	Analog Mobile (E)	CDMA Mobile (E)	Total Erlangs (E)	Analog Subscribers	CDMA Subscribers	Total Subscribers
57	0	0	0	12.34 × 3 = 37.0	0	37.0	1850	0	1850
48	1	66	60.24%	9.83 × 3 = 29.5	14.9 × 3 = 44.7	74.2	1475	2235	3710
42	2	132	80.91%	8.2 × 3 = 24.6	34.68 × 3 = 104.0	128.6	1230	5200	6430
36	3	198	89.3%	6.62 × 3 = 19.9	55.33 × 3 = 166	185.9	995	8300	9295
30	4	264	93.73%	5.084 × 3 = 15.3	76.38 × 3 = 229.1	244.4	765	11,455	12,220
24	5	330	96.43%	3.627 × 3 = 10.9	97.69 × 3 = 293.1	304.0	545	14,655	15,200
18	6	396	98.13%	2.277 × 3 = 6.8	119.2 × 3 = 357.6	364.4	340	17,880	18,220
12	7	462	99.22%	1.092 × 3 = 3.3	140.7 × 3 = 422.2	425.5	165	21,110	21,275
6	8	528	99.86%	0.223 × 3 = 0.7	162.4 × 3 = 487.3	488	35	24,365	24,400
0	9	594	100%	0	184.2 × 3 = 552.5	552.5	0	27,625	27,625

The impact on introducing CDMA in an analog system results in reducing the cell analog capacity by 20%, but it doubles the total capacity of the cell. This means that at least 20% of the offered traffic load must come from dual-mode mobiles before activating CDMA. To realize 100% capacity increase in the cell, about 60% of the offered traffic load in the cell must be from dual-mode mobiles.

12.13 The Transition from an Analog System to a Digital System

The transition from an analog system to the CDMA system, as offered by cellular system equipment vendors and envisioned by the cellular system operators, generally falls into three basic modes:

- An overlay design of two separate, independent systems—one is analog and the other is CDMA
- A completely integrated system where CDMA and analog service is offered everywhere
- A partially integrated system in which CDMA and analog coverage is provided in part of the overall service area and only analog coverage exists in the remainder of the service area

12.13.1 Overlay Design

In the overlay scenario, a new CDMA system is superimposed over the existing analog system. The CDMA overlay could require a one-to-one digital base station for each analog base station, but it may use a smaller number of larger coverage area cells than the existing analog system to reduce the cost of base stations. Fig. 12-6 shows a CDMA base station coverage area that is three times larger than the analog base station coverage area. Thus, it requires one-third as many base stations to provide CDMA coverage over the service area. Such an overlay design could be operated as two separate, independent systems that are operated by different vendors from separate mobile switching centers (MSC) or as a single system in which a single MSC controls both types of base stations. Handoffs between the CDMA and analog system are not possible, and handoffs from analog to CDMA are not allowed by the IS-95 standard. The advantages and disadvantages of the full overlay design are

- **Advantages**
    - ◆ It allows independent analog and CDMA systems in the market.
    - ◆ There can be a different vendor rather than the analog system vendor for the CDMA system if this is desired.
    - ◆ The service provider will be able to advertise that CDMA (digital) service is available throughout the service area.
    - ◆ Since a smaller number of base stations can be used, the investment in digital equipment should be lower than for designs that require digital equipment in a larger number of base stations.

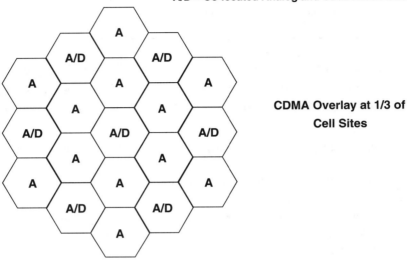

A = Analog Base Stations

A/D = Co-located Analog and CDMA Base Stations

CDMA Overlay at 1/3 of Cell Sites

Figure 12-6   Overlay Design

- **Disadvantages**

  - There is capacity loss due to segmentation of the cellular spectrum since there will be fewer radio frequencies for the analog subscribers.
  - The grade of service to analog subscribers, whose numbers may increase if analog terminal dumping occurs, will require investing in additional analog base station infrastructure.
  - There is increased system operational complexity. The engineering, Operation, Administration, and Maintenance (OA&M) of a two-system CDMA/analog overlay is much more complex than for a single system.
  - The analog-only base stations may require additional RF filters to reduce the probability that a nearby CDMA mobile will overload the analog base station receiver.

12.13.2 Integrated Design

In this scenario (Fig. 12-7), the system is designed to support both analog and CDMA customers everywhere in the service area. As with the other designs, this would be a transitional approach containing the capability for high CDMA capacity in the core and lower CDMA capacity in the noncore areas. Over time, the area with higher CDMA capacity would expand to include more and more of the system. The advantages and disadvantages of this approach are

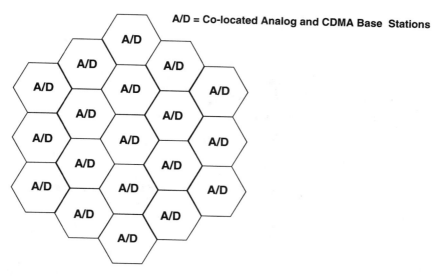

Figure 12-7 Integrated Design

- **Advantages**

  - The entire system would have complete digital coverage.
  - This scenario avoids receiver overload and other radio problems.
  - There is no cellular spectrum segmentation.
  - Full-spectrum efficiency is achieved through high reuse (N = 1) everywhere.
  - Dual-mode terminals and handoffs between CDMA and analog channels are not required (except for roaming into analog coverage areas).
  - OA&M is simplified from that of the overlay system.
  - The system operator can advertise digital service everywhere in the service area.

- **Disadvantages**

  - It requires digital equipment everywhere in the system. Although the investment costs can be reduced in areas that need only a low-capacity digital service, it is still a larger investment than for the partial digital system option.

12.13.3 Partial CDMA Coverage, Integrated System

In the partially integrated system, only a part of the system is converted to support analog and digital traffic (usually the core of the system, most in need of traffic relief from the digital design). Surrounding the core area of the system, a transition or buffer zone is required to avoid interference between cochannel analog and digital channels. Cochannels are not allowed in the buffer zone. Beyond the buffer zone base stations can be assigned analog channels that are cochannel with digital channels within the core. A dual-mode mobile assigned a digital channel

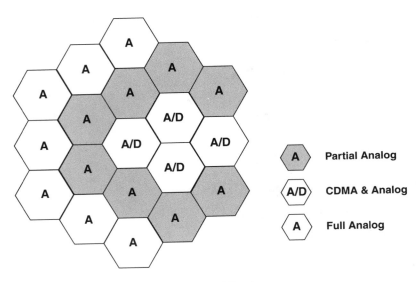

Figure 12-8   Partial CDMA Coverage—Integrated Design

in the core would be handed off to an analog channel as the mobile approaches the edge of the digital coverage area, and then it would be handed to a cell in the buffer zone. Mobiles assigned an analog channel in the analog-only cells would not transition to a digital channel once inside the digital coverage area.

The transition zone can gradually be moved outward from the core, and digital coverage can be expanded. The rate of expansion can be determined by the mix of terminals in the system, the need for capacity relief, the strategy for moving the customer base to digital terminals, and economics of the system operator. Fig. 12-8 shows an example of a simplified model with a uniform cell size. The buffer zone between the CDMA/analog cells and analog-only cells may require two tiers of cells depending on the propagation and relative sizes of the actual cells. If the cells in the buffer zone are at full capacity, then adding CDMA will reduce the capacity of the buffer zone cells since frequencies required by CDMA channels cannot be used in the buffer zone.

Fig. 12-9 shows the case where a CDMA base station interferes with the analog mobile. The performance criterion for analog mobiles is that the carrier signal from the analog base station is 17 dB greater than the total interference on the channel. If the analog system is designed for

$$\frac{S_{anal}}{I_{coch}} = 18 \text{ dB} \qquad (12.46)$$

and if

$$\frac{S_{anal}}{I_{coch} + I_{CDMA}} \geq 17 \text{ dB} \qquad (12.47)$$

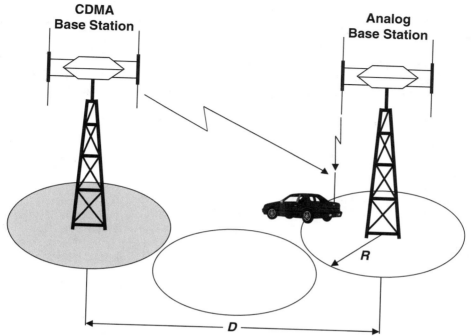

Figure 12-9 CDMA Base Station Interference with Analog Mobile

provides adequate voice quality, then

$$\frac{S_{\text{anal}}}{I_{\text{CDMA}}} > 24 \text{ dB} \tag{12.48}$$

Table 12-8 lists the relative sizes of CDMA and analog base stations to achieve the criterion as a function of separation between the CDMA and analog center frequencies. With the recommended 9 guard channels, the CDMA base station has to be only slightly farther away than the

Table 12-8 Required D/R^a Ratios vs. Center Frequency Separation between CDMA and Analog Center Frequencies: CDMA BS Interferes with Analog MS

Center Frequency Separation (f_s)	Required D/R
$f_s < 900$ kHz	≥ 2.90
$900 \leq f_s < 1980$ kHz	≥ 1.33
$f_s \geq 1980$ kHz	≥ 1.14

[a] D = distance between analog and CDMA base stations
R = radius of the analog base station

radius of the analog cell to have adequate analog voice-quality performance. If the analog base station assigns frequencies within the CDMA channel, then the CDMA base station must be about 3 times the analog cell radius away, or a one cell buffer zone for equal-size analog and CDMA cells. (R = radius of the analog base station, and D = distance between analog and CDMA base stations.)

Fig. 12-10 shows a case where the CDMA mobile interferes with the analog base station. The performance criterion for the analog base station is that the received signal be 17 dB greater than the total received noise. If the analog system is designed for

$$\frac{S_{anal}}{I_{coch}} = 18 \text{ dB} \tag{12.49}$$

and if

$$\frac{S_{anal}}{I_{coch} + I_{CDMA}} \geq 17 \text{ dB} \tag{12.50}$$

provides adequate voice quality, then

$$\frac{S_{anal}}{I_{CDMA}} > 24 \text{ dB} \tag{12.51}$$

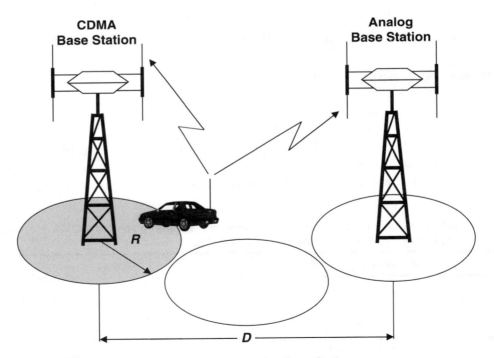

Figure 12-10   CDMA Mobile Interference with Analog Base Station

Table 12-9 Required D/R^a Ratio vs. Frequency Separation between CDMA and Analog Center Frequencies: CDMA MS Interferes with Analog BS

Frequency Separation f_s	Required D/R
$f_s < 900$ kHz	≥ 1.23
$900 \leq f_s < 1980$ kHz	≥ 1.05
$f_s \geq 1980$	≥ 1.02

[a] D = distance between the analog and CDMA base stations
R = radius of the CDMA base station

Table 12-9 lists the required D/R ratios as a function of the separation between the CDMA and analog center frequencies to achieve the criterion, where D = distance between the analog and CDMA base station and R = radius of the CDMA base station. With the recommended guard 9 channels, the CDMA mobile can be close to the analog base station before the interference becomes a problem. This situation is less critical than the previous case.

Both rules (Tables 12-8 and 12-9) must be satisfied when deploying CDMA base stations to prevent interference to the existing analog cells. Meeting these criteria will insure that the interference from the analog transmitters to the CDMA receivers is not a problem as well.

The following are the advantages and disadvantages of the partial integrated design:

- **Advantages**

  - The CDMA capacity advantage can be placed where it is needed most, i.e., in the core system. Investment only for the core system will be needed.
  - OA&M is simpler than the overlay approach.
  - The design avoids the receiver overload problem of the overlay design.

- **Disadvantages**

  - The system operator cannot advertise digital-everywhere service.
  - Handoff is required between CDMA and analog coverage areas.
  - IS-95 does not provide analog-to-CDMA handoff, so a call initiated in the analog area will not provide any digital features available in the digital coverage area.
  - Voice-quality changes may be perceived during CDMA-to-analog handoffs.

EXAMPLE 12.6

Consider a small-city cellular system that is growing at the predicted growth rate for the next 7 years (see Table 12.10). The start-up system required 9 omnidirectional-coverage base stations and has grown to 29 directional analog base stations to provide service for 36,000 Busy Hour Call Attempts (BHCA). Based on the predictions, this system must expand to provide capacity for 100,000 BHCA at the end of 7 years. The service provider has chosen to provide CDMA service over the complete coverage area by overlaying the coverage with 10 minicells. The service provider reduces the analog subscribers gradually as indicated in Table 12.10. The traffic per subscriber during the busy hour is 0.02 Erlangs.

Year 0 (all analog) (see Fig. 12-11)

- Total traffic during busy hour: $36,000 \times 0.02 = 720.0$ Erlangs
- Traffic per sector: $\dfrac{720.0}{29 \times 3} = 8.276$ Erlangs
- Number of voice channels per sector to provide 2% blocking: 14

Table 12-10   Prediction for Analog and Digital Subscribers

End of Year	Analog Subscribers	CDMA Subscribers	Total Subscribers
0	36,000	0	36,000
1	29,000	13,000	42,000
2	24,000	22,000	46,000
3	16,000	39,000	55,000
4	10,000	52,000	62,000
5	4,000	66,000	70,000
6	0	82,000	82,000
7	0	100,000	100,000

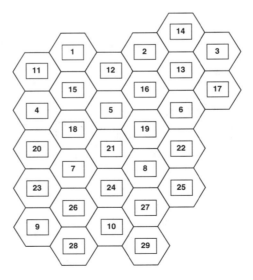

Figure 12-11   Configuration at Year 0—all analog cells

Table 12-11 summarizes the results at the end of year 0.

Year 1 (see Fig. 12-12)

- CDMA traffic: $13,000 \times 0.02 = 260$ Erlangs
- Analog traffic: $29,000 \times 0.02 = 580.0$ Erlangs

To provide full CDMA coverage, CDMA minicell equipment is added to 10 base stations.

- CDMA traffic per sector: $\dfrac{260.0}{10 \times 3} = 8.67$ Erlangs

One CDMA channel provides 14.9 Erlangs per sector with 2% system blocking to serve the dual-mode mobiles, so the service provider decides to eliminate all analog channels in the expanded spectrum and to use the spectrum for one CDMA channel in the minicell equipment. This reduces the analog capacity by 3 channels per sector in all 29 base stations.

- Analog traffic per sector: $\dfrac{29,000 \times 0.02}{29 \times 3} = 6.67$ Erlangs
- Number of analog channels per sector: $14 - 3 = 11$

Table 12-11 Number of Analog and CDMA BSs and Capacity at the End of Year 0

No. of BSs		No. of RF Channels/Sector		Capacity Erlangs/Sector		
Analog	CDMA	Analog	CDMA	Analog	CDMA	BHCA
29	0	14	0	8.2	0	36,000

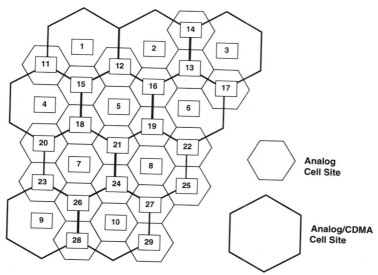

Figure 12-12 Configuration at the End of Year 1 and Year 2

These 11 channels will carry 6.83 Erlangs of traffic at 4% blocking for the analog subscribers. The single CDMA channel will carry all the offered traffic load with virtually no blocking for the CDMA subscribers. If the grade of service for the analog subscribers is an unacceptable business strategy, then additional analog capacity will have to be provided by adding additional channels in the limited spectrum band, possibly resulting in additional cochannel interference. The results are summarized in Table 12-12. In the calculations, we assume an average of 22 CDMA calls per sector with 2% blocking. As can be seen, the offered CDMA traffic load (8.67 Erlangs per sector) is much lower than the 2% blocking capacity (14.9 Erlangs per sector). CDMA subscribers will experience virtually no blocking while analog subscribers will experience 4% blocking.

Year 2 (see Fig. 12-12)

- CDMA traffic per sector: $\dfrac{22,000 \times 0.02}{10 \times 3} = 14.67$ Erlangs

One CDMA channel on 10 base stations with 2% blocking provides 14.9 Erlangs per sector.

- Analog traffic per sector: $\dfrac{24,000 \times 0.02}{29 \times 3} = 5.52$ Erlangs

- Number of analog channels per sector: $14 - 3 = 11$

These 11 channels will carry 5.84 Erlangs of traffic at 2% blocking for analog subscribers. Table 12-13 summarizes the results.

Year 3 (see Fig. 12-13)

- CDMA traffic per sector: $\dfrac{39,000 \times 0.02}{10 \times 3} = 26.0$ Erlangs

Two CDMA channels on 10 base stations with 2% blocking gives 34.68 Erlangs per sector.

- Analog traffic per sector: $\dfrac{16,000 \times 0.02}{23 \times 3} = 4.64$ Erlangs

We remove 6 analog base stations, leaving 23 base stations to carry analog traffic.

- Number of analog channels per sector: $14 - 5 = 9$

Table 12-12 Number of Analog and CDMA BSs and Capacity at the End of Year 1

No. of BSs		No. of RF Channels/Sector		Capacity Erlangs/Sector		
Analog	CDMA	Analog	CDMA	Analog	CDMA	BHCA
29	10	11	1	6.83[a]	14.9	42,000

[a] 4% blocking.

Table 12-13 Number of Analog and CDMA BSs and Capacity at the End of Year 2

No. of BSs		No. of RF Channels/Sector		Capacity Erlangs/Sector		
Analog	CDMA	Analog	CDMA	Analog	CDMA	BHCA
29	10	11	1	5.84	14.9	46,200

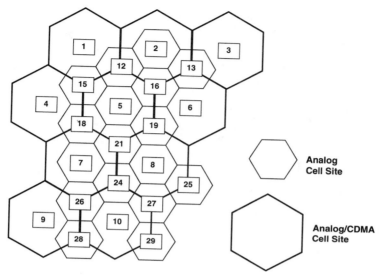

Figure 12-13 Configuration at the End of Year 3

These 9 channels will carry 4.35 Erlangs per sector of traffic at 2% blocking for the analog subscribers. Since the offered traffic load is slightly more than the capacity at 2% blocking, the analog subscribers will experience about 2.5% blocking, whereas the CDMA subscribers will experience almost no blocking since offered load is less than the capacity. The results are summarized in Table 12-14.

Year 4 (see Fig. 12-14)

- CDMA traffic per sector: $\dfrac{52,000 \times 0.02}{10 \times 3} = 34.67$ Erlangs

Two CDMA channels on 10 base stations with 2% blocking gives 34.68 Erlangs per sector.

- Analog traffic per sector: $\dfrac{10,000 \times 0.02}{20 \times 3} = 3.33$ Erlangs
- Number of analog channels per sector: $14 - 5 = 9$

We remove 3 analog base stations, leaving 20 base stations to carry analog traffic.

These 9 channels will carry 4.35 Erlangs per sector of traffic at 2% blocking for the analog subscribers. The results are summarized in Table 12-15.

Table 12-14 Number of Analog and CDMA BSs and Capacity at the End of Year 3

No. of BSs		No. of RF Channels/Sector		Capacity Erlangs/Sector		
Analog	CDMA	Analog	CDMA	Analog	CDMA	BHCA
23	10	9	2	4.35	34.68	55,000

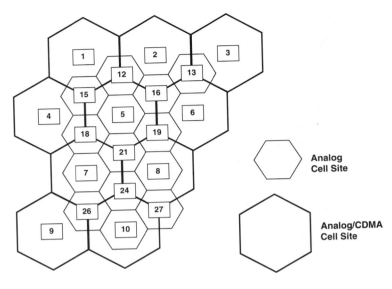

Figure 12-14   Configuration at End of Year 4

Table 12-15   Number of Analog and CDMA BSs and Capacity at the End of Year 4

No. of BSs		No. of RF Channels/Sector		Capacity Erlangs/Sector		
Analog	CDMA	Analog	CDMA	Analog	CDMA	BHCA
20	10	9	2	4.35	34.68	62,000

Year 5 (see Fig. 12-15)

- CDMA traffic per sector: $\dfrac{66,000 \times 0.02}{10 \times 3} = 44.0$ Erlangs

Three CDMA channels on 10 base stations with 2% blocking gives 55.33 Erlangs per sector.

- Analog traffic per sector: $\dfrac{4,000 \times 0.02}{10 \times 3} = 2.67$ Erlangs

We remove 10 analog base stations, leaving 10 base stations to carry analog traffic.

- Number of analog channels per sector: $14 - 7 = 7$

These 7 channels will carry 2.94 Erlangs per sector at 2% blocking. The results are summarized in Table 12-16.

Year 6 (all digital) (see Fig. 12-16)

- CDMA traffic per sector: $\dfrac{82,000 \times 0.02}{10 \times 3} = 54.67$ Erlangs

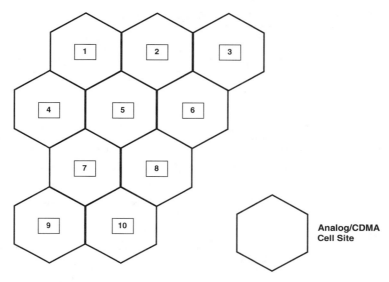

Figure 12-15   Configuration at End of Year 5

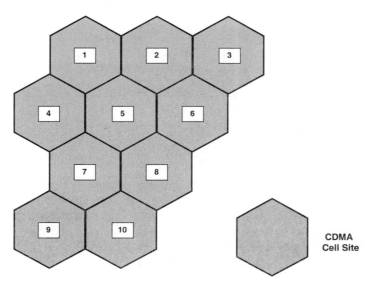

Figure 12-16   Configuration at End of Year 6 and Year 7

Table 12-16   Number of Analog and CDMA BSs and Capacity at the End of Year 5

No. of BSs		No. of RF Channels/Sector		Capacity Erlangs/Sector		
Analog	CDMA	Analog	CDMA	Analog	CDMA	BHCA
10	10	7	3	2.94	55.33	70,000

Table 12-17   Number of Analog and CDMA BSs and Capacity at the End of Year 6

No. of BSs		No. of RF Channels/Sector		Capacity Erlangs/Sector		
Analog	CDMA	Analog	CDMA	Analog	CDMA	BHCA
0	10	0	4	0	76.38	82,000

Table 12-18   Number of Analog and CDMA BSs and Capacity at the End of Year 7

No. of BSs		No. of RF Channels/Sector		Capacity Erlangs/Sector		
Analog	CDMA	Analog	CDMA	Analog	CDMA	BHCA
0	10	0	4	0	76.38	100,000

We eliminate all remaining analog base stations and use 4 CDMA channels on 10 base stations with 2% blocking to give 76.38 Erlangs per sector. The results are summarized in Table 12-17.

Year 7 (see Fig. 12-16)

- CDMA traffic per sector: $\dfrac{100,000 \times 0.02}{10 \times 3} = 66.67$ Erlangs

Four CDMA channels per sector provides 76.38 Erlangs of traffic at 2% blocking. The results are given in Table 12-18.

12.14 Facilities Engineering

CDMA technology offers a significant capacity improvement with respect to analog and other digital technologies. However, in order to fully realize these capacity improvements, the service provider must properly engineer the CDMA system. This section discusses the engineering of facilities and its relationship to call capacity.

Facilities encompass terrestrial facilities, radio facilities, transcoders (vocoders), and network facilities. As discussed earlier, transcoders can be physically located at the base station (see Fig. 12-17) or at the MSC (see Fig. 12-18).

Only the radio facility is related to the capacity of the radio link. Also the service provider must configure a sufficient number of CDMA channels to support the maximum number of simultaneous calls (CDMA channels include the pilot channel, sync channel, access channels,

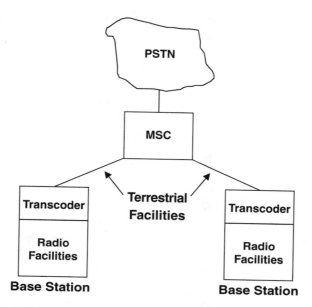

Figure 12-17 Basic CDMA Facilities Configuration (Transcoder at Base Station)

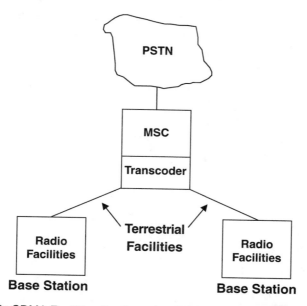

Figure 12-18 Basic CDMA Facilities Configuration (Transcoder at MSC)

paging channels, and traffic channels). Once this is done, the required frequency spectrum can be determined. This section provides an analytical discussion of the capacity of the reverse radio link and a qualitative discussion of the capacity of the forward radio link.

Unlike analog and TDMA technology, CDMA technology does not impose a definite limit on the radio capacity. Rather, CDMA technology exercises a *soft limit*, in which mobile subscribers experience a level of degradation that is related to the total interference and the thermal noise. The capacity is *soft* since the number of mobile subscribers can be increased if the service provider is willing to lower the grade of service and thus decrease customer satisfaction. With a greater number of simultaneous CDMA calls, the noise floor increases. If the noise floor increases, the probability of receiving a correct frame decreases, i.e., the FER increases. Interference is generated by other CDMA mobile stations occupying the same radio spectrum on the same cell or on a different cell. Moreover, for 800-MHz operation, interference is generated if a mobile station operating in the analog mode is occupying a portion of the frequency spectrum that is used by the CDMA channel. However, in a properly engineered CDMA system, analog operation is restricted to cells sufficiently separated from the cell serving the CDMA mobile. This significantly reduces the interference to the CDMA mobile.

In order for a CDMA system to achieve the expected capacity enhancement, it is imperative that power control be properly functioning in both the forward and reverse directions of the radio channel (refer to chapter 10). Power control is executed only on the traffic channels. However, in the following analysis, it is not assumed that power control is perfect. This fact is reflected by assuming that the instantaneous power varies about the desired E_b/I_t level with a log-normal distribution having a standard deviation σ_c. Typical values of σ_c are on the order of 1.5 to 2.5 dB [12].

One method of estimating the capacity is to determine the probability that a CDMA channel does not have sufficient bandwidth to accommodate a mobile station for a given frame interval and still satisfy the interference constraints. This event is called an *outage*. During an outage of the reverse radio channel, the FER can exceed the desired maximum limit. This situation is not catastrophic but does lead to degraded service. In the following analysis, the desired interference is given by $I_0/N_0 < 1/\eta$. Typically, η is between 0.25 and 0.1 which corresponds to power ratios of I_0/N_0 between 6 dB and 10 dB.

The explicit formula [12,13] for the normalized average user occupancy (λ/μ in terms of Erlangs per sector) is given by

$$\frac{\lambda}{\mu}v(1 + f) = \Delta'_r \times F(B, \sigma_c) \tag{12.52}$$

where: λ = average call rate for the entire CDMA system,

$\dfrac{1}{\mu}$ = average call duration,

v = voice activity factor,

B = $\dfrac{[Q^{-1}(P_{out})]^2}{\Delta'_r}$,

$$\Delta'_r = \frac{\dfrac{B_w}{R}}{\dfrac{E_{b0}}{N_t}} \times (1 - \eta), \text{ and}$$

$$F(B, \sigma_c) = \frac{1}{\alpha_c}\left[1 + \frac{\alpha^3_c \cdot B}{2}\left(1 - \sqrt{1 + \frac{4}{\alpha^3_c \times B}}\right)\right]$$

in which

$$\alpha_c = e^{(\beta\sigma_c)^2/2}$$
$$\beta = (\ln[10])/10, \text{ and}$$

$$Q(z) = \left(\int_z^\infty e^{[(-x^2)/2]}dx\right)/(\sqrt{2\pi})$$

P_{out} is the probability of outage,
f is the mean interference from neighboring cells, and
E_{b0}/I_t is the median of the desired E_b/I_t.

Using Eq. (12.52), the following configurations are analyzed:

1. Hard handoff only for a CDMA system with a frequency bandwidth of 1.25 MHz and equipped with 8-kbps transcoders (see Table 12-19), B_w = 1.25 MHz, R = 9.6 kbps, and B_w/R = 130.
2. Soft handoffs with a maximum of 2 cells with a 1.25-MHz bandwidth and equipped with 8-kbps transcoders (see Table 12-20).
3. Soft handoffs with a maximum of 3 cells with a 1.25-MHz bandwidth and equipped with 8-kbps transcoders (see Table 12-21).
4. Hard handoffs only with a 1.25-MHz bandwidth and equipped with 13-kbps transcoders (see Table 12-22), B_w = 1.25 MHz, R = 14.4 kbps, and B_w/R = 87.
5. Soft handoffs with a maximum of 2 cells with a 1.25-MHz bandwidth and equipped with 13-kbps transcoders (see Table 12-23).
6. Soft handoffs with a maximum of 3 cells with a 1.25-MHz bandwidth and equipped with 13-kbps transcoders (see Table 12-24).
7. Hard handoffs only with a 10-MHz bandwidth and equipped with 8-kbps transcoders (see Table 12-25), B_w = 10 MHz, R = 9.6 kbps, and B_w/R = 1042.
8. Soft handoff with a maximum of 2 cells with a bandwidth of 10-MHz and equipped with 8-kbps transcoders (see Table 12-26).
9. Soft handoffs with a maximum of 3 cells with a 10-MHz bandwidth and equipped with 8-kbps transcoders (see Table 12-27).

Configurations 1, 2, and 3 correspond to wireless systems supporting standards TIA IS-95A and TIA IS-96A 8-kbps vocoder; configurations 4, 5, and 6 correspond to wireless systems supporting standards TIA IS-95A and Qualcomm's proprietary 13-kbps vocoder; configurations 7, 8, and 9 correspond to TIA SP-2977 and TIA IS-96A. Furthermore, in all these configurations, it is assumed that attenuation due to propagation losses decreases as the fourth power of distance and that the log-normal component has a standard deviation of 8 dB (refer to Eq. [12.2]). The voice activity factor v is assumed to equal 0.4, and the standard deviation for power control σ_c is assumed to be 2.5 dB. In Tables 12-19 through 12-27, we use f = 2.38 for hard handoffs, f = 0.77 for soft handoffs with a maximum of 2 cells, and f = 0.57 for soft handoffs with a maximum of 3 cells.

Comparing Tables 12-19 to 12-27, we can make several observations:

1. Hard handoffs reduce the capacity on the reverse radio link by approximately 50% with respect to two-way soft handoffs (Table 12-19 vis-à-vis Table 12-20).
2. If a system is equipped with 13-kbps transcoders rather than 8-kbps transcoders, the capacity is reduced approximately 40% (Table 12-21 vis-à-vis Table 12-24). This observation is consistent with our expectations since the rate is increased from 9.6 kbps to 14.4 kbps (50% increase).

Table 12-19   Configuration 1—Hard Handoffs Only (f = 2.38),
Bandwidth = 1.25 MHz, 8-kbps Transcoders (B_w/R = 130)

η	E_{b0}/I_t	P_{out}	λ/μ (Erlangs)
0.10	2.51 (4 dB)	0.01	21.63
0.25	2.51 (4 dB)	0.01	17.38
0.10	3.98 (6 dB)	0.01	12.38
0.25	3.98	0.01	9.86
0.10	2.51	0.05	24.18
0.25	2.51	0.05	19.63
0.10	3.98	0.05	14.23
0.25	3.98	0.05	11.48

Table 12-20   Configuration 2—Soft Handoffs (f = 0.77),
Bandwidth = 1.25 MHz, 8-kbps Transcoders (B_w/R = 130)

η	E_{b0}/I_t	P_{out}	λ/μ (Erlangs)
0.10	2.51 (4 dB)	0.01	41.30
0.25	2.51	0.01	33.19
0.10	3.98 (6 dB)	0.01	23.63
0.25	3.98	0.01	18.83
0.10	2.51	0.05	46.17
0.25	2.51	0.05	37.49
0.10	3.98	0.05	27.17
0.25	3.98	0.05	21.92

Table 12-21   Configuration 3—Soft Handoffs (f = 0.57), Bandwidth = 1.25 MHz, 8-kbps Transcoders ($B_w/R = 130$)

η	E_{b0}/I_t	P_{out}	λ/μ (Erlangs)
0.10	2.51 (4 dB)	0.01	46.56
0.25	2.51	0.01	37.42
0.10	3.98 (6 dB)	0.01	26.64
0.25	3.98	0.01	21.23
0.10	2.51	0.05	52.05
0.25	2.51	0.05	42.27
0.10	3.98	0.05	30.63
0.25	3.98	0.05	24.71

Table 12-22   Configuration 4—Hard Handoffs Only (f = 2.38), Bandwidth = 1.25 MHz, 13-kbps Transcoders ($B_w/R = 87$)

η	E_{b0}/I_t	P_{out}	λ/μ (Erlangs)
0.10	2.51 (4 dB)	0.01	13.25
0.25	2.51	0.01	10.46
0.10	3.98 (6 dB)	0.01	7.43
0.25	3.98	0.01	5.86
0.10	2.51	0.05	15.17
0.255	2.51	0.05	12.25
0.10	3.98	0.05	8.79
0.25	3.98	0.05	7.04

Table 12-23   Configuration 5—Soft Handoffs (f = 0.77), Bandwidth = 1.25 MHz, 13-kbps Transcoders ($B_w/R = 87$)

η	E_{b0}/I_t	P_{out}	λ/μ (Erlangs)
0.10	2.51 (4 dB)	0.01	25.29
0.25	2.51	0.01	20.17
0.10	3.98 (6 dB)	0.01	14.18
0.25	3.98	0.01	11.19
0.10	2.51	0.05	28.97
0.255	2.51	0.05	23.39
0.10	3.98	0.05	16.79
0.25	3.98	0.05	13.45

Table 12-24 Configuration 6—Soft Handoffs (f = 0.57),
Bandwidth = 1.25 MHz, 13-kbps Transcoders (B_w/R = 87)

η	E_{b0}/I_t	P_{out}	λ/μ (Erlangs)
0.10	2.51 (4 dB)	0.01	28.52
0.25	2.51	0.01	22.74
0.10	3.98 (6 dB)	0.01	15.99
0.25	3.98	0.01	12.62
0.10	2.51	0.05	32.66
0.255	2.51	0.05	26.37
0.10	3.98	0.05	18.93
0.25	3.98	0.05	15.17

Table 12-25 Configuration 7—Hard Handoffs Only (f = 2.38),
Bandwidth = 10 MHz, 8-kbps Transcoders (B_w/R = 1042)

η	E_{b0}/I_t	P_{out}	λ/μ (Erlangs)
0.10	2.51 (4 dB)	0.01	221.51
0.25	2.51	0.01	182.21
0.10	3.98 (6 dB)	0.01	134.93
0.25	3.98	0.01	110.62
0.10	2.51	0.05	230.52
0.255	2.51	0.05	190.34
0.10	3.98	0.05	141.87
0.25	3.98	0.05	116.86

Table 12-26 Configuration 8—Soft Handoffs (f = 0.77),
Bandwidth = 10 MHz, 8-kbps Transcoders (B_w/R = 1042)

η	E_{b0}/I_t	P_{out}	λ/μ (Erlangs)
0.10	2.51 (4 dB)	0.01	423.00
0.25	2.51	0.01	347.93
0.10	3.98 (6 dB)	0.01	257.66
0.25	3.98	0.01	211.24
0.10	2.51	0.05	440.21
0.255	2.51	0.05	363.48
0.10	3.98	0.05	270.91
0.25	3.98	0.05	223.16

Table 12-27 Configuration 9—Soft Handoffs (f = 0.57), Bandwidth = 10 MHz, 8-kbps Transcoders (B_w/R = 1042)

η	E_{b0}/I_t	P_{out}	λ/μ (Erlangs)
0.10	2.51 (4 dB)	0.01	476.88
0.25	2.51	0.01	392.28
0.10	3.98 (6 dB)	0.01	290.48
0.25	3.98	0.01	238.15
0.10	2.51	0.05	496.29
0.255	2.51	0.05	409.79
0.10	3.98	0.05	305.42
0.25	3.98	0.05	251.59

3. If the bandwidth increases from 1.25 MHz to 10 MHz (eightfold increase), the capacity increases approximately 10 times (Table 12-20 vis-à-vis Table 12-26).

We investigate the effect on the reverse radio link capacity if the power control is perfect, i.e., $\sigma_c = 0$. In this case, $F(B, \sigma_c)$ in Eq. (12.52) simplifies to

$$F(B, \sigma_c) = 1 + \frac{B}{2}\left(1 - \sqrt{1 + \frac{4}{B}}\right)$$

(12.53)

As an example of applying Eq. (12.53), we can compare the results in Table 12-20 in which $\eta = 0.10$, $E_{b0}/I_t = 6$ dB, and $P_{out} = 0.01$. In this case, $\lambda/\mu = 23.63$ Erlangs. If the power control is perfect, i.e., $\sigma_c = 0$, λ/μ increases to 27.14 Erlangs or a 15% increase. However, Eq. (12.53) is based upon Eq. (12.2), which may not adequately model all radio environments. In such cases, the engineer needs to use either a more complicated mathematical model or to execute a computer simulation.

The power of the reverse link channels for a specific user is adjusted at a rate of 800 times per second in order for the received power at the base station to provide the same minimum E_b/I_t as required for the specified link quality. The accuracy of the power control affects the reverse link capacity (see Fig. 12-19).

When determining the capacity of the reverse radio link, one must also include the capacity needed to support the access channels. Both Call Setup and Registration messages are transmitted on the access channel. The decrease of the capacity on the reverse radio link is small due to supporting the access channels, typically about a 1% reduction of the supportable Erlang traffic. Thus this reduction is ignored in the determination of the radio link capacity.

When designing a CDMA system for a given call load, the engineer must determine the number of CDMA Channel Modem (CM) circuits that must be supported at each cell. A CM operates at baseband rather than at RF frequencies. It demodulates the CDMA signal for a given mobile station and combines signals from multiple sectors of a given cell during softer handoffs (see Fig. 12-20). The number of CMs is affected by the number of simultaneous soft handoffs

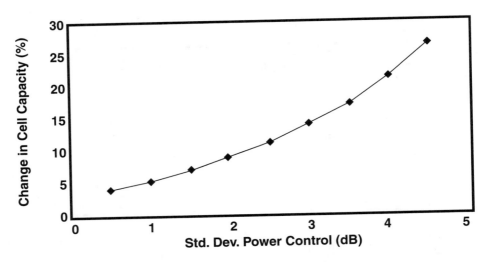

Figure 12-19   Effect of Power Control on Reverse Link Capacity

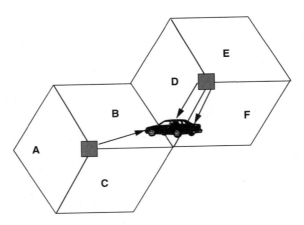

Figure 12-20   Soft-Softer Handoff Configuration

but not upon softer handoffs. In our analysis, we assume that a CM is not dedicated to a particular sector or CDMA carrier, although this assumption is dependent upon the actual manufacturer's implementation.

The exact distribution of two-way soft handoffs, three-way soft handoffs, softer handoffs, and soft-softer handoffs is very dependent upon the radio environment and radio configuration. The service provider needs to tune the system in order to optimize the radio link capacity. This tuning process includes adjusting the transmitted power of the pilot channels and the threshold levels that trigger handoffs. TIA IS-95A defines several thresholds: *T_ADD*, *T_DROP*,

Table 12-28 Handoff Distribution—CDMA Formal Field Test, November 18–23, 1991

Handoff Type	Mean (%)	Standard Deviation (%)
Softer	62.80	19.14
Two-way soft	6.52	19.05
Soft-softer	4.42	7.90
Three-way soft	0.40	0.82
Three-way softer	0.20	0.42
No handoff	25.66	10.80

T_TDROP, and T_COMP. The base station sends the values of these thresholds in the Systems Parameter message (which is transmitted on the paging channel) and the Handoff Direction message or the Extended Handoff Direction message (which is transmitted on the traffic channel). During a call, the mobile station measures the strength of the pilot channel of the serving cells (sectors) and potential candidates. The mobile station sends a Pilot Strength Measurement message to the base station in order to initiate a possible handoff for one of the following reasons:

1. Pilot's signal strength of a serving cell drops below *T_DROP* for a duration equal to *T_TDROP.*[*] *T_TDROP* is determined by the service provider and ranges from 0 to 15 seconds.
2. Pilot's signal strength of a candidate exceeds *T_ADD.*[†]
3. Pilot's signal strength of a candidate exceeds that of a serving cell by *T_COMP.*[‡]

As an example, Table 12-28 shows the handoff distribution exhibited during CDMA trials in San Diego, California.

As a general rule, a CDMA system is tuned so that calls are in some form of soft handoff, i.e., two-way soft, soft-softer, or three-way soft, for approximately 40–60% of the time. Percentages greater than this often are not justified by the improvement of the call quality. It is interesting to note that the aggregated results of the CDMA Formal Field Test (Table 12-28) indicate the total soft handoff (the sum of two-way soft, three-way soft, and soft-softer handoffs) is 12% of the time. This observation may be rationalized by the fact that calls were in total softer handoff (the sum of softer, soft-softer, or three-way softer) 67% of the time. Also note that the standard deviation of the handoff distributions is large, particularly for two-way soft and soft-softer handoffs. There are two reasons for this:

- The handoff distribution varies with the cells that are serving the call. Cells are tuned differently in order to optimize the call capacity.

---

[*] As defined in Section 6.6.6.2.5.2 of TIA IS-95A, this pilot is contained in the Active Set.
[†] TIA IS-95A refers to the pilot as being in the Neighbor Set or in the Remaining Neighbor Set.
[‡] TIA IS-95A refers to the pilot as being in the Candidate Set.

- The handoff distribution varies with the environment of the mobile station. Even though the mobile station may be served by the same cell, the terrain within the cell's domain varies sufficiently to significantly affect the handoff distributions.

When a call is in a two-way soft, soft-softer, or three-way softer handoff, two CMs are required to support the call. When a call is in a three-way soft handoff, three CMs are needed. Only one CM is needed for the call during a softer handoff or when no handoff configuration occurs.

EXAMPLE 12.7

Assume an equal call distribution across all cells and sectors with the same handoff distribution. The handoff distribution is as follows: 40% softer handoff, 20% two-way soft handoff, 10% soft-softer handoff, 29% no handoff, and 1% three-way soft handoff. There are two CDMA carriers, each having a bandwidth of 1.25 MHz. The system is equipped with transcoders that conform to TIA IS-96A, i.e., 8-kbps voice coding and 9.6 kbps on the physical layer. Assume $\eta = 0.25$, $E_{b0}/I_0 = 6$ dB, and $P_{out} = 0.01$. The system is configured with 10 cells, each having 3 sectors. The average call duration is 90 seconds, and each mobile subscriber generates 0.03 Erlang during the busy hour.

Determine the number of calls that can be supported per hour by the system. Determine the number of mobile subscribers that can be provided service if 2% blocking is acceptable. Also, determine the number of CMs that must be equipped to support the calculated number of subscribers.

From Table 12-21, the capacity of the reverse radio link per sector is:

$$21.23\frac{\text{Erlangs}}{\text{carrier}} \times 2 \text{ carriers} = 42.46 \text{ Erlangs}$$

Thus, each sector can simultaneously support 42 CDMA channels. This does not equal the number of simultaneous calls since some of the channels are assigned as the second and third channels for calls in softer and soft handoff. A CDMA channel is required for each sector configured in the call (refer to Fig. 12-21).

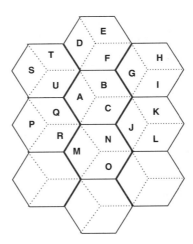

Figure 12-21   Sector Configuration for Example 12.7

In order to determine the number of simultaneous calls that can be supported by the CDMA system, we must associate each call with one sector even though the given call is being served by multiple sectors. In this example, we assume that the call is associated with the oldest serving sector, although there may be other assignments. If complete homogeneity is assumed, we can determine the total number of simultaneous calls supported by the entire system by multiplying the number of sectors ($10 \times 3 = 30$) by the number of simultaneous calls supported by each sector. To illustrate this point, let us determine the number of simultaneous calls supported by sector B as shown in Fig. 12-21. Softer handoffs are served by sectors B-C and B-A; two-way soft handoffs by B-G and B-F; three-way soft handoffs by B-G-F; soft-softer handoffs by B-C-G, B-C-J, B-C-F, B-C-N, B-A-U, B-A-F, B-A-Q, and B-A-G; no handoffs by B. Let the number of simultaneous calls supported by the sector be x. Then x can be determined by

$$x + 0.40x + 0.2x + 0.1x + 0.01x = 42$$

where 40% of the channels are supporting other sectors in softer handoff, 20% of the channels are supporting other cells in two-way soft handoff, 10% of the channels are supporting other cells in soft-softer handoff, and 1% of the channels are supporting cells in three-way soft handoff.

Thus, x equals 24 simultaneous calls per sector.

Therefore, the entire CDMA system can serve

$$10 \text{ cells} \times 3 \frac{\text{sectors}}{\text{cell}} \times 24 \frac{\text{calls}}{\text{sector}} = 720 \text{ calls}$$

In order to determine the number of subscribers that can be supported by this system by each sector during the busy hour, we use the Erlang-B formula. Each sector can support 24 simultaneous calls, which is equivalent to *trunks* or *radios*. From Erlang-B tables, we find that 16.63 Erlangs per sector can be supported during the busy hour. Thus, the number of mobile subscribers that the system supports is determined by

$$0.03 \frac{\text{Erlang}}{\text{subscriber}} \times N(\text{subscribers}) = 16.63 \frac{\text{Erlang}}{\text{sector}} \times 3 \frac{\text{sector}}{\text{cell}} \times 10 \text{ cells}$$

$$N = \frac{16.63 \times 3 \times 10}{0.03} \text{ subscribers}$$

$$N = 16{,}630 \text{ subscribers}$$

Next, we calculate the number of CMs that need to be equipped at each sector. We have already determined that each sector can support 42 CDMA channels. If a call is in a softer or no handoff, one CM is needed; for two-way soft handoff and softer-soft handoff, two CMs are needed; for a three-way soft handoff, three CMs are needed. The number of CMs that need to be equipped at each sector for supporting the traffic channels is

$$x + 0.2x + 0.1x + 0.01x = 42 \text{ CM}$$

where 20% of the CMs are supporting other cells in soft handoff, 10% of the CMs are supporting other cells in soft-softer handoff, and 1% of the CMs are supporting other cells in three-way soft handoff.

$$x = 32 \text{ CM}$$

In addition, CMs must be equipped for the access channel. Even though the access channel has a negligible effect upon the reverse radio channel ($16.31 \times 0.01 = 0.02$ Erlangs), a CM must be equipped to support this channel.

12.15 Design Considerations at the Boundary of a CDMA System

Standards do not currently support soft handoffs between CDMA systems operated by different service providers. Thus, if a mobile station moves between these CDMA systems, a hard hand-off will occur. This results in the reduction of capacity. Comparing Table 12-19 with Table 12-20, we see that the capacity is degraded by approximately 50%. The service provider has several options in order to mitigate this problem. First, the boundary cells (sectors) can be located at low traffic areas. If this is not practical, additional spectrum can be allocated at the boundary sectors. Neither option is very appealing to the service provider. Consequently, TIA TR45 has developed an intervendor soft handoff.

12.16 Interfrequency Handoff

The CDMA network may have multiple frequency carriers in each cell, and a hot-spot cell could have a larger number of frequencies than neighboring cells. Also, in hierarchical cell structures, microcells have a different frequency than the macrocell overlying the microcells. Therefore, an efficient scheme is required to hand off between different frequencies. A blind handoff used by an IS-95A CDMA network does not provide an adequate call quality. The mobile station should be able to determine the signal strength and quality of another carrier frequency while still maintaining the connection with the current carrier frequency.

A CDMA transmission is continuous; there are no idle slots for the interfrequency measurements. Therefore, compressed mode and dual receiver have been proposed as a solution to interfrequency handoff. In the compressed (slotted) mode, measurement slots are created by transmitting the data frame with a lower spreading ratio during a shorter period, and the rest of the frame is used for measurements on other carriers. The dual receiver can measure other frequencies without affecting the reception of the current frequency.

The dual-receiver approach is considered suitable if the mobile terminal uses antenna diversity. During the interfrequency measurements, one receiver branch is switched to another frequency for measurements, while the other branch keeps receiving from the current frequency. The loss of diversity gain during measurements needs to be compensated for with higher forward link transmission power. The advantage of the dual-receiver approach is that there is no break in the current frequency connection. Fast closed-loop power control is running all the time.

The slotted-mode approach shown in Fig. 12-22 is considered attractive for the mobile station without antenna diversity. The information transmitted during a frame is compressed in time. For a compressed mode, there are a number of alternatives for implementations: variable spreading factor, code rate increase, multicode, and higher modulation. Variable rate spreading and code rate increment (puncturing) cause 1.5 to 2.5 dB loss in E_b/I_t, and higher-order modulation causes even higher loss (~ 5 dB). This is due to a break in power control and less coding. A further drawback of the variable spreading factor is that simple terminals have to be able to operate with different spreading ratios.

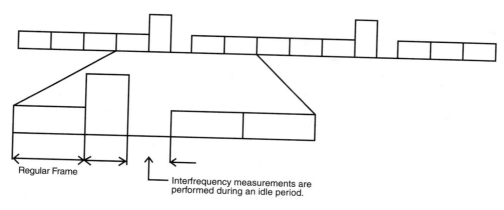

Regular Frame

Interfrequency measurements are
performed during an idle period.

Figure 12-22 Interfrequency Slotted Mode

12.17 Summary

This chapter discussed principles for engineering a CDMA system including propagation models, link budgets, and facilities engineering. However, extensive RF measurements and RF modeling are needed in order to plan a real commercial system. The chapter's intent was to provide some tools and a better understanding toward achieving this goal. We also observed that the calculated path losses with the Okumura/Hata model are about 13 to 15 dB lower than the path losses obtained using the Walfisch-Ikegami (COST 231) model. This is because the Okumura/Hata model neglects several important parameters such as street width and street orientation.

The chapter also covered the concepts of the delay spread, coherence bandwidth, and Doppler spread. When the signal bandwidth is much larger than the coherence bandwidth, the signal is called wideband signal and the fading is frequency selective. This means that only a portion of the signal bandwidth fades at any instant of time. With signal bandwidth much smaller than the coherence bandwidth, flat fading of the entire signal occurs.

With the signal symbol interval being much larger than the coherence time, the channel fades rapidly compared to the symbol rate. This is called fast fading relative to symbol time, and frequency dispersion occurs, causing signal distortion. With the signal symbol interval being much smaller than the coherence time, the channel does not change during the symbol interval; it is referred to as the slow-fading channel, relative to the symbol time. The chapter concludes with a discussion of various schemes to transition from an analog system to a digital system.

12.18 References

1. Forney, G. D., "The Viterbi Algorithm," *Proceedings of IEEE* 61(3), March 1978, pp. 268–78.

2. Garg, V. K., and Wilkes, J. E., *Wireless and Personal Communications Systems*, Prentice Hall PTR, Upper Saddle River, NJ, 1996.

3. Hata, M., "Empirical Formula for Propagation Loss in Land Mobile Radio Services," *IEEE Transactions on Vehicular Technology* 29(3), 1980.

4. Jakes, W. C., ed., *Microwave Mobile Communications*, John Wiley, New York, 1974.

5. Motley, A. J., and Keenan, J. M., "Radio Coverage in Buildings," *British Telecom Tech. Journal*, Special Issue on Mobile Communications 8(1), January 1990, pp. 19–24.

6. Okumura, Y., Ohmori, E., Kawano, T., and Fukuda, K., "Field Strength and Its Variability in VHF and UHF Land-Mobile Radio Service," *Rev. Elec. Communication Lab.* 16, 1968, pp. 825–73.

7. Rappaport, T. S., *Wireless Communications*, Prentice Hall PTR, Upper Saddle River, NJ, 1996.

8. Sampei, Seiichi, *Applications of Digital Wireless Technologies to Global Wireless Communications*, Prentice Hall PTR, Upper Saddle River, NJ, 1997.

9. Seidel, S. Y., and Rappaport, T. S., "914 MHz Path Loss Prediction Models for Indoor Wireless Communications in Multifloored Buildings," *IEEE Trans., Antennas & Propagation* 40(2), February 1992.

10. Sklar, B., "Rayleigh Fading Channels in Mobile Digital Communication Systems—Part I: Characterization," *IEEE Communication Magazine* 35(9), September 1997, pp. 136–46.

11. Sklar, B., "Rayleigh Fading Channels in Mobile Digital Communication Systems—Part II: Mitigation," *IEEE Communication Magazine* 35(9), September 1997, pp. 148–55.

12. Viterbi, Andrew J., "CDMA Principles of Spread Spectrum Communication," Addison-Wesley Publishing Company, Reading, MA, 1995.

13. Viterbi, Audrey M., and Viterbi, Andrew J., "Erlang Capacity of a Power Controlled CDMA System," *IEEE Journal on Selected Areas in Communications*, August 1993.

14. Walfisch, J., and Bertoni, H. L., "A Theoretical Model of UHF Propagation in Urban Environment," *IEEE Trans., Antennas & Propagation*, to be published.

15. Weissberger, M. A., "An Initial Critical Summary of Models for Predicting the Attenuation of Radio Waves by Trees," ESD-TR-81-101, Electromagnet Compat. Analysis Center, Annapolis, MD, July 1982.

Reverse and Forward Link Capacity of IS-95 CDMA System

13.1 Introduction

The CDMA system is an interference-limited system in which link performance depends on the ability of the receiver to detect a signal in the presence of interference. In order for a CDMA link to perform satisfactorily, the designer must specify a frame error rate (FER). Field trials help the designer to establish the required E_b/I_t values on the reverse link and various channels of the forward link that will maintain the specified FER. The key issue in CDMA network design is to minimize multiple access interference. Power control is critical to reduce multiaccess interference. Designers must include the interference from other cells in the system to determine the actual reuse factor in the CDMA system.

In the forward direction (BS to MS) a pilot signal is used by the mobile demodulator to provide a coherent reference that is effective even in a fading environment because the desired signal and pilot fade together. In the reverse direction (MS to BS), no pilot is used for power efficiency considerations because, unlike the forward direction, an independent pilot would be required for each signal. A modulation consistent with, and relatively efficient for, noncoherent reception is used for the reverse link.

The maximum number of mobiles that can be supported on the forward link of a CDMA system is different from the maximum number that can be supported on the reverse link. Normally, the capacity of a CDMA system depends upon the reverse link capacity. The forward link capacity is governed by the total transmitted power of the cell site and its distribution to traffic channels and other overhead channels including the pilot, paging, and sync channels. If the power amplifier cannot provide enough power to the forward traffic channels, system capacity may become forward link limited. Soft handoffs improve the capacity of the reverse link; however, they also affect the capacity of the forward link—the forward link capacity is reduced by the number and types of soft handoffs.

This chapter presents procedures for calculating the capacity of the reverse and forward links of a CDMA system. We establish the relationship for determining the *pole point* or asymptotic cell capacity that can be achieved as the power received at the base station from a mobile approaches infinity. We also relate the ratio of the received cell power to cell noise with cell loading. The chapter also covers a procedure to develop a *link safety margin parameter* for each of the forward link channels.

13.2 Reverse Link Capacity

Consider an omnidirectional cell site serving a given set of mobiles. We divide mobiles into two groups—the mobiles that are powered up, and those that are not powered up. The mobiles that are powered up are further divided into four subgroups:

- Active and transmitting (i.e., mobiles in conversational mode)
- Active, but not transmitting (mobiles in nonconversational mode)
- Idle and transmitting (mobiles in access mode)
- Idle and not transmitting (mobiles in nonaccess mode)

We assume that interference at the cell site by mobiles in the access mode is typically too small to worry about—it can be accounted for as a source of some degradation in system quality and system capacity. We focus only on the active mobiles in our analysis. We assume there are M mobiles that are transmitting at a given time in a cell. In a CDMA environment, for each mobile, there are $(M - 1)$ cochannel interferers. At the cell site, the average signal power received from the ith mobile is S_{ri}. This signal power provides a bit energy equal to E_b, i.e.,

$$E_b = \frac{S_{ri}}{R} \tag{13.1}$$

where R = mobile transmission rate in bps.

The thermal noise power is $N_0 B_w$, where N_0 is the thermal noise power spectral density (psd) and B_w is the spreading bandwidth. The average cochannel interference psd at the base station is given as

$$I_0 = \frac{1}{B_w} \sum_{i=1}^{M-1} v_f \cdot S_{ri} \tag{13.2}$$

where v_f = voice activity factor.

In Eq. (13.2) we assume a perfect power control on the reverse link, and the signals transmitted from all the mobiles arrive at the base station with the same received power; in other words, $S_{ri} = S_r$ for all values of i (i.e., $1 \le i \le M - 1$). The total interference and thermal noise psd will be

$$I_t = I_0 + N_0 = \frac{1}{B_w} \cdot \sum_{i=1}^{M-1} v_f \cdot S_{ri} + N_0 \tag{13.3}$$

Recognizing that $S_{ri} = S_r$, from Eq. (13.3) we get

$$I_t = \frac{(M-1) \cdot v_f \cdot S_r}{B_w} + N_0 \tag{13.4}$$

E_b/I_t will be given as

$$\frac{E_b}{I_t} = \left(\frac{B_w}{R}\right) \cdot \frac{S_r}{[N_0 B_w + (M-1) \cdot v_f \cdot S_r]} = G_p \cdot \frac{S_r}{[N_0 B_w + (M-1) \cdot v_f \cdot S_r]} \tag{13.5}$$

where G_p = processing gain = B_w/R

Next we express the signal strength S_r in dB as

$$S_r = P_m + G_m + G_b + G_{dv} + G_{sho} + L_p + M_{fade} + L_b + L_{pent} + L_c \tag{13.6}$$

where G_m = transmit antenna gain of the mobile (dB)
G_b = receive antenna gain of the base station (dB)
G_{dv} = base station antenna diversity (dB)
G_{sho} = soft-handoff gain
L_b = body loss (dB)
L_c = cable connection loss (dB)
L_p = path loss (dB)
L_{pent} = penetration loss through a vehicle or building
M_{fade} = log-normal shadow margin (dB)
P_m = transmit power of the mobile (dB)

Solving Eq. (13.5) for M, we get

$$M = 1 + G_p \cdot \left[\frac{1}{(E_b/I_t) \cdot v_f}\right] - \frac{N_0 \cdot B_w}{S_r \cdot v_f} \tag{13.7}$$

and solving Eq. (13.5) for S_r, we get:

$$S_r = \frac{(E_b/I_t) \cdot N_0}{\dfrac{1}{R} - \dfrac{(M-1)v_f(E_b/I_t)}{B_w}} \tag{13.8}$$

If we include an interference factor f (see section 13.4) from the other cells, we can rewrite Eq. (13.5) as

$$\frac{E_b}{I_t} = G_p \cdot \frac{S_r}{N_0 \cdot B_w + (M-1) \cdot v_f \cdot S_r(1+f)} \tag{13.9}$$

We also include an imperfect power control factor, η_c, and rewrite Eq. (13.9) as

$$\frac{E_b}{I_t} = G_p \cdot \frac{S_r}{B_w \cdot N_0 + (M-1) \cdot v_f \cdot (S_r/\eta_c) \cdot (1+f)} \tag{13.10}$$

Solving Eq. (13.10) for M, we get

$$M = 1 + G_p \cdot \left[\frac{\eta_c}{(E_b/I_t) \cdot v_f \cdot (1+f)} \right] - \frac{N_0 \cdot B_w \cdot \eta_c}{S_r \cdot v_f \cdot (1+f)} \qquad (13.11)$$

Solving Eq. (13.10) for S_r, we get

$$S_r = \frac{(E_b/I_t) \cdot N_0}{\dfrac{1}{R} - \dfrac{(M-1)(v_f)(1+f)(E_b/I_t)}{B_w \cdot \eta_c}} \qquad (13.12)$$

From Eq. (13.11) the maximum value of M is given as

$$M_{max} = 1 + G_p \cdot \left[\frac{\eta_c}{(E_b/I_t) \cdot (v_f) \cdot (1+f)} \right] \qquad (13.13)$$

M_{max} is called the *pole point* or *asymptotic cell capacity* that is achieved when $S_r \to \infty$. For simplification we neglect 1 and rewrite Eq. (13.13) as

$$M_{max} \approx G_p \cdot \left[\frac{\eta_c}{(E_b/I_t) \cdot (v_f) \cdot (1+f)} \right] \qquad (13.14)$$

To gain further insight into the capacity dynamic, we can rewrite Eq. (13.12) as

$$\frac{S_r/\eta_c}{N_0 B_w} = \frac{1}{M_{max} \cdot v_f \cdot (1+f) \cdot (1-\rho)} \qquad (13.15)$$

where $\rho = M/M_{max}$ = cell loading factor

Equation (13.15) is plotted in Fig. 13-1. The required SNR per mobile increases in a nonlinear fashion with ρ. The power per mobile is much larger for a heavily loaded cell than it is for the lightly loaded cell. Practical capacity limits may be set at a point where the slope becomes too steep (such as $\rho = 0.8$). Also, for the smallest practical value of $M_{max} - M$ to be equal to 1, Eq. (13.15) indicates that the maximum received power per mobile is close to the cell-site noise level.

We further adjust Eq. (13.15) to reflect the total received power (interference) equal to $P_{rec} = v_f \cdot (1+f) \cdot (S_r/\eta_c) \cdot M$. The ratio of P_{rec} to cell-site noise can be expressed in terms of the loading factor ρ. Eq. (13.16) shows that the ratio of interference and noise also rises in a nonlinear fashion with ρ (see Fig. 13-2).

$$\frac{P_{rec}}{N_0 B_w} = \frac{\rho}{1-\rho} \qquad (13.16)$$

Since total power P_{total} is equal to $P_{rec} + N_0 B_w$, we can rewrite Eq. (13.16) as

$$\frac{P_{total}}{N_0 B_w} = \frac{P_{rec} + N_0 B_w}{N_0 B_w} = \frac{\rho}{1-\rho} + 1 = \frac{1}{1-\rho} \qquad (13.17)$$

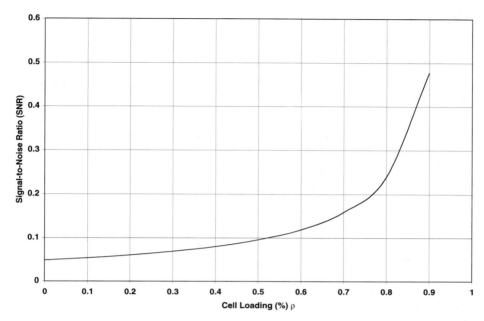

Figure 13-1 SNR vs. Cell Loading

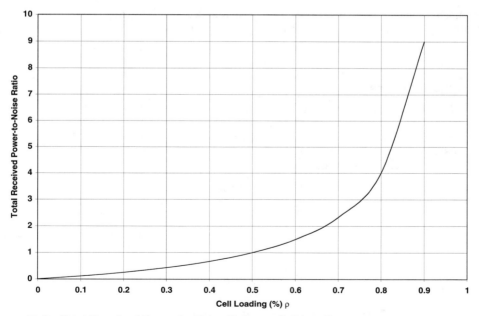

Figure 13-2 Total Received Power-to-Noise Ratio vs. Cell Loading

Typical average values of different parameters are

$v_f = 0.5$ $(0.4 \rightarrow 0.6)$

$E_b/I_t = 6 \rightarrow 7$ dB (4 to 5)

$f = 0.67$ $(0.56 \rightarrow 1.28)$ based on path-loss exponent $\gamma = 4$ and standard deviation of path loss of 6 to 10 dB

$\eta_c = 0.8$ $(0.7 \rightarrow 0.85)$

Using these average values, we estimate the cell capacity as

$$M_{max} = \frac{1.23 \times 10^6}{9600 \times 4} \cdot \frac{0.8}{0.5 \times (1.67)} \approx 31 \text{ (with } E_b/I_t = 6 \text{ dB)}$$

and

$$M_{max} = \frac{1.23 \times 10^6}{9600 \times 5} \cdot \frac{0.8}{0.5 \times (1.67)} \approx 25 \text{ (with } E_b/I_t = 7 \text{ dB)}$$

With a three-sector antenna, a practical gain of 2.55 can be achieved; the capacity range per sector will be

$$M_{sector} = 31 \times \frac{2.55}{3} \approx 26 \text{ or } M_{sector} = 25 \times \frac{2.55}{3} \approx 21$$

The capacity range of a sector will be 21 to 26. In practice, sector loading is often limited to 0.5–0.7 of the calculated values, giving an average number of mobiles per sector equal to 13–16.

EXAMPLE 13.1

Calculate the required E_b/I_t using the following parameters: $B_w = 1.23$ MHz; $R = 9.6$ kbps; $P_m = 63$ mW (18 dBm); $L_c = -2$ dB; $G_m = 0$ dB; $L_p = -135$ dB; $M_{fade} = -8$ dB; $G_b = 9$ dB; F (noise figure) = 5 dB; $T = 290$ degrees Kelvin (K); k_b = Boltzmann's constant = 1.380662×10^{-23}; $v_f = 0.4$; $M = 20$. Assume all other parameters to be 0.

The received signal power will be

$$S_r = P_m + L_c + G_m + G_b + L_p + M_{fade}$$

$$S_r = 18 + (-2) + (0) + (9) + (-135) + (-8) = -118 \text{ dBm}$$

$$N_0 = FTk_b = 3.16228 \times 290 \times 1.380662 \times 10^{-23} = 1.266 \times 10^{-20}W = 1.266 \times 10^{-17}\text{mW}$$

$$\frac{E_b}{I_t} = \left(\frac{B_w}{R}\right)\frac{S_r}{[N_0 B_w + (M-1)v_f S_r]}$$

$$\frac{E_b}{I_t} = \frac{1.23 \times 10^6}{9.6 \times 10^3} \times \frac{10^{-11.8}}{[1.266 \times 10^{-17} \times 1.23 \times 10^6 + 19 \times 0.4 \times 10^{-11.8}]} \text{ dB}$$

$$\frac{E_b}{I_t} = \frac{128 \times 10^{-11.8}}{10^{-11}[1.5557 + 1.2045]} = \frac{128 \times 10^{-0.8}}{2.7602} = 7.345 = 8.66 \text{ dB}$$

EXAMPLE 13.2

For the IS-95 CDMA system, a chip rate of 1.2288 Mcps is specified for the data rate of 9.6 kbps (i.e., 8-kbps vocoder). The required E_b/I_t is specified as 7.0 dB. Calculate the pole capacity. What is the average number of mobiles that can be supported by a sector of the three-sector cell?. Assume: interference factor from the neighboring cells $f = 0.55$; the voice activity factor $v_f = 0.5$; the power control accuracy factor $= 0.80$; the gain due to sectorization $= 2.55$. Assume all other parameters to be 0. How much reduction in sector capacity will occur with a 13.0-kbps vocoder provided all other things remain unchanged?

$$M_{max} \approx G_p \cdot \left[\frac{\eta_c}{(E_b/I_t) \cdot (v_f) \cdot (1 + f)} \right]$$

$$G_p = \frac{B_w}{R} = \frac{1.23 \times 10^6}{9.6 \times 10^3} = 128, \frac{E_b}{I_t} = 7 \text{ dB} = 5.0$$

$$M_{max} = \frac{128 \times 0.80}{5.0 \times 0.5 \times (1 + 0.55)} = 26.42$$

$$\text{Average subscriber/sector} = \frac{26.42 \times 2.55}{3} = 22.46 \approx 22$$

With a 13.0-kbps vocoder (data rate $R = 14.4$ kbps), the processing gain will be

$$G_p = \frac{1.23 \times 10^6}{14.4 \times 10^3} = 85.4$$

The reduction in sector capacity will be

$$= \frac{128 - 85.4}{128} \times 100 = 33.28\%$$

EXAMPLE 13.3

A total of 36 equal-power mobiles share a frequency band through a CDMA system. Each mobile transmits information at 9.6 kbps with a DSSS BPSK-modulated signal. Calculate the minimum chip rate of the PN code in order to maintain a bit error probability of 10^{-3}. Assume: interference factor from other cells $f = 0.60$; voice activity factor $v_f = 0.5$; power control accuracy factor $= 0.8$.

$$\text{Bit error probability for BPSK } P_b = Q\left(\sqrt{\frac{2E_b}{I_t}}\right) \approx \frac{e^{-E_b/I_t}}{2\sqrt{\pi(E_b/I_t)}} = 10^{-3}$$

$$\text{Required } \frac{E_b}{I_t} \approx 4.8 = 6.8 \text{ dB}$$

$$M = 36 = \frac{G_p}{E_b/I_t} \times \frac{1}{1+f} \times \frac{1}{v_f} \times 0.8$$

$$\frac{G_p}{4.8} \times \frac{1}{1.6} \times \frac{1}{0.5} \times 0.8 = 36$$

$$\therefore G_p = 172.8$$

$$\text{Chip rate} = 172.8 \times 9.6 \times 10^3 = 1.6588 \text{ Mcps}$$

EXAMPLE   13.4

Calculate the Erlang capacity of a sector for a three-sector cell using the following parameters: carrier bandwidth, $B_w = 1.23$ MHz; RS2 $R = 14.4$ kbps; required $E_b/I_t = 7$ dB; voice activity factor $v_f = 0.4$; interference due to other cells $f = 0.6$; three-sector antenna gain $\alpha = 2.61$; cell loading factor $\rho = 0.54$; outage or call blocking probability $P_{out} = 2\%$.

$$G_p = \frac{B_w}{R} = \frac{1.23 \times 10^6}{14.4 \times 10^3} = 85.4$$

$$M_{max} = 1 + \frac{G_p}{(E_b/I_t)_{reqd}} \cdot \frac{1}{v_f} \cdot \frac{1}{1+f} \cdot \alpha = 1 + \frac{85.4}{5.012} \cdot \frac{1}{0.4} \cdot \frac{1}{1+0.6} \cdot 2.61 = 70.4$$

$$M_{max} \text{ per sector} = 70.4/3 = 23.48 \approx 24$$

$$M = 0.54 \times 24 \approx 13$$

From the Erlang B table at 2% blocking probability for $M = 13$ channels, the capacity = 7.4 Erlangs.

13.3 Multicell Network

In a CDMA multiple-cell (multicell) network the same frequency band is used in all cells unlike other access technologies in which the bandwidth used in a given cell is reused only in cells that are sufficiently far away to avoid cochannel interference. To compare CDMA with other multiple access schemes, capacity is determined (or measured) as the total number of users in the multicell network rather than the number of users per bandwidth or per isolated cell.

The CDMA system is an interference-limited system. CDMA link performance depends on the ability of the receiver to discern a signal in the presence of interference. In order for the performance of a CDMA link to be satisfactory, an FER of about 1% is recommended. Field trials are conducted to establish the required E_b/I_t value on the reverse link and various channels of the forward link to maintain the recommended FER. The link budget is established to achieve the value of E_b/I_t. The required value of E_b/I_t depends upon the propagation environment and the speed of the mobile. Based on field trials, the following values of E_b/I_t are suggested:

- **Low-speed mobiles, speed ≤ 5 mph: 5 dB.** In this case, the duration of fades is much larger than the time between power control updates for a mobile. Thus, the effect of any fade is compensated by a quick response of the power control mechanism.
- **Medium-speed mobiles, speed ≈ 30 mph: 7 dB.** The advantages of high or low speed are not applicable; therefore, the required E_b/I_t is somewhat higher.
- **High-speed mobiles, speed ≥ 60 mph: 6 to 6.5 dB.** In this case, the fade duration is smaller compared to chip length. Thus, only burst errors occur on the links that are corrected by interleaving and Viterbi decoding. Therefore the required E_b/I_t is low.

The key issue in a CDMA network design is to minimize multiple access interference. Power control is critical to multiaccess interference. Each cell controls the transmit power of its own mobiles. However, a serving cell is unable to control the power of mobiles in the neighboring cells. The mobiles in the neighboring cells introduce additional interference, thereby reducing the capacity of the reverse link. In Eq. (13.10) we include this effect by a factor f. The interference from other cells determines the actual reuse factor of the CDMA system. CDMA networks are designed to tolerate a certain amount of interference and, therefore, have a capacity advantage over TDMA or FDMA in this regard.

13.4 Intercell Interference

The intercell interference factor f is difficult to evaluate because the serving cell does not have control over the power received from mobiles in other cells. The f depends on the geometry of the serving cell and neighboring cells. It will be small if the serving cell radius is large, if the path-loss slope has a higher value, or if the standard deviation of path loss is small.

For $\gamma = 4$ and a standard deviation of path loss $\sigma = 8$ dB, the upper bound on f is 0.77. Table 13-1 lists the value of f with two-way and three-way soft handoff for different values of σ and $\gamma = 4$ [5].

The cell loading is a measure of the total interference I_0 allowed in the system in reference to thermal noise.

Table 13-1 Intercell Interference factor f for $\gamma = 4$

σ dB	Other Cell Interference Factor f	
	Two-Way Soft Handoff	Three-Way Soft Handoff
0	0.44	0.44
2	0.43	0.43
4	0.47	0.45
6	0.56	0.49
8	0.77	0.57
10	1.28	0.75
12	2.62	1.17

$$\rho = \frac{M}{M_{max}} \approx \frac{I_0}{I_0 + N_0} \tag{13.18}$$

where M = active number of mobiles in the cell, and
M_{max} = maximum possible mobiles in the cell.

ρ equal to 0.5 implies that the interference in the system is equal to the thermal noise level. $\rho < 0.5$ implies that the system is *noise limited*, whereas $\rho > 0.5$ indicates that the system is *interference limited*. Typically, a value of ρ between 0.5 and 0.7 is used.

13.5 Erlang Capacity of a Single Cell

To calculate the Erlang capacity of a single cell in a CDMA system, we assume that the number of active users M can be modeled by Poison distribution.

$$p_m = \frac{(\lambda/\mu)^M}{M!} \cdot e^{-\lambda/\mu} \tag{13.19}$$

where λ/μ = offered average traffic load in Erlang,
λ = average arrival rate of users, and
$1/\mu$ = average time per call.

The call service time τ per user is assumed to be exponentially distributed, so that the probability that τ exceeds T is given as

$$p_r(\tau > T) = e^{-\mu T} \qquad T > 0 \tag{13.20}$$

Using these assumptions, it has been shown in reference [5] that the blocking or outage probability p_{out} is

$$p_{out} = e^{-(\lambda v_f)/\mu} \cdot \sum_{K \lfloor \Delta_r' \rfloor}^{\infty} \left(\frac{v_f \lambda}{\mu}\right)^K \cdot \frac{1}{K!} \approx Q\left(\frac{\Delta_r' - (v_f \lambda)/\mu}{\sqrt{(v_f \lambda)/\mu}}\right) \tag{13.21}$$

where $\Delta_r' = \dfrac{G_p(1-\eta)}{(E_b/I_t)_{sp}}$, and

$\dfrac{1}{\eta}$ = the ratio of total interference plus thermal noise power to thermal noise power.

By taking into account the interference from other cells and an imperfect power control, Eq. (13.21) can be modified as

$$p_{out} \approx Q\left[\frac{\Delta_r' - v_f(\lambda/\mu)(1+f)e^{(\beta\sigma_c)^2/2}}{\sqrt{v_f \cdot (\lambda/\mu) \cdot (1+f)} \cdot e^{(\beta\sigma_c)^2}}\right] \tag{13.22}$$

where $\beta = (\ln 10)/10$, and
    σ_c = standard deviation of power control.

We may invert the approximate expression for blocking probability Eq. (13.22) by solving a quadratic equation, to obtain the explicit formula for normalized average user occupancy, λ/μ, in terms of Erlangs per sector [3,6] as

$$(\lambda/\mu) \cdot v_f \cdot (1 + f) = \Delta_r' \cdot F(B, \sigma_c) \qquad (13.23)$$

where $B = \dfrac{[Q^{-1}(P_{\text{out}})]^2}{\Delta_r'}$, and

$$F(B, \sigma_c) = \frac{1}{\alpha_c} \cdot \left[1 + \frac{\alpha_c^3 B}{2} \left(1 - \sqrt{1 + \frac{4}{\alpha_c^3 B}} \right) \right] \text{ in which}$$

$\alpha_c = e^{(\beta \sigma_c)^2/2}$; $\beta = (\ln 10)/10 = 0.2303$

EXAMPLE 13.5

Find the Erlang capacity of a CDMA cell assuming

- Blocking or outage probability $(P_{\text{out}}) = 1\%$
- Log-normal shadowing margin $(M_{\text{fade}}) = 8$ dB
- Path-loss exponent $(\gamma) = 4$
- Voice activity factor $(v_f) = 0.4$
- Other cell interference factor $(f) = 0.55$
- Spreading bandwidth $(B_w) = 1.23$ MHz
- Data rate $(R) = 9.6$ kbps
- $(E_b/I_t)_{\text{sp}} = 7$ dB = 5.0
- $1/\eta = 10$
- $\sigma_c = 2$ dB = 1.5849

$$Q^{-1}(0.01) = 2.33; \ \Delta_r' = \frac{G_p}{(E_b/I_t)_{\text{sp}}} \cdot (1 - \eta) = \frac{1.23 \times 10^6}{9.6 \times 10^3} \cdot \frac{(1 - 0.1)}{5} = 23.04$$

$$\alpha_c = e^{(0.2322 \times 1.5849)^2/2} = 1.0701$$

$$B = \frac{(2.33)^2}{23.04} = 0.2356$$

$$F(B, \sigma_c) = \frac{1}{1.0701} \left[1 + \frac{(1.0701)^3 \times 0.2356}{2} \left(1 - \sqrt{1 + \frac{4}{(1.0701)^3 \times 0.2356}} \right) \right] = 0.5494$$

$$\frac{\lambda}{\mu} = \frac{23.04 \times 0.5494}{0.4 \cdot (1 + 0.55)} = 20.42 \text{ Erlangs}$$

The number of users from the Erlang B table at 1% blocking ~ 30.

13.6 Forward Link Capacity

An important feature of CDMA that contributes to the added capacity on the reverse link is *soft handoff* [2]. In a CDMA network, a mobile can be served by multiple cells simultaneously. However, that same feature puts an additional burden on the forward link. Since multiple cells have to provide service to the same mobile, additional resources are allocated on the forward link. Forward link performance differs vastly from that of the reverse link because

- Access is one to many instead of many to one.
- Synchronization and coherent detections are facilitated by use of a common pilot channel.
- The interference is received from a few concentrated large sources (cells) rather than many distributed small ones (mobiles).

To maximize the capacity of the forward link, it is essential to control the power of the cell so that power can be allocated to individual mobiles according to their needs. More power is provided to those mobiles that receive the highest interference from neighboring cells. Mobiles on the boundaries may be in soft handoff, in which case they also receive signal power from two or more cells. Power control on the forward link is accomplished by measuring the mobile power received from its serving cell and the total received power. The information about these two power values is transmitted to the serving cell.

For the forward link a *figure of merit* is defined for various channels. The figure of merit is the difference between the received (rec) and specified (sp) E_b/I_t. The link safety margin parameter for each of the channels on the forward link is defined as

$$M_{\text{pilot}} = (E_c/I_t)_{\text{rec}} - (E_c/I_t)_{\text{sp}} > 0 \tag{13.24a}$$

$$M_{\text{traffic}} = (E_b/I_t)_{\text{rec}} - (E_b/I_t)_{\text{sp}} > 0 \tag{13.24b}$$

$$M_{\text{sync}} = (E_b/I_t)_{\text{rec}} - (E_b/I_t)_{\text{sp}} > 0 \tag{13.24c}$$

$$M_{\text{paging}} = (E_b/I_t)_{\text{rec}} - (E_b/I_t)_{\text{sp}} > 0 \tag{13.24d}$$

Note for the pilot channel E_c/I_t is used instead of E_b/I_t—this is because the pilot channel does not carry any information. Energy per chip, E_c, is used, the chip rate being 1.2288 Mcps.

The forward link budget is used to confirm that quantities in Eqs. (13.24a–d) are positive and that there is sufficient margin for the forward link to perform efficiently. Of M_{pilot}, M_{traffic}, M_{sync}, and M_{paging}, the first two are more critical. If these two are positive, then the other two are also likely to be positive. For perfect link balance, all margin parameters should be 0, particularly M_{pilot} and M_{traffic}. The suggested values for the specified E_b/I_t and E_c/I_t parameters are

- Pilot channel: $(E_c/I_t)_{\text{sp}} = -15$ dB
- Traffic channel: $(E_b/I_t)_{\text{sp}} = 7$ dB
- Sync channel: $(E_b/I_t)_{\text{sp}} = 7$ dB
- Paging channel: $(E_b/I_t)_{\text{sp}} = 7$ dB

We use the following assumptions for the CDMA forward link budget:

1. All mobiles are

   ◆ at the cell edge
   ◆ at least in two-way soft handoff
   ◆ traveling at a medium speed
   ◆ $(E_b/I_t) = 7$ dB for 1% FER

2. Power control is working perfectly for all mobiles.
3. Total forward link traffic channels' power is equally divided among all mobiles.

Forward link capacity depends on the power that is available for the traffic channels. The power allocation to each overhead channel (i.e., P_{pilot}, P_{sync}, and P_{paging}) is determined from field tests. The suggested power allocations for the forward link channels are

- $P_{pilot} = 15\text{--}20\%\ P_{cell\text{-}site}$
- $P_{sync} = 10\%$ of $P_{pilot} = 1.5\text{--}2\%\ P_{cell\text{-}site}$
- $P_{paging} = 30\text{--}40\%$ of $P_{pilot} = 7\%\ P_{cell\text{-}site}$
- $P_{traffic} = [1 - (0.2 + 0.02 + 0.07)] = 71\text{--}76.5\%\ P_{cell\text{-}site}$

Note that P_{paging} and $P_{traffic}$ represent the total allocated power for all the paging and traffic channels, respectively, and $P_{cell\text{-}site}$ is the total transmit power of the cell site.

$$P_{(traffic)/(mobile)} = P_{traffic}/(M_{total} \cdot \alpha_{chan}) \tag{13.25}$$

$$M_{total} = M(1 + \xi_{co}) \tag{13.26}$$

where M = number of active mobiles per sector,
  ξ_{co} = channel overhead factor for extra traffic channels required for mobiles in different types of soft handoffs (see Table 13-2), and
  α_{chan} = channel activity factor.

$$P_{(paging)/(channel)} = P_{paging}/N_p \tag{13.27}$$

where N_p = number of paging channels.

$P_{(traffic)/(mobile)}$ is a nominal value. Actual power allocated for each mobile can be up to ± 4 dB around this value depending on the forward link power control for each mobile. On the forward link, extra traffic channels are required for the mobiles in various types of soft handoffs. The percentage of the coverage area in handoff is a design criterion. The extra number of traffic channels in handoff can be related to the area in handoff. Table 13-2 provides the suggested values.

13.6.1 Pilot Channel

The mobile measures the E_c/I_t of the pilot channel continuously and compares it against threshold values of the handoff parameters T_ADD and T_DROP (E_c is the energy per chip and

Table 13-2 Channel Overhead Factor for Various Types of Soft Handoffs

Type of Handoff	% Area in Handoff	ξ_{co}
Soft	25%	0.25
Softer	20%	0.20
Soft-soft	10%	0.20
Soft-softer	10%	0.20
Total ξ_{co}	—	0.85

I_t is the interference plus noise density measured on the pilot channel). The mobile reports the results of these comparisons to the serving cell. The serving cell decides whether or not the mobile needs handoff. The E_c/I_t of the pilot channel is important for determining whether or not the mobile is within the coverage area of the particular cell. The pilot signal from a cell is transmitted at a relatively higher power than those of other forward link logical channels (i.e., paging, sync, traffic). In order to set up a call, the mobile must receive the pilot signal successfully. The pilot channel acts as a coherent carrier phase reference for demodulation of other logical channels on the forward link. Since the E_c/I_t effectively determines the coverage area of a cell or sector, it is essential that the E_c/I_t be sufficiently large.

13.6.2 Traffic Channel

Let $(S_1)_m$ equal the power received by the mth mobile from the cell/sector providing maximum power (i.e., serving cell), and let $(S_2)_m \cdots (S_Q)_m$ equal the power received by the mth mobile from neighboring cells.

Thus

$$(S_1)_m > (S_2)_m \cdots > (S_Q)_m > 0 \tag{13.28}$$

We assume that the power received from Q cells or sectors is significant and that all other cells' power is negligible. We assume that all cell sites beyond the second ring around a serving cell contribute negligible received power, so that $Q \leq 18$. The received bit energy-to-interference plus thermal noise for the mth mobile will be [4]

$$\left(\frac{E_b}{I_t}\right)_m \geq \left(\Phi_t \cdot \frac{B_w}{R} \cdot \frac{\omega_m(S_1)_m}{\displaystyle\sum_{j=1}^{Q(S_j)_m} (S_j)_m + N_0 B_w} \right) \tag{13.29}$$

where Φ_t = fraction of total cell-site power assigned to traffic channels,

$(1 - \Phi_t)$ = fraction of total cell power that is assigned to transmission of overhead channels (pilot, sync, and paging channels),

N_0 = thermal noise density,

B_w = spreading bandwidth,

R = data rate,

G_p = processing gain = B_w/R,

ω_i = fraction of total power allocated to the ith mobile,

         Note: the weighting factor ω_i is proportional to the total sum of other base station powers, $S_2, S_3, \ldots, S_Q$, relative to the mobile's own base station power S_1

M = number of users in mobile's own cell or sector.

$$\sum_{i=1}^{M} \omega_i \le 1 \tag{13.30}$$

From Eq (13.29) the weighting factor ω_m is given as

$$\omega_m \le \frac{(E_b/I_t)_m}{\Phi_t G_p} \left[1 + \left(\frac{\sum_{j=2}^{Q} (S_j)}{(S_1)} \right)_m + \frac{\sigma_n^2}{(S_1)_m} \right] \tag{13.31}$$

where σ_n^2 = thermal power.

Since $\Phi_t S_1$ is the maximum total power allocated to the cell/sector containing the given mobile and M is the total number of mobiles in the cell/sector, we define the relative received cell power as

$$f_m = 1 + \left(\frac{\sum_{j=2}^{Q} S_j}{S_1} \right)_m \tag{13.32}$$

Next we combine Eq. (13.30) and Eq. (13.31) to get

$$\sum_{i=1}^{M} f_i \le \frac{G_p \Phi_t}{E_b/I_t} - \sum_{i=1}^{M} \frac{\sigma_n^2}{(S_1)_i} = \Delta f \tag{13.33}$$

Generally the background noise is well below the total largest received cell-site signal power and the second term in Eq. (13.33) is typically negligible relative to the first term. The capacity can be estimated from the outage or blocking probability, defined as

$$P_{out} = p_r[\mathrm{BER} > (\mathrm{BER})_{sp}] \tag{13.34}$$

where $(\mathrm{BER})_{sp}$ = specified bit error rate for which E_b/I_t is equal to $(E_b/I_t)_{sp}$.

We compute Δ_f' for $(E_b/I_t)_{sp}$ from Eq. (13.33) and express the outage or blocking probability as

$$p_{\text{out}} = p_r\left[\sum_{i=1}^{M} f_i > \Delta_f'\right] \tag{13.35}$$

The distribution of $\sum_{i=1}^{M} f_i$ cannot be expressed in a closed form. The simulation results for
the blocking probability for the forward link of IS-95 are shown in Fig. 13-3 for $G_p = 128$, with
20% of the transmitted power in the cell/sector to the pilot channel and with the required $E_b/I_t = $
5 dB for the traffic channel to ensure BER $\leq 10^{-3}$ [2]. The reduction of 2 dB relative to reverse
link is justified by the coherent reception using the pilot as a reference, as compared to the noncoherent detection in the reverse link. In the simulation, powers from base stations were represented as the product of the fourth order of distance and a log-normally distributed attenuation.
With these parameters, the forward link can support the BER of 10^{-3} for more than 99% of the
time for 38 mobiles per sector or 114 mobiles per cell.

EXAMPLE 13.6

Using the following parameters of an IS-95 CDMA system, estimate the sector capacity based on
reverse and forward link performance.

- Spreading bandwidth (B_w) = 1.23 MHz
- Data rate (R) = 9.6 kbps
- $1/\eta = 10$
- Outage or blocking probability (P_{out}) = 1%
- $(E_b/I_t)_{\text{sp}}$ for the reverse link = 7 dB = 5.0
- $(E_b/I_t)_{\text{sp}}$ for the forward link = 5 dB

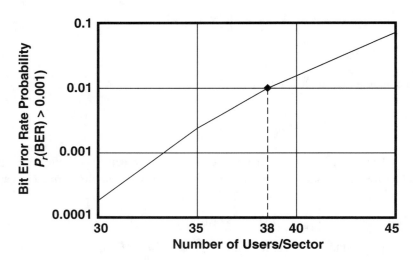

Figure 13-3 Forward Link Capacity of a Cellular CDMA System

- Standard deviation of power control (σ_c) = 2.5 dB = 1.7783
- Voice activity factor (v_f) = 0.5
- Interference from other cells (f) = 0.65
- Path-loss exponent (γ) = 4
- Standard deviation for shadow margin = 8 dB
- Channel overhead factor for soft handoffs (ξ_{co}) = 0.85

Reverse Link

$$Q^{-1}(0.01) = 2.33$$

$$\Delta_r' = \frac{1.23 \times 10^6}{9.6 \times 10^3} \times \frac{(1 - 0.1)}{5} = 23.04$$

$$B = (2.33)^2/23.04 = 0.2356$$

$$\alpha_c = e^{(0.2303 \times 1.7783)^2/2} = 1.8075$$

$$F(B, \sigma_c) = \frac{1}{1.0875}\left[1 + \left(\frac{(1.0875)^3 \times 0.2356}{2}\right)\left(1 - \sqrt{1 + \frac{4}{(1.0875)^3 \times 0.2356}}\right)\right] = 0.5339$$

For perfect power control $\alpha_c = e^0 = 1$

$$F(B, \sigma_c) = 1 + \frac{0.2356}{2}\left(1 - \sqrt{1 + \frac{4}{0.2356}}\right) = 0.6183$$

Efficiency of power control: $\eta_c = \dfrac{0.5339}{0.6183} = 0.8635$

$$\frac{\lambda}{\mu} = \frac{23.04 \times 0.5339}{0.5 \times (1 + 0.65)} = 14.91 \text{ Erlangs}$$

$$\left(\frac{\lambda}{\mu}\right)_{perfect} = \frac{14.91}{0.8635} = 17.27 \text{ Erlangs}$$

From the Erlang B table at 1% blocking with 14.91 Erlangs, $\approx$ 23 mobiles can be supported per sector, whereas with perfect power control we can support $\approx$ 27 mobiles. We lose about 15% of the sector capacity due to imperfect power control. With a loading factor of about 70%, the reverse link capacity will be about 16 mobiles per sector.

Forward Link

From Fig. 13-3 we can see that the forward link can support 38 mobiles per sector with perfect power control. If we assume the same accuracy of the power control as on the reverse link, the sector capacity will be reduced to

$$\text{No. of mobiles per sector} \approx 38 \times 0.8635 = 33$$

Next we consider the effect of soft handoffs on forward link capacity. If the channel overhead factor is ξ_{co} = 0.85, the sector capacity based on the performance of the forward link will be

$$\text{Sector capacity} = \frac{33}{1.85} \approx 18$$

In this example, the sector capacity is controlled by forward link performance. In practice, a loading factor of about 70% is often suggested; this will give the sector capacity about 13 mobiles. Note that the sector capacity of the forward link is significantly affected by total percentage and distribution of soft handoffs.

13.7 CDMA Cell Size

For CDMA systems with AWGN channels, the signal-to-interface ratio (SIR)[*] is an accurate measure of performance [1]. The SIR, the number of users in the cell/sector M, and the maximum mobile transmit power establish the size of the cell on the reverse link.

In a noise- (coverage-) limited system, cell size is the main concern rather than capacity. Receiver sensitivity is used to calculate the size of the cell. Fading margin, propagation path loss, and the minimum received signal level needed to achieve an acceptable performance are the key factors in establishing the cell size. Propagation path losses are determined either from statistical propagation models or from actual measurements.

In an interference- (capacity-) limited system, the size of the cell is determined mainly by the level of the interference from other users. On the reverse link, the maximum path loss that the cell can tolerate is determined by the SIR, the number of simultaneous users, and the maximum power that the mobile can transmit. The maximum cell size should be such that the mobile can close the link.

In a noise-limited system, the minimum SIR can be translated to a minimum signal strength requirement on the forward link. The forward link cell boundary is defined by pilot E_c/I_t. The maximum cell size should be such that, within the coverage area, the received pilot E_c/I_t should be above a predefined threshold.

- **Reverse link cell size.** At the cell site, the SIR per antenna can be given by

$$\text{SIR}(r) = \frac{p_m \cdot L_p(r) \cdot G'_b \cdot G'_m}{[N_0 B_w]_{\text{cell}} + [M/f_r - 1](v_f \cdot p_m) L_p(r) G'_b G'_m} \tag{13.36}$$

where p_m = mobile's power amplifier output,
  G'_b = cell antenna gain including cable losses,
  G'_m = mobile antenna gain including cable losses,
  $L_p(r)$ = reverse link path loss,
  M = number of users in a cell/sector,
  v_f = voice activity factor,
  f_r = frequency reuse factor, and
  $[N_0 B_w]_{\text{cell}}$ = thermal noise of the cell.

The quantity $[N_0 B_w]_{\text{cell}} + [M/f_r - 1] v_f p_m(r) L_p(r) G'_b G'_m\}$ depends only on system loading. It has been shown that [7]

---

[*] SIR includes both interference and thermal noise.

$$1 + \frac{[M/f_r - 1]([v_f \cdot p_m] \cdot L_p(r)G'_b G'_m)}{N_0 B_w} = \frac{1}{1 - \rho} \tag{13.37}$$

where ρ = system loading factor.

The maximum path losses that the mobile can tolerate are given by

$$L_p(r) = \frac{(\text{SIR})[N_0 B_w]_{\text{cell}}\left(\frac{1}{1 - \rho}\right)}{p_m G'_b G'_m} \tag{13.38}$$

We express Eq. (13.38) in dB to get

$$L_p(r) = (\text{SIR})_{\text{min}} + [N_0 B_w]_{\text{cell}} - p_m - G'_b - G'_m - 10\log(1 - \rho) \tag{13.39}$$

The maximum transmission loss will be

$$T(r) = L_p(r) + G'_b + G'_m \tag{13.40}$$

- **Forward link cell size.** On the forward link, the parameter that determines the cell size is the pilot E_c/I_t which is given as

$$\frac{E_c}{I_t} = \frac{\phi_p \cdot p_c \cdot L_p(r) \cdot G'_b \cdot G'_m}{[N_0 B_w]_{\text{mob}} + I_{oc}(r)B_w + I_0(r)B_w} = \frac{\phi_p \cdot p_c \cdot T(r)}{[N_0 B_w]_{\text{mob}} + I_0(r)B_w \cdot (1 + \xi)}$$

$$= \frac{\phi_p \cdot p_c \cdot T(r)}{[N_0 B_w]_{\text{mob}} + p_c T(r)(1 + \xi)} \tag{13.41}$$

Solving for $T(r)$ we get

$$T(r) = \frac{(E_c/I_t)(N_0 B_w)_{\text{mob}}}{p_c[\phi_p - (E_c/I_t)(1 + \xi)]} \tag{13.42}$$

We express Eq. (13.42) in dB to get

$$T(r) = (E_c/I_t)_{\text{min}} + (N_0 B_w)_{\text{mob}} - p_c - 10\log[\phi_p - (10^{(E_c/I_t)_{\text{min}}/10})(1 + 10^{\xi/10})] \tag{13.43}$$

$$L_p(r) = (E_c/I_t)_{\text{min}} + (N_0 B_w)_{\text{mob}} - p_c - 10\log[\phi_p - (10^{(E_c/I_t)_{\text{min}}/10})(1 + 10^{\xi/10})]$$
$$- G'_b - G'_m \tag{13.44}$$

where ϕ_p = fraction of the cell power allocated to the pilot,
        p_c = cell output,
        G'_b = cell antenna gain including cable losses,
        G'_m = mobile antenna gain including cable losses,
        $I_{oc}(r)$ = other cell interference power spectral density,

$I_o(r)$ = serving cell interference power spectral density,

ξ = I_{oc}/I_o,

$[N_0 B_w]_{mob}$ = thermal noise at the mobile, and

$(E_c/I_t)_{min}$ = minimum required value for pilot.

EXAMPLE 13.7

Calculate transmission loss vs. cell loading for a 200-mW mobile unit. Assume: $(E_b/I_t)_{min} = 7$ dB; processing gain at the cell site = 21 dB; cell noise figure = 5 dB.

$$(SIR)_{min} + \text{processing gain} = (E_b/I_t)_{min}$$

$$(SIR)_{min} = 7 - 21 = -14 \text{ dB}$$

$$(N_0 B_w)_{cell} = [3.1622 \times 290 \times 1.38066 \times 10^{-23} \times 1.2288 \times 10^6] \times 10^3 \text{ mW} = -108 \text{ dBm}$$

$$T(r) = (SIR)_{min} + (N_0 B_w)_{cell} - P_m - 10\log(1-\rho) = -14 - 108 - 23 + 10\log(1-\rho)$$

$$= -145 - 10\log(1-\rho) \text{ dB}$$

Fig. 13-4 shows a plot of $T(r)$ vs. ρ.

EXAMPLE 13.8

Plot maximum transmission loss (dB) vs. % of power allocated to the pilot channel. Assume: $(E_c/I_t)_{min} = -15$ dB; mobile noise figure = 8 dB; $I_{oc}/I_o \sim 2.5$ dB; cell-site output = 44 dBm.

$$T(r) = (E_c/I_t)_{min} - P_c + (N_0 B_w)_{mob} - 10\log[\phi_p - (10^{(E_c/I_t)_{min}/10})(1 + 10^{I_{oc}/I_o})]$$

$$T(r) = -15 - 44 - 105 - 10\log[\phi_p - 0.03162 \times (1 + 1.7783)]$$

$$T(r) = -164 - 10\log[\phi_p - 0.08785] \text{ dB}$$

Fig. 13-5 shows a plot of $T(r)$ vs. ϕ_p.

13.8 Forward and Reverse Link Balance

A more powerful forward link results in extra interference to mobiles in other cells. On the other hand, a more powerful reverse link sacrifices capacity. It is desirable then to design the system so that the two boundaries coincide. Balanced links minimize interference and eliminate associated handoff problems. The boundary of the cell on the reverse link is determined by cell loading, and the boundary on the forward link is obtained by a minimum pilot (E_c/I_t). In order to keep the two boundaries close, we need to equate the path loss on both links. From Eqs (13.39) and (13.44) we define the balance factor B_f as

$$B_f = [(SIR)_{min} - (E_c/I_t)_{min}] + [N_0 F_{cell} - N_0 F_{mob}] \cdot B_w + [p_c - p_m]$$

$$+ 10\log\left(\frac{(\phi_p - 10^{(E_c/I_t)_{min}/10})(1 + 10^{(I_{oc}/I_o)/10})}{1 - \rho}\right) \quad (13.45)$$

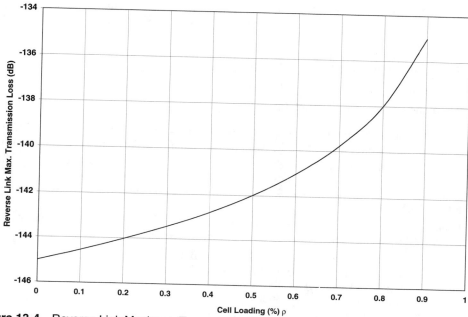

Figure 13-4 Reverse Link Maximum Transmission Loss vs. Cell Loading

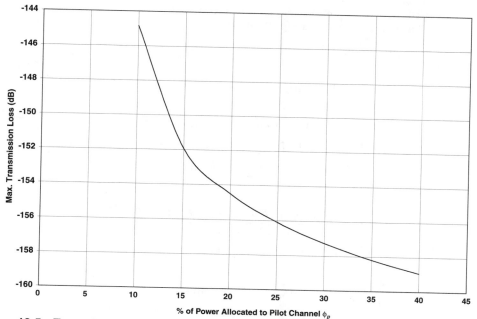

Figure 13-5 Transmission Loss vs. Percent of Power Allocated to the Pilot Channel

Based on B_f, the system designer can decide which link is the limiting factor. If $B_f < 0$, the system is forward link limited; $B_f = 0$, links are balanced; if $B_f > 0$ the system is reverse link limited. A more realistic rule is

$B_f < -\delta$: the system is forward link limited

$|B_f| \le \delta$: the two links are balanced

$B_f > \delta$: the system is reverse link limited

δ is the parameter that takes into account the tolerance in all factors involved in calculating B_f. A good system design should insure that the two links are balanced. This makes handoff transition smoother and reduces the amount of interference.

EXAMPLE 13.9

Using the following data, calculate the power allocated to the pilot channel. What is the allocated power for the pilot channel to balance the forward and reverse link?

- Maximum mobile power (p_m) = 200 mW (23 dBm)
- Maximum cell-site power (p_c) = 10 W (40 dBm)
- Voice activity factor (v_f) = 0.4
- Processing gain at the base station = 21 dB
- Number of users per cell (M) = 20
- Cell loading factor (ρ) = 0.5
- Cell noise figure (F_{cell}) = 5 dB
- Mobile noise figure (F_{mob}) = 8 dB
- Cell-site antenna gain including cable loss (G_b') = 6 dB
- Mobile antenna gain including cable loss (G_m') = 0 dB
- I_{oc}/I_o = 2.5 dB
- $(E_c/I_t)_{min} = -15$ dB
- $(E_b/I_t)_{min} = 7$ dB

$$(\text{SIR})_{min} + \text{processing gain} = (E_b/I_t)_{min}$$

$$(\text{SIR})_{min} = -21 + 7 = -14 \text{ dB}$$

Using Eq. (13.39), the path loss (cell site) on the reverse link is

$$L_p(r) = -14 - 108 - 23 - 6 - 10\log(1 - 0.5) = -148 \text{ dB}$$

$$T(r) = -142 \text{ dB}$$

From Eq. (13.42)

$$\phi_p = \frac{(E_c/I_t)_{min}[(N_0 B_w)_{mob} + p_c \cdot T(r)(1 + 10^{(I_{oc}/I_o)/10})]}{p_c \cdot T(r)}$$

$$p_c = 10^4$$

$$T(r) = 10^{-14.2}$$

$$(N_0 B_w)_{mob} = -105 \text{ dB} = 10^{-10.5}$$

$$(E_c/I_t)_{min} = -15 \text{ dB} = 10^{-1.5} = 0.03163$$

$$\phi_p = \frac{0.03163 \cdot [10^{-10.5} + 10^4 \cdot 10^{-14.2} \cdot (1 + 1.7783)]}{10^4 \cdot 10^{-14.2}} = 0.1078 = 10.78\%$$

For the balanced condition, we use Eq. (13.45) with $B_f = 0$ and solve for ϕ_p

$$0 = [-14 - (-15)] + [-108 - (-105)] + [40 - 23] + \left[10\log\left(\frac{\phi_p - 10^{-1.5}(1 + 1.7783)}{0.5}\right)\right]$$

$$\therefore \phi_p = 0.1037 = 10.37\%$$

EXAMPLE 13.10

Consider a minicell in which the maximum transmission loss is equal to -112 dB. The other data include

- $[N_0B_w]_{mob} = -105$ dB
- $[N_0B_w]_{cell} = -108$ dB
- $[E_b/I_t]_{min} = 7$ dB
- $[E_c/I_t]_{min} = -15$ dB
- Cell loading = 50%
- $I_{oc}/I_o = 2.5$ dB

Find the output of the mobile and cell site.

From Eq. (13.39)

$$-112 = -14 - 108 - p_m - 10\log(1 - 0.5)$$

$$p_m = -7 \text{ dBm } (= 0.2 \text{ mW})$$

For the minicell, the pilot strength can be determined as

$$\left(\frac{E_c}{I_t}\right)_{min} = \frac{\phi_p}{1 + I_{oc}/I_o}$$

$$\therefore \phi_p = \left(\frac{E_c}{I_t}\right)_{min}(1 + I_{oc}/I_o) = 10^{-1.5} \cdot (1 + 10^{0.25}) = 0.0879 \sim 9\%$$

From Eq. (13.43)

$$T(r) = \left(\frac{E_c}{I_t}\right)_{min} - p_c' + [N_0B_w]_{mob} - 10\log[\phi_p - (10^{(E_c/I_t)_{min}/10})(1 + 10^{(I_{oc}/I_o)/10})]$$

$$\therefore -112 = -15 - p_c + (-105) - 10\log[0.09 - (10^{-1.5})(1 + 10^{0.25})]$$

$$p_c = -15 + 112 - 105 + 26.7 = 18.7 \text{ dBm } (\sim 74 \text{ mW})$$

13.9 Forward Link Budget

The forward link budget is calculated to confirm that the link safety margin parameters (Eq. 13.46) are positive and that there is sufficient margin for the forward link to work efficiently. M_{pilot} and $M_{traffic}$ are more critical. If both of these are positive, then the other two are also generally positive.

$$M_{\text{pilot}} = \left(\frac{E_c}{I_t}\right)_{\text{rec}} - \left(\frac{E_c}{I_t}\right)_{\text{sp}} \qquad (13.46a)$$

$$M_{\text{traffic}} = \left(\frac{E_b}{I_t}\right)_{\text{rec}} - \left(\frac{E_b}{I_t}\right)_{\text{sp}} \qquad (13.46b)$$

$$M_{\text{sync}} = \left(\frac{E_b}{I_t}\right)_{\text{rec}} - \left(\frac{E_b}{I_t}\right)_{\text{sp}} \qquad (13.46c)$$

$$M_{\text{paging}} = \left(\frac{E_b}{I_t}\right)_{\text{rec}} - \left(\frac{E_b}{I_t}\right)_{\text{sp}} \qquad (13.46d)$$

where *rec* means received value and *sp* means required value.

We use the following procedure to determine link safety margins.

The total cell-site power P_{total} will be

$$P_{\text{total}} = 10\log(10^{0.1P_{\text{traffic}}} + 10^{0.1P_{\text{pilot}}} + 10^{0.1P_{\text{sync}}} + 10^{0.1P_{\text{paging}}}) \text{ dBm} \qquad (13.47)$$

where P_{total} = total cell-site effective radiated power (ERP) (dBm),
 P_{sync} = ERP of sync channel (dBm),
 P_{pilot} = ERP of pilot channel (dBm),
 P_{paging} = ERP of paging channel (dBm), and
 P_{traffic} = ERP of all traffic channels (dBm).

$$P_{(\text{traffic})/(\text{user})} = \frac{P_{\text{traffic}}}{(M_{\text{total}} \cdot \alpha_{\text{chan}})} = P_{\text{traffic}} - 10\log\alpha_{\text{chan}} - 10\log M_{\text{total}} \text{ dBm} \qquad (13.48)$$

where $P_{(\text{traffic})/(\text{user})}$ = ERP of a traffic channel (dBm),
 α_{chan} = channel activity factor,
 M_{total} = $M(1 + \eta_{co})$, and
 η_{co} = traffic channel overhead percentage due to soft handoff.

The received power at the mobile on each channel from the serving cell site will be

$$P_{r,\text{total}} = P_{\text{total}} + GL \qquad (13.49)$$

$$P_{r,\text{pilot}} = P_{\text{pilot}} + GL \qquad (13.50)$$

$$P_{r,(\text{traffic})/(\text{user})} = P_{(\text{traffic})/(\text{user})} + GL \qquad (13.51)$$

$$P_{r,\text{sync}} = P_{\text{sync}} + GL \qquad (13.52)$$

$$P_{r,\text{paging}} = P_{\text{paging}} + GL \qquad (13.53)$$

where $GL = G_m + L_c + L_b + L_{\text{pent}} + M_{\text{fade}} + L_p + G_b,$

in which L_p = average propagation path loss between cell site and mobile (dB),
L_{pent} = penetration loss (dB),
L_b = body/orientation loss (dB),
L_c = cell-site feeder loss (dB),
M_{fade} = margin for log-normal shadowing (dB),
G_m = mobile antenna gain (dB), and
G_b = cell-site antenna gain (dB).

In-cell interference is caused by other users radiated from the same cell and is given as

$$I_{\text{ss-ch}} = 10\log(10^{0.1P_{r,\text{total}}} - 10^{0.1P_{r,\text{ch}}}) - 10\log B_w \text{ dBm/Hz} \qquad (13.54)$$

where ch is the pilot, paging, sync, or traffic/user, and
B_w = bandwidth.

Out-of-cell interference is caused by users in other cells and is given as

$$I_{\text{os-ch}} = I_{\text{ss-ch}} + 10\log(1/f_r - 1) \text{ dBm/Hz} \qquad (13.55)$$

where f_r = reuse factor.

Total interference will be

$$I_{\text{ch}} = 10\log(10^{0.1I_{\text{ss-ch}}} + 10^{0.1I_{\text{os-ch}}}) \text{ dBm/Hz} \qquad (13.56)$$

The thermal noise density will be

$$N_0 = 10\log(290 \times 1.38 \times 10^{-23}) + N_f + 30 \text{ dBm/Hz} \qquad (13.57)$$

where N_f = noise figure for the mobile.

The energy per bit for a channel will be

$$E_{\text{bch}} = P_{r,\text{ch}} - 10\log R_{\text{ch}} \qquad (13.58)$$

where R_{ch} = data rate for the channel.

The E_b/I_t for a channel can be calculated as

$$\frac{E_{\text{bch}}}{N_0 + I_{\text{ch}}} = P_{r,\text{ch}} - 10\log R_{\text{ch}} - 10\log(10^{0.1N_0} + 10^{0.1I_{\text{ch}}}) \text{ dB} \qquad (13.59)$$

Using Eq. (13.59) we can write

$$\left(\frac{E_b}{I_t}\right)_{\text{rec, pilot}} = P_{r,\text{pilot}} - 10\log B_w - 10\log(10^{0.1N_0} + 10^{0.1I_{\text{pilot}}}) \text{ dB} \qquad (13.60)$$

$$\left(\frac{E_b}{I_t}\right)_{rec,\,paging} = P_{r,\,paging} - 10\log R_{paging} - 10\log(10^{0.1N_0} + 10^{0.1I_{paging}})\,dB \quad (13.61)$$

$$\left(\frac{E_b}{I_t}\right)_{rec,\,sync} = P_{r,\,sync} - 10\log R_{sync} - 10\log(10^{0.1N_0} + 10^{0.1I_{sync}})\,dB \quad (13.62)$$

$$\left(\frac{E_b}{I_t}\right)_{rec,\,traffic} = P_{r,\,(traffic)/(user)} - 10\log R_{traffic} - 10\log(10^{0.1N_0} + 10^{0.1I_{traffic}})\,dB \quad (13.63)$$

EXAMPLE 13.11

Use the following data to calculate link safety margin parameters for the forward link channels of a CDMA system.

- Pilot channel ERP (P_{pilot}) = 33.8 dBm
- Sync channel ERP (P_{sync}) = 23.8 dBm
- Paging channel ERP (P_{paging}) = 29.5 dBm
- Traffic channels ERP ($P_{traffic}$) = 41.0 dBm
- Number of users per sector on the reverse link = 13
- Channel overhead due to soft handoff (η_{co}) = 0.85
- Path loss between cell site and mobile (L_p) = −130.2 dB
- Penetration loss (L_{pent}) = −15 dB
- Body/orientation loss (L_b) = −2 dB
- Fade margin (M_{fade}) = −10.3 dB
- Mobile antenna gain (G_m) = 2 dB
- Cell-site antenna gain (G_b) = 13 dB
- Cable losses (L_c) = −1.5 dB
- Channel activity factor (α_{chan}) = 0.42
- Bandwidth (B_w) = 1.2288 MHz
- Traffic channel rate = 9600 bps
- Sync channel rate = 1200 bps
- Paging channel rate = 4800 bps
- Cell reuse factor (f_r) = 0.65

$$P_{total} = 10\log[10^{4.1} + 10^{3.38} + 10^{2.95} + 10^{2.38}] = 42.0734\,dBm$$

$$M_{total} = 13\,(1 + 0.85) = 24$$

$$P_{(traffic)/(user)} = 41.0 - 10\log(0.42) - 10\log 24 = 30.9654\,dBm$$

$$GL = 2 + 13 - 1.5 - 2.0 - 15 - 10.3 - 130.2 = -144\,dB$$

$$P_{r,\,total} = 42.0734 - 144 = -101.9266\,dBm$$

$$P_{r,\,pilot} = 33.8 - 144 = -110.2\,dBm$$

$$P_{r,\,sync} = 23.8 - 144 = -120.2\,dBm$$

$$P_{r,\,paging} = 29.5 - 144 = -114.5\,dBm$$

$$P_{r,\,(\text{traffic})/(\text{user})} = 30.9654 - 144 = -113.0346 \text{ dBm}$$

$$I_{\text{ss-pilot}} = 10\log(10^{-10.19266} - 10^{-11.02}) - 10\log(1.2288 \times 10^6) = -163.5211 \text{ dBm/Hz}$$

$$I_{\text{os-pilot}} = -163.5211 + 10\log(1/0.65 - 1) = -166.2096 \text{ dBm/Hz}$$

$$I_{\text{pilot}} = 10\log(10^{-16.35211} + 10^{-16.62096}) = -161.65 \text{ dBm/Hz}$$

$$I_{\text{ss-sync}} = 10\log(10^{-10.19266} - 10^{-12.02}) - 60.9 = -162.8865 \text{ dBm/Hz}$$

$$I_{\text{os-sync}} = -162.8865 - 2.6885 = -165.575 \text{ dBm/Hz}$$

$$I_{\text{sync}} = 10\log(10^{-16.28865} + 10^{-16.5575}) = -161.0157 \text{ dBm/Hz}$$

$$I_{\text{ss-paging}} = 10\log(10^{-10.19266} - 10^{-11.45}) - 60.9 = -163.0736 \text{ dBm/Hz}$$

$$I_{\text{os-paging}} = -163.0736 - 2.6885 = -165.7621 \text{ dBm/Hz}$$

$$I_{\text{paging}} = 10\log(10^{-16.30736} + 10^{-16.57621}) = -161.2030 \text{ dBm/Hz}$$

$$I_{\text{ss-(traffic)/(user)}} = 10\log(10^{-10.19266} - 10^{-11.30346}) - 60.9 = -163.1769 \text{ dBm/Hz}$$

$$I_{\text{os-(traffic)/(user)}} = -163.1769 - 2.6885 = -165.8654 \text{ dBm/Hz}$$

$$I_{\text{(traffic)/(user)}} = 10\log(10^{-16.31769} + 10^{-16.58654}) = -161.306 \text{ dBm/Hz}$$

$$N_0 = 10\log(290 \times 1.38 \times 10^{-23}) + 30 + 8 = -165.9772 \text{ dBm/Hz}$$

$$\left(\frac{E_c}{I_t}\right)_{\text{rec, pilot}} = -110.2 - 10\log(1.2288 \times 10^6) - 10\log(10^{-16.597} + 10^{-16.165}) = -10.82 \text{ dB}$$

$$\left(\frac{E_b}{I_t}\right)_{\text{rec, sync}} = -120.2 - 10\log(1200) - 10\log(10^{-16.597} + 10^{-16.10157}) = 8.83 \text{ dB}$$

$$\left(\frac{E_b}{I_t}\right)_{\text{rec, paging}} = -114.5 - 10\log(4800) - 10\log(10^{-16.597} + 10^{-16.12}) = 8.64 \text{ dB}$$

$$\left(\frac{E_b}{I_t}\right)_{\text{rec, traffic}} = -113.0346 - 10\log(9600) - 10\log(10^{-16.597} + 10^{-16.1306}) = 7.18 \text{ dB}$$

$$M_{\text{pilot}} = -10.82 - (-15) = 4.18$$

$$M_{\text{traffic}} = 7.18 - 7.0 = 0.18$$

$$M_{\text{paging}} = 8.64 - 7.0 = 1.64$$

$$M_{\text{sync}} = 8.83 - 7.0 = 1.83$$

Since all the link safety parameters are positive, the power allocations on forward link channels are satisfactory.

Note that link budget calculations can be easily performed using Microsoft Excel or any other program as shown in Table 13-3.

Table 13-3 Link Budget Calculations

channel	total	pilot	sync	paging	fwd traffic
total trfc pwr					57.00
voice activ					0.42
# channels					22.00
% users in soft handoff					25.00
power output	58.49	51.50	41.50	46.94	46.37
path loss	146.00	146.00	146.00	146.00	146.00
fade margin	6.20	6.20	6.20	6.20	6.20
rx ant gain	−3.00	−3.00	−3.00	−3.00	−3.00
rx power	−96.71	−103.70	−113.70	−108.26	−108.83
in cell I		−158.58	−157.69	−157.92	−157.88
freq reuse factor		0.65	0.65	0.65	0.65
out cell I		−161.26	−160.38	−160.61	−160.57
total I		−156.71	−155.82	−156.05	−156.01
noise figure		8.00	8.00	8.00	8.00
noise		−165.98	−165.98	−165.98	−165.98
I/N		9.27	10.15	9.93	9.97
N+I		−156.22	−155.42	−155.63	−155.59
data rate		1.23E+06	1200	9600	9600
eb/(I+N)		−8.37	10.93	7.55	6.95
channel					rvse trfc
power output					23.00
path loss					146.00
fade margin					6.20
rx ant gain					14.00
bs cable loss					2.50
rx power					−117.70
voice activ					0.40
in cell I					−169.35
freq reuse factor					0.60
out cell I					−171.11
total I					−167.13
noise figure					5.00

Table 13-3   Link Budget Calculations (Continued)

channel	total	pilot	sync	paging	fwd traffic
noise					−168.98
I/N					1.84
N+I					−164.95
data rate					9600
eb/(I+N)					7.43

13.10 Summary

In this chapter, we developed necessary equations to calculate the reverse and forward link capacity of an IS-95 CDMA system. We found that the maximum number of mobiles that can be supported on the forward and reverse links of a CDMA system is different. Reverse link capacity improves with soft handoffs; however, soft handoffs affect the capacity of the forward link.

In a noise-limited system, cell size rather than capacity is the main concern. Receiver sensitivity is used to calculate the size of the cell. In an interference-limited system, the size of the cell is determined mainly by the level of the interference from other users.

The chapter includes several numerical examples to illustrate the procedure and demonstrate the importance of several system parameters in capacity calculations. It concludes with link safety margin parameters for the forward link channels.

13.11 References

1. Borth, D. E., and Pursley, M. B., "Analysis of Direct Sequence Spread Spectrum Multiple Access Communication over Rician Fading Channels," *IEEE Transactions on Communications* COM-27 (10), October 1979, pp. 1566–77.

2. Gilhousen, K., Jacobs, I., Padovani, R., Viterbi, A., Weaver, L., and Whearley, C., III, "On the Capacity of a Cellular CDMA System," *IEEE Transactions on Vehicular Technology* 40 (2), May 1991, pp. 303–12.

3. Gilhousen, K. S., Jacobs, I. M., Padovani, R., and Weaver, L. A., "Increased Capacity Using Satellite Communications," *IEEE Transactions on Select Areas in Communications* JSAC-8 (4), May 1990, pp. 503–14.

4. Glisic, S., and Vucctic, B., *Spread Spectrum CDMA Systems for Wireless Communications*, Artech House, Inc., Boston, 1997.

5. Viterbi, A. J., *CDMA*, Addison-Wesley Publishing Company, New York, 1995.

6. Viterbi, A. M., and Viterbi, A. J., "Erlang Capacity of a Power Controlled CDMA System," *IEEE Journal of Selected Areas in Communications* 11 (6), 1993, pp. 892–900.

7. Weber, C. L., et al., "Performance Considerations of CDMA Systems," *IEEE Transactions on Vehicular Technology*, February 1981, pp. 3–9.

13.12 Problems

1. A chip rate of 1.2288 Mcps is used for IS-95 RS2 (14.4 kbps with a 13-kbps vocoder). The required E_b/I_t is 7 dB. Calculate the average number of mobiles supported by a sector of the three-sector cell; also calculate the Erlang capacity of a sector at 2% blocking. Assume: interference from other cells $f = 0.6$; cell loading $\rho = 0.6$; voice activity factor $v_f = 0.4$; power control accuracy factor $\eta_c = 0.90$; gain due to sectorization $\alpha = 2.61$.

2. Calculate the pole capacity of the IS-95 CDMA system with a chip rate of 1.2288 Mcps and RS2 (14.4 kbps). Assume: interference factor due to neighboring cells $f = 0.67$; voice activity factor $v_f = 0.6$; power control accuracy factor $\eta_c = 0.80$; gain due to three-sector antenna $\alpha = 2.55$.

3. A total of 20 equal-power mobiles share a frequency band through a CDMA system. Each mobile transmits data at 16 kbps with a DSSS BPSK-modulated signal. Calculate the minimum chip rate of the PN sequence in order to maintain a bit error probability of 10^{-6}. Assume: interference factor due to other cells $f = 0.6$; power control accuracy factor $\eta_c = 0.8$; gain due to three-sector antenna $\alpha = 2.55$.

4. Calculate the Erlang capacity of a sector for the three-sector cell and the number of users per sector using the following parameters: carrier bandwidth = 1.23 MHz, RS1, $R = 9.6$ kbps; $(E_b/I_t)_{min} = 7$ dB; voice activity factor $v_f = 0.5$; interference due to other cells $f = 0.67$; cell loading factor $\rho = 0.5$; call blocking probability, $P_{out} = 1\%$; three-sector antenna gain $\alpha = 2.55$; $1/\eta = 10$; standard deviation of power control $\sigma_c = 2$ dB.

5. Calculate total transmission loss in dB for the forward link of a CDMA system using the following data:

   ♦ Cell output power = 40 dBm
   ♦ Allocated power for pilot channel = 15% of cell output
   ♦ Mobile noise figure = 8 dB
   ♦ $(E_c/I_t)_{min} = -13$ dB
   ♦ $I_{oc}/I_o = 2.5$ dB

6. Calculate total transmission loss in dB for the reverse link of a CDMA system using the following data:

   ♦ Mobile output = 200 mW (23 dBM)
   ♦ $(E_b/I_t) = 7$ dB
   ♦ Cell noise figure = 5 dB
   ♦ Cell loading $\rho = 60\%$

7. Using the data given in Problems 5 and 6, find the allocated power of the pilot channel for balancing the forward and reverse links.

CHAPTER 14

Wireless Data

14.1 Introduction

This chapter covers data communication services and OSI upper layers and presents wireless data systems including wide-area systems and high-speed Wireless Local Area Networks (WLANs). We describe activities for wireless data standards and outline the error-control methods used by the standards. Also included are packet radio protocols and their channel efficiency formulas. The contention function of packet radio models the mechanism where mobile stations access the network on the access channel. Packet services are one of four data services supported in CDMA. The other three are asynchronous data, facsimile, and short message services (similar to paging).

We discuss the standards for data services supported by CDMA cellular/PCS systems and present highlights of the TIA IS-99, TIA IS-637, and TIA IS-657 standards. The chapter includes the architecture for each of the four data services and the protocol stacks that are supported by the services.

We include both sets of standards (CDMA and non-CDMA) for two reasons: first, the WLANs all use some form of spread spectrum communications, either frequency hopping or direct sequence spreading; second, the two methods (WLANs and CDMA) are part of a larger wireless network that many companies are constructing. With the phenomenal growth of laptop personal computers and the Internet, wireless data is no longer limited to just e-mail or faxes. It encompasses the ability to send and receive data any time from any place in the world. It gives a user at a remote location full access to all of the desktop services that would normally be available at an office PC. Data services are delivering the same promise that voice services have recently delivered: any time, any where communications.

14.2 Data Communication Services

End-to-end communication services are classified as either *synchronous* (sync) or *asynchronous* (async). A sync communication service delivers a bit stream with a fixed delay and a given bit error rate. Voice communication is an example of sync communication service. Sync delivery of a 64-kbps voice bit stream can be implemented by dividing the bit stream into packets that are received with random delays and are stored in a buffer to hold the bits until they are delivered. This implementation of a sync transmission service is called *packetized voice*. In packetized voice, a buffer is used to absorb the random fluctuations in the packet transmission delays. Another implementation of the sync transmission of the bit stream is to use a dedicated coaxial cable that propagates the bits one after the other, all with the same delay.

In an async communication service, the bit stream to be transferred is divided into packets. The packets are received by the destination with varying delays, and a fraction of them may not be received correctly at the destination. An async communication service is evaluated by its Quality of Service (QoS). QoS deals with parameters, such as the packet error rate, delay, throughput, reliability, and security of the communication.

There are two classes of async communication services: *connection oriented* and *connectionless*. A connection-oriented communication service delivers the packet in sequence—i.e., in correct order—and confirms the delivery. Depending on the QoS requirements, the delivery may be guaranteed to be free of errors. Thus, connection-oriented service looks from end to end like a dedicated link, which may be noiseless or noisy. A connectionless communication service delivers the packets individually. The packets can be delivered out of order, and some may contain errors while others may be lost. Some connectionless services provide an acknowledgment (ACK) of correctly delivered packets. Thus connectionless services are similar to mail service provided by the post office: letters may be delivered out of order; normal mail delivery does not guarantee the delivery. Yet another class of communication service called *expedited data* is used in some applications. It corresponds to a potentially faster delivery of packets, usually by making them jump to the head of the queues of packets that are waiting to be transmitted.

Communication services are implemented by transporting bits over the network. One essential objective of the bit transport is connectivity, where one network user should be able to exchange information with many other users. It should be possible to *route* the bits of one user to any one of a large number of other users. The property to vary the path followed by the bits is called *switching*. There are three basic methods used for switching bits in communication networks:

- Circuit switching
- Virtual-circuit packet switching
- Datagram packet switching

In circuit switching, the switch connects transmission paths to establish a circuit between the transmitter and receiver. Circuit switching is quite suitable for continuous data transmission services.

A packet-switched network uses another scheme. The nodes of the network, *packet-switching nodes*, play a role similar to that of switches in a circuit-switched network. Packet-switched networks can use two different methods for selecting the path followed by packets: *virtual circuit* (VC) and *datagram*. In the VC transport, the different packets that are part of the same information transfer are sent along the same path. The packets follow one another as if they were using a dedicated circuit even though they may be interleaved with other packet streams. Some implementations of VC perform an error control on each link between successive nodes. Thus, not only are the packets delivered in sequence by each node to the next node along the path, but they are also transmitted without errors. This is implemented by each node checking the correctness of the packets it receives and asking the previous node along the path to retransmit incorrect packets. VC packet switching does not need a buffer at the destination.

Since multiple virtual circuits may exist between the source-destination pair, routing cannot be done on the basis of source-destination address only. Data packets must carry an indication of VC identification as well. Routing is done based on explicit route number and destination address. An explicit routing table at each node associates an appropriate outgoing transmission group with the destination address and explicit route number. Changing the explicit route number for a given destination will cause a new path to be followed. This introduces alternative route capability. If a link or node along the path becomes inoperative, any session using that path can be reestablished on an explicit route by bypassing the failed element. Explicit routes can also be assigned on the basis of type of traffic, type of physical media along the path (satellite or terrestrial, for example), or other criteria. Routes could also be listed on the basis of cost, the lowest-cost route being assigned first, then the next-lowest-cost route, and so forth.

In datagram packet switching, the bits are grouped as packets. Each packet is labeled with the address of its destination. The packets are routed independently of one another and arrive at destination out of sequence. Datagram packet switching requires buffers at the source and the destination. In datagram packet switching networks, each network node keeps a complete (global) topological database that is updated regularly as topological changes occur. Generally, the routing philosophy of datagram networks is to route packets (datagram) along paths of minimum time delay.

14.3 OSI Upper Layers

This section summarizes the role of the OSI upper layers (transport, session, presentation, and application).

The *transport layer* segments the messages into packets of acceptable size and performs the reassembly at the destination. It may multiplex many low-rate transmissions onto one virtual circuit or divide a high-rate transmission into parallel virtual circuits. The transport layer controls transmission errors and requests retransmission of packets corrupted by transmission errors. In addition, the flow may be controlled by some mechanism to prevent one host from sending data faster than the destination host can handle.

The *session layer* sets up the call and takes care of the authentication of the user and of billing. The session layer supervises the synchronization (packet numbering) and the recovery in case of failures. It also closes the session at the end of the transmission.

The *presentation layer* asks the session layer to set up a call. It specifies the destination's name and the type of transmission (e.g., datagram, high priority). The presentation layer translates between the local syntax used by the application process and the transfer syntax, as well as performing the required encryption and data compression.

The *application layer* provides information transfer services for user application programs. The user interacts with the application layer through a user interface. The application layer is composed of Specific Application Service Elements (SASEs) that use the services of Common Application Service Elements (CASEs). A CASE establishes the association between SASEs and may include an Association Control Service Element (ACSE), a Remote Operation Service Element (ROSE), and a Commitment Concurrency and Recovery (CCR) element.

14.4 Wireless Data Systems

We can classify wireless data systems into two basic categories: wide-area wireless data systems, and high-speed wireless local area networks. WLANs and wide-area wireless data systems serve different categories of user applications and, therefore, have different system design objectives. Wireless data services are used for transaction processing and for interactive, broadcast, and multicast services. *Transaction processing* is used for credit card verification, paging, taxi calls, vehicle theft reporting, and notice of voice or electronic mail. *Interactive services* include database access and remote LAN access. *Broadcast services* are general information services, weather and traffic advisory services, and advertising. *Multicast services* are similar to subscribed information services, law enforcement communications, and private bulletin boards.

In the following sections, we briefly describe wide-area wireless data systems and WLANs that have been deployed in the United States.

14.4.1 Wide-Area Wireless Data Systems

Wide-area wireless data systems are designed to provide high mobility, wide-area coverage, and low data rate digital data communications to both vehicles and pedestrians. The technical challenge is to design a system that efficiently uses the available bandwidth to serve large numbers of users distributed over wide geographical areas. Table 14-1 gives the details of wide-area wireless packet data systems deployed in the United States—there are Specialized Mobile Radio Services (SMRS) allocations centered around 450 MHz and 900 MHz.

The ARDIS data network was developed by Motorola as a joint venture between Motorola and IBM to support IBM field service repair people. It is now a public service offering and is solely owned by Motorola. RAM Mobile Data is another public offering that uses the Ericsson Mobitex technology. Both the ARDIS and RAM networks are evolving to data rates of 19.2 kbps. They have been designed to make use of standard, two-way voice, land mobile-radio channels, with 12.5 or 25 kHz channel spacing.

Table 14-1   Wide-Area Wireless Packet Data Systems

	RAM Mobile (Mobitex)	ARDIS (KDT)	Metricom (MDN)	CDPD
Data rate	19.2 kbps	19.2 kbps	76 kbps	19.2 kbps
Channel spacing	12.5 kHz	25 kHz	160 kHz	30 kHz
Access	slotted ALOHA CSMA		FHSS (ISM)	unused AMPS channels
Frequency (MHz)	$f_c \sim 900$	$f_c \sim 800$	$f_c \sim 915$	$f_c \sim 800$
Transmit power (W)	0.16 to 10 under power control	40	1	1.6
Modulation	GMSK[a]	GMSK	GMSK	GMSK BT[b] = 0.5

[a] GMSK = Gaussian minimum shift keying.
[b] BT = channel width × bit duration.

The CDPD technology shares the 30-kHz spaced 800-MHz voice channels used by the AMPS systems. The data rate is 19.2 kbps. The CDPD base station equipment shares cell sites with the voice cellular radio system. The aim is to reduce the cost of providing packet data service by sharing the resources with voice cellular systems. This strategy is similar to one that has been used by nationwide fixed wireline packet data networks to provide an economically viable data service by using a small portion of the capacity of the networks designed mainly for voice traffic.

Another approach used in wide-area wireless packet data networks is based on the micro-cell concept of providing coverage in smaller areas. The microcell data networks are designed for stationary or low-speed users. The basic aim is to reduce the cost of providing wireless data service by using small and inexpensive base stations that can be installed on utility poles, the sides of buildings, and inside buildings. The strategy is similar to the one being proposed for Personal Communications Networks (PCNs). BS-to-BS wireless links are used to reduce the cost of interconnecting in a data network. A large microcell network of small, inexpensive base stations has been installed in the lower San Francisco Bay area by Metricom. The slow Frequency Hopping Spread Spectrum (FHSS) in the 902–928 MHz U.S. Industrial Scientific Medical (ISM) band has been used. Transmitter power is 1 W maximum. Power control is used to minimize interference and maximize battery life.

14.4.2   High-Speed Wireless Local Area Networks

A WLAN typically supports a limited number of users in a well-defined indoor area. System aspects such as bandwidth efficiency and product standardization are not crucial. The maximum achievable data rate is an important consideration in the selection of a WLAN. The transmission channel characteristics and signal processing techniques are important.

WLANs are used to extend wired LANs for convenience and mobility. Three different approaches have been used for connectivity of WLANs. The first approach includes access to

Wide-Area Networks (WANs) and Metropolitan Area Networks (MANs). In the wide area, the network transmission systems use the cellular arrangement and the wired long-distance network. The data is packetized to meet the immediate demands of the users' community. Data must be in a proper form and format to prevent excessive overhead and consequent latency in transport. The second approach deals with localized communications services for the added convenience of connections between building floors and desktops in a dynamic environment. Flexibility to provide quick connections for moves, additions, and changes gives the organization significant improvement over the basic wired LAN. The third approach is the flexible mobile LAN arrangement to access a company intranet. This form of connectivity is becoming important in all walks of life and business communities. As the workforce becomes more mobile, the need to provide untethered connectivity is increasing exponentially.

Two different technical approaches exist with the WLANs. These are based on radio and optical technologies. In the radio-based technology, there are two solutions: the licensed microwave radio frequency range (18–23 GHz) or the unlicensed radio frequency range (902–928 MHz, 2.4–2.4835 GHz, and 5.75–5.825 GHz). In the unlicensed radio frequency, there are two options. The first option uses Frequency Hopping Spread Spectrum (FHSS) technology, whereas the second option uses Direct Sequence Spread Spectrum (DSSS) technology. The 902–928 MHz frequency band is an unlicensed ISM band that allows manufacturers to supply products with very limited constraints. Newer products are also emerging that use the 2.4-GHz band. The following are the major limitations of the unlicensed frequency-band WLANs:

- The system is restricted to 100 mW of output.
- The system must not interfere with other radio-frequency equipment in the same area.
- The system must go through an FCC-type acceptance process (in the international sector, this is called homologation or type acceptance, and the frequencies may be different) using either 902–928 MHz, 2.4–2.4835 GHz, or 5.75–5.825GHz frequencies in various ISM bands.

14.4.2.1 Spread Spectrum Radio-Based WLANs

WLANs use spread spectrum techniques to allow flexibility and minimize interference while not being license bound. WLANs have been produced using FHSS and DSSS approaches with different speeds. The motivation to use spread spectrum for packet radio systems comes from improved multipath resistance, the ability to coexist with other systems, and the antijamming nature of the code. In an office environment, spread spectrum is a promising choice because it reduces the effects of multipath caused by reflections from the walls and increases the mobility of the terminals within the office environment. The low spectral power density per user of spread spectrum permits an overlay with certain existing systems and reduces the concerns about health-related issues in high-power transmission. Spread spectrum offers the potential for greater range and higher data rates compared with optical technology. It improves interception resistance and provides data privacy.

Table 14-2 provides a partial list of WLANs available in the United States, two of which we will describe briefly—AT&T WaveLAN, based on the FHSS technology, and Telesystem ARLAN, based on the DSSS technology [1].

- **AT&T WaveLAN.** This system supports speeds of up to 20 Mbps and works with various network operating systems. WaveLAN uses a DS Quadrature Phase Shift Keying (QPSK) multiplexing scheme to transmit across the entire broadband at higher signal rates. Through multiplication of the original narrowband signal with PN sequence, the code is spread across several frequencies. WaveLAN offers better security because the conventional radio receiver cannot decode the signal without knowing the actual spreading pattern. WaveLAN can operate up to 800 feet with a power output of 250 mW. It works in any laptop, notebook, or palmtop PC that is equipped for a Personal Computer Memory Card International Association (PCMCIA) card. WaveLAN allows users to operate in a cellular network for LANs. Each WaveLAN is assigned its own identification code and can receive data only if its code corresponds to that of the cell it occupies. Users can move anywhere within their assigned cell and still be able to communicate intracell. If users need to move between cells, they must first stop the application from running, then reconfigure their address ID to match with the cell they are moving into. With roaming this is automatic. WaveLAN is capable of interfacing directly with the backbone cable systems at standard LAN cable speeds.

- **Telesystems Advanced Radio LAN (ARLAN).** ARLAN uses DSSS technology. Using a conventional cable system, ARLAN devices called *access points* are attached to the cable to allow for a full range of interconnections. A microcell can be configured from the backbone network by setting an access point to act like a wireless repeater. Telesystems Micro-cellular Architecture (TMA) allows the network to cover various applications and various-sized facilities. With multiple base station antennas, the network can be extended to create microcells, each with its own operating area and devices. TMA is supported by firmware in each of the ARLAN devices. It supports multiple overlapping cells, creating a seamless network within the building. Handoff from cell to cell is a part of the network concept that allows for LAN connectivity of users who need to move freely throughout departments or floors within the building. Using SS technology, the system can select various center frequencies and allows for the coexistence of multiple devices operating within the same area but serving different needs. ARLAN 600 was designed for high-noise, industrial applications and uses a spreading ratio of up to 100. It offers a full range of interfaces for async and sync data transfer from terminals and hosts. The system operates in the 915-MHz and 2.4-GHz frequency ranges and uses packet burst duplex transmission capabilities. Access to the ARLAN network is packet-switched Carrier-Sense Multiple Access with Collision Avoidance (CSMA/CA) (see section 14.6.2.3 for details on CSMA). Power output for these devices is up to 1 W for distances of up to 500 feet diameter in an office environment and up to 3000 feet diameter in factories or open-plan offices indoors. For line-of-

Table 14-2 Partial List of WLAN Products

Product	Freq.	Link Rate	User Rate	Protocol	Access	No. of Chan. or Spread Factor	Mod/Coding	Power (mW)	Network Topol.
Altair Plus Motorola	18–19 GHz	15 Mbps	5.7 Mbps	Ethernet			4-level FSK[a]	25 peak	8 devices per radio
WaveLAN AT&T	902–928 MHz	2 Mbps	1.6 Mbps	Similar to Ethernet	DSSS		DQPSK[b]	250	peer to peer
AirLAN Solectek	902–928 MHz		2 Mbps	Ethernet	DSSS		DQPSK	250	PCMCIA with antenna
Freeport Windata	902–928 MHz	16 Mbps	5.7 Mbps	Ethernet	DSSS	32 chips per bit	16 PSK[c]/Trellis	650	hub
Intersect Persoft Inc.	902–928 MHz		2 Mbps	Ethernet; Token-ring	DSSS		DQPSK	250	hub
LAWN O'Neill Comm.	902–928 MHz		38.4 kbps	AX.25	SS	20 users per channel; max. 4 channels		20	peer to peer
WiLAN WiLAN Inc	902–928 MHz	20 Mbps	1.5 Mbps per channel	Ethernet; Token-ring	CDMA/TDMA	3 channels; 10–15 links each	Unconv.	30	peer to peer
Radio Port ALPS Electric	902–928 MHz		242 kbps	Ethernet	SS			100	peer to peer
ARLAN 600 Telesys.	902–928 2.4 GHz		1.35 Mbps	Ethernet	DSSS			100 Max.	PCs with antennas
Radio Link Cal. Microwave	902–928 2.4 GHz	250 kbps	64 kbps		FHSS	250 ms/hop 500-kHz space			hub
RangeLAN Proxim, Inc.	902–928 MHz		242 kbps	Ethernet; Token-ring	DSSS	3 channels		100	
RangeLAN 2 Proxim Inc.	2.4 GHz	1.6 Mbps	50 kbps	Ethernet; Token-ring	FHSS	10 channels @ 5kbs; 15 subchannels each		100	peer to peer bridge
Netwave Xircom	2.4 GHz	1 Mbps per adopter		Ethernet; Token-ring	FHSS	82 1-MHz channel or hops			hub
Freelink Cabletron System	2.4 and 5.8 GHz		5.7 Mbps	Ethernet	DSS	32 chips per bit	16 PSK trellis	100	100

[a] FSK = frequency-shift keying.
[b] DQPSK = differential quadratic phase-shift keying.
[c] PSK = phase-shift keying.

sight building-to-building communications, the system can achieve distances of 6 miles. With microcell architecture, each cell is capable of handling up to 1 Mbps. The ARLAN 655 and 670 are complete wireless network interface cards that are mounted inside a PC, workstation (WS), or other device. They provide the same functionality as a conventional LAN adapter card and can support multiple topologies in conjunction with the network operating systems.

14.5 WLAN Standards

All standards for WLANs employ unlicensed bands. There are two approaches that can be used to regulate an unlicensed band. One approach is based on a standard to allow different vendors to communicate with one another using a set of interoperable rules. This approach is taken by IEEE 802.11 and ETSI's RES 10, HIPERLAN. In the second approach, a minimum set of rules or "spectrum etiquette" is established to allow terminals designed by different vendors to have a fair share of the available channel frequency/time resources and to coexist in the same band. This approach does not preclude the first approach. The second approach has been pursued by WINForum. In a coexisting environment, a vendor can be interoperable with another vendor by using the same protocol and transmission scheme.

The three major standard activities for WLANs are IEEE 802.11, HIPERLAN, and WIN-Forum. IEEE 802.11 developed a standard for DSSS, FHSS, and infrared light technology using the ISM bands as the radio channel. The HIPERLAN standard is aimed at the 5.2- and 17.1-GHz bands in European countries. WINForum's goal is to obtain a PCS band for unlicensed data and voice applications and to develop spectrum etiquette for them.

14.5.1 IEEE 802.11

IEEE 802.11 addresses the physical and media access (MAC) protocol layers for peer-to-peer and peer-to-centralized communications topologies using DSSS or FHSS over radio or infrared light technology. Both SS systems operate in the 2.4–2.4835 GHz ISM band. This band has been selected over the 902–928 MHz and 5.725–5.85 GHz ISM bands because it is widely available in most countries. In the 2.4–2.4835 GHz band, more than 80 MHz of bandwidth is available that is suitable for high-speed data communication. Also, implementation in this band is more cost effective as compared with the implementation in the higher frequencies. IEEE 802.11 supports DSSS with Binary Phase-Shift Keying (BPSK) and Quadrature Phase-Shift Keying (QPSK) modulation for data rates of 1 and 2 Mbps, respectively, as well as FHSS with Gaussian Frequency-Shift Keying (GFSK) modulation and two hopping patterns with data rates of 1 and 2 Mbps. For DSSS the band is divided into five overlapping 26-MHz subbands centered at 2.412, 2.442, 2.470, 2.427, and 2.457 GHz, with the last two overlapping the first three. This set-up provides five orders of frequency selectivity for the user. It is quite cost effective in improving the transmission reliability in the presence of interference or severe frequency-selective multipath fading. For FHSS, the channel is divided into 79 subbands, each with a 1-MHz bandwidth, and three patterns of 22 hops are user options. A minimum hop rate of 2.5 hops/second is assigned to provide slow frequency hopping in which each packet is sent in one hop and, if it is destroyed, the following packet is sent from another hop for which the channel condition would be different. This approach provides a very effective time-frequency diversity and takes advantage of a retransmission scheme to provide a robust transmission. The IEEE 802.11 standard avoids rigid requirements and leaves room for vendors to maneuver in the following areas:

1. **Multiple physical media**—FHSS and DSSS radio, as well as infrared light; additional media as approved in the future.
2. **Common MAC layer regardless of physical layer**—all IEEE 802.11-compliant WLANs use CSMA/CA algorithm similar to Ethernet's Carrier-Sense Multiple Access with Collision Detection (CSMA/CD) MAC layer.
3. **Common frame format**—frames including headers and error protection fields are the same, regardless of whether the attached wired LAN is 802.3 Ethernet or 802.5 token ring; the access point handles conversion of 802.11 frames to wireline frame format.
4. **Multiple on-air data rates**—1 or 2 Mbps, with the possibility of higher rates in the future.
5. **Power limit**—a maximum power of 1 W (or +30 dBm), as mandated by the FCC; there is no minimum power requirement, which leaves open the possibility of low-power implementations.

The standard defines the basic media and configuration issues, transmission procedures, throughput requirements, and range characteristics for WLAN technology, focusing more on access applications that involve the use of personal digital assistants (PDAs) and portable PCs rather than trunk applications (see Figs. 14-1 and 14-2). Trunk applications use wireless as part of the enterprise backbone for transmitting data from building to building, whereas access applications allow users of portable PCs, PDAs, and other wireless devices to tap into corporate LANs from anywhere in an office or on a factory floor.

The radio transmitter in each user end-station is always listening for activity on the WLAN. If one end-station is transmitting, another will not. The system has a preset time-out to block a user from dominating the network, to avoid unnecessary transmission collisions, and to allow priority traffic through. This is the function of the CSMA/CA access control mechanism. Once it is determined that the network is free, the end-station ramps up to full power and sends a *preamble* (a standard signaling message) to the access point. The preamble is a repeated bit pattern followed by a special bit sequence. It allows the access point to lock onto the signal before the data is sent. After the link is established, the end-station sends address and protocol information. The header is followed by the data, which is transmitted at the on-air data rate. After the error-check word is sent, the end-station listens for acknowledgment from the destination. If no acknowledgment is received, the data is re-sent. The sequence is repeated until all the data has been sent and acknowledged.

The IEEE 802.11 committee has specified that data rates for wireless systems must be either 1 or 2 Mbps. Either the user chooses the rate, or the system selects the best one according to the conditions. The on-air data rate includes message headers, retransmissions, and latency

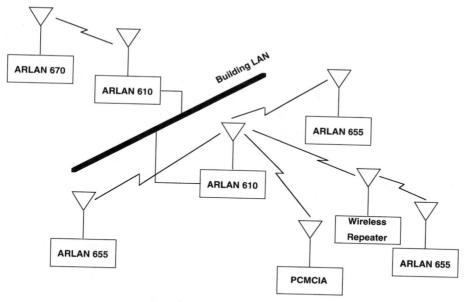

Figure 14-1 Access Application for WLAN

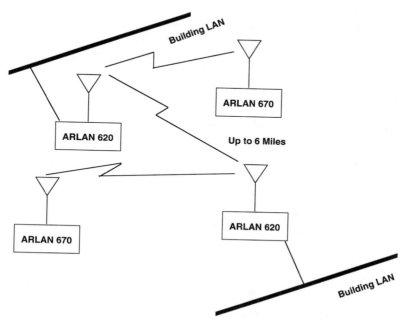

Figure 14-2 Access Application for WLAN between Two Buildings with Line of Sight

(the time between when a network station begins to seek access to a transmission channel and when that access is granted). Header overhead and retransmissions primarily affect performance of large data transfers, whereas latency has the greatest effect on short, bursty data transfers; this is because the latency involved in setting up a transmission introduces more delay than the transmission of message overhead or the retransmissions. Therefore, throughput on a WLAN is lower for short messages than for longer messages. The actual throughput of an IEEE 802.11 system on an on-air data rate of 2 Mbps is about 1.0–1.5 Mbps for long messages and 0.5–1.0 Mbps for short messages. Throughput is also affected by the range of the system. In a typical office environment, the range of an IEEE 802.11 WLAN is 200–300 feet, which is sufficient to cover most partitioned areas and an outside rim of walled offices.

Sensitivity of the system is crucial because signal power can be affected drastically by obstacles. Sensitivity figures are the smallest amounts of received power that the radio can use. The IEEE 802.11 standard requires a sensitivity of less than –80 dBm. One issue that is not addressed by the standard is roaming capability. Roaming is made possible with overlapping WLAN cells in a configuration similar to that used for analog cellular phones. Roaming is considered to be part of the application- or driver-level technology, so vendors will likely resort to different schemes to achieve it.

14.5.2 Wireless Information Networks Forum

The Wireless Information Networks Forum (WINForum) addresses WLAN and wireless Private Branch Exchange (WPBX) services and focuses on spectrum etiquette to provide fair access to an unlicensed band widely used for different applications and devices. The etiquette does not preclude any common air interface standards or access technologies. It demands Listen-Before-Talk (LBT); thus, a device may not transmit if the spectrum it will occupy is already in use within its range. The power is limited to keep the range short and allow operation in densely populated office areas. The power and connection time are related to the occupied bandwidth to equalize the interference and provide a fair access to frequency/time resources. In the view of WINForum, the asynchronous transmission used in WLAN applications is bursty, begins transmission within milliseconds, uses short bursts that contain large amounts of data, and releases the link quickly. On the other hand, the isochronous transmission, typified by voice services such as a WPBX, uses long holding time, periodic transmission, and flexible link access times that may be extended up to a second. The asynchronous subbands may range from 50 kHz to 10 MHz, whereas the isochronous subbands may be divided into 1.25-MHz segments. The two types are technically contrasting and cannot share the same spectrum.

14.5.3 High-Performance Radio Local Area Network

ETSI's subtechnical committee, RES 10, has been assigned the task of developing a standard for High-Performance Radio Local Area Network (HIPERLAN). The committee secured two bands at 5.12–5.30 GHz and 17.1–17.3 GHz for HIPERLAN to operate at a minimum useful bit rate of 20 Mbps for point-to-point applications with a range of 50 m. It is expected that, at this rate and range, a data rate of 500–1000 Mbps—comparable to FDDI—can be achieved for a standard building floor of approximately 1000 m square. RES 10 is responsible for defining a radio transmission technique, including type of modulation, coding, and channel access, as well as the specific protocols.

14.5.4 U.S. Advanced Research Project Agency

The U.S. Advanced Research Project Agency (ARPA) has sponsored WLAN projects at the University of California at Berkeley (UCB) and University of California at Los Angeles (UCLA). The UCB Infopad project is based on a coordinated network architecture with fixed coordinating nodes and DSSS (CDMA), whereas the UCLA project is for peer-to-peer networks and uses FHSS. Both ARPA-sponsored projects are concentrated on the 900-MHz ISM band.

14.6 Access Methods

14.6.1 Fixed-Assignment Access Methods

In the fixed-assignment access method, a fixed allocation of channel resources (frequency or time, or both) is made on a predetermined basis to a single user. The three basic access methods—FDMA, TDMA, and CDMA—are examples of the fixed-assignment access method. In this section, we discuss only the CDMA method. With CDMA, multiple users operate simultaneously over the entire bandwidth of the frequency/time signal domain, and the signals are kept separate by their distinct user-signal codes. As we discussed in chapter 2, the number of the users that can be supported simultaneously by a DS-CDMA system is

$$ M = \left[\frac{G_p}{E_b/N_0} \right] \times \frac{1}{1+f} \times \eta_c \times \frac{1}{v_f} \times \alpha \qquad (14.1) $$

where G_p = processing gain = B_w/R_s,
 B_w = bandwidth,
 R = information rate,
 E_b/N_0 = bit energy-to-noise density,
 f = interference from other cells,
 η_c = power control factor,
 v_f = voice activity factor (= 1 for data service), and
 α = gain due to sector antenna.

EXAMPLE 14.1

Consider a CDMA system that uses QPSK modulation and convolutional coding. The system has a bandwidth of 1.25 MHz and transmits data at 9.6 kbps. Find the number of users that can be supported by the system and the bandwidth efficiency. Assume a three-sector antenna with an effective gain = 2.6, η_c = 0.9, and an interference factor f = 0.5. A BER of 10^{-3} is required.

$$ G_p = \frac{1.25 \times 10^6}{9.6 \times 10^3} = 130.2 $$

$$P_b = 10^{-3} = \frac{1}{2}\text{erfc}\sqrt{\frac{E_b}{N_0}}$$

$$\frac{E_b}{N_0} \approx 7 \text{ dB (5)}$$

$$M = \frac{130.2}{5} \times \frac{1}{1+0.5} \times 2.6 \times 0.9 = 40.6 \approx 40$$

$$\eta_{bw} = \frac{40 \times 9.6}{1.25 \times 10^3} = 0.307 \text{ bit/sec/Hz}$$

EXAMPLE 14.2

Consider a QPSK/DSSS WLAN that is designed to transmit in the 902–928 MHz ISM band. The symbol transmission rate is 0.1 mega symbols per second (Msps). An orthogonal code with 4 symbols is used. A BER of 10^{-5} is required. How many users can be supported by the WLAN? A three-sector antenna with a gain = 2.6 is used. Assume an interference factor $f = 0.5$ to account for the interference from users in other cells and $\eta_c = 0.9$. What is the bandwidth efficiency of the system?

$$\text{Band width } B_w = 928-902 = 26 \text{ MHz}$$

$$\text{Data rate } R = R_s\log_2 4 = 0.1\log_2 2^2 = 0.2 \text{ Mbps}$$

$$G_p = \frac{B_w}{R} = \frac{26}{0.2} = 130$$

$$P_b = 10^{-5} = \frac{1}{2}\text{erfc}\sqrt{\frac{E_b}{N_0}}$$

$$\frac{E_b}{N_0} \approx 10 \text{ dB (10)}$$

$$M = \frac{130}{10} \times \frac{1}{1+0.5} \times 2.6 \times 0.9 = 20.3 \approx 20$$

$$\eta_{bw} = \frac{20 \times 0.2}{26} = 0.154 \text{ bit/sec/Hz}$$

14.6.2 Random-Access Methods

When each user has a steady flow of information to transmit (for example, a data file transfer or a facsimile transmission), fixed-assignment access methods are useful because they make efficient use of communication resources. However, when the information to be transmitted is bursty in nature, fixed-assignment access methods result in a waste of communication resources. Furthermore, in a cellular system where subscribers are charged based on channel connection time, fixed-assignment access methods may be too expensive to use for transmitting short messages. Random-access protocols provide flexible and efficient methods for managing channel access when transmitting short messages. Random-access methods give each user freedom to gain access to the network whenever the user has information to send. Because of this freedom, these schemes result in contention among users trying to access the network simultaneously. Contention may cause collisions resulting in the need to retransmit the information. The commonly used random-access protocols are pure ALOHA, slotted ALOHA, and CSMA/CD. The following sections briefly describe details of each of these protocols and provide necessary throughput expressions.

14.6.2.1 Pure ALOHA

In the pure ALOHA scheme, users transmit information whenever they have information to send. A user sends information in packets. After sending a packet, the user waits a length of time equal to the round-trip delay for an acknowledgment (ACK) of the packet from the receiver. If no ACK is received, the packet is assumed to be lost in a collision, and it is retransmitted with a randomly selected delay to avoid repeated collisions.* The normalized throughput S (average packet arrival rate divided by the maximum throughput) of the pure ALOHA protocol is given as

$$S = Ge^{-2G}$$

$$(14.2)$$

where G = normalized offered traffic load.

Note from Eq. (14.2) that the maximum throughput occurs at traffic load $G = 50\%$ and is $S = 1/2e$. This is about 0.184. Thus, the best channel utilization with pure ALOHA protocol is only 18.4%.

14.6.2.2 Slotted ALOHA

In the slotted ALOHA system, the transmission time is divided into time slots. Each time slot is made exactly equal to the packet transmission time. Users are synchronized to the time slots so that, whenever a user has a packet to send, the packet is held and transmitted in the next time slot. With the synchronized time slots scheme, the interval of a possible collision for any packet is reduced to one packet time from two packet times, as in the pure ALOHA scheme. The normalized throughput S for the slotted ALOHA protocol is given as

$$S = Ge^{-G}$$

$$(14.3)$$

where G = normalized offered traffic load.

The maximum throughput for slotted ALOHA occurs at $G = 1.0$ (Eq. [14.3]), and it is equal to $1/e$ or about 0.368. This implies that, at the maximum throughput, 36.8% of the time slots carry the successfully transmitted packets, whereas the same percentage of the time slots remain empty.

---

* Note that the protocol on CDMA access channels as implemented in TIA IS-95A is based upon the pure ALOHA approach. The mobile station randomizes its attempt for sending a message on the access channel and may retry if an acknowledgment is not received from the base station. For further details, reference Section 6.6.3.1.1.1 of TIA IS-95A.

14.6.2.3   Carrier-Sense Multiple Access

CSMA protocols have been widely used in both wired and wireless LANs. These protocols provide enhancements over pure and slotted ALOHA protocols. The enhancements are achieved through use of additional capability at each user station to sense the transmissions of other user stations. The carrier-sense information is used to minimize the length of collision intervals. For carrier sensing to be effective, propagation delays must be less than packet transmission times. There are two general classes of CSMA protocols: nonpersistent and p-persistent. Each of these classes can be used with the slotted or unslotted operation.

- **Nonpersistent CSMA.** A user station does not sense the channel continuously while it is busy. Instead, after sensing the busy condition, it waits for a randomly selected interval of time before sensing again. The algorithm works as follows: if the channel is found to be idle, the packet is transmitted; if the channel is sensed to be busy, the user station backs off to reschedule the packet to a later time. After backing off, the channel is sensed again, and the algorithm is repeated.

- **p-persistent CSMA.** The slot length is typically selected to be the maximum propagation delay. When a station has information to transmit, it senses the channel. If the channel is found to be idle, it transmits with probability p. With probability $q = 1 - p$, the user station postpones its action to the next slot, where it senses the channel again. If that slot is idle, the station transmits with probability p or postpones again with probability q. The procedure is repeated until either the frame has been transmitted or the channel is found to be busy. When the channel is detected busy, the station then senses the channel continuously and, when it becomes free, it starts the above procedure again. If the station initially senses the channel to be busy, it simply waits one slot and applies the above procedure.

  - 1-persistent CSMA: 1-persistent CSMA is the simplest form of p-persistent CSMA. It signifies the transmission strategy, which is to transmit with probability 1 as soon as the channel becomes idle. After sending the packet, the user station waits for an ACK, and, if it is not received within a specified amount of time, the user station waits for a random amount of time and then resumes listening to the channel. When the channel is again found to be idle, the packet is retransmitted immediately.

For more details on CSMA, refer to references [3, 21].

The throughput expressions for the CSMA protocols are as follows:

- Unslotted nonpersistent CSMA

$$S = \frac{Ge^{-aG}}{G(1 + 2a) + e^{-aG}} \tag{14.4}$$

- Slotted nonpersistent CSMA

$$S = \frac{aGe^{-aG}}{1 - e^{-aG} + a} \qquad (14.5)$$

- Unslotted 1-persistent CSMA

$$S = \frac{G[1 + G + aG(1 + G + (aG)/2)]e^{-G(1+2a)}}{G(1+2a) - (1 - e^{-aG}) + (1 + aG)e^{-G(1+a)}} \qquad (14.6)$$

- Slotted 1-persistent CSMA

$$S = \frac{Ge^{-G(1+a)}[1 + a - e^{-aG}]}{(1+a)(1 - e^{-aG}) + ae^{-G(1+a)}} \qquad (14.7)$$

where S = normalized throughput,
G = normalized offered traffic load,
a = $\dfrac{\tau}{T_p}$,
τ = propagation delay, and
T_p = packet transmission time.

EXAMPLE 14.3

Consider a WLAN installation in which the maximum propagation delay is 0.4 μsec. The WLAN operates at a data rate of 10 Mbps, and packets have 400 bits. Calculate the throughput with: (1) an unslotted nonpersistent, (2) a slotted persistent, and (3) a slotted 1-persistent CSMA protocol.

$$T_p = \frac{400}{10} = 40 \text{ μsec}$$

$$a = \frac{\tau}{T_p} = \frac{0.4}{40} = 0.01$$

$$G = \frac{40 \times 10^{-6} \times 10 \times 10^6}{400} = 1$$

- Unslotted nonpersistent

$$S = \frac{1 \times e^{-0.01}}{(1 + 0.02) + e^{-0.01}} = 0.493$$

- Slotted nonpersistent

$$S = \frac{0.01 \times 1e^{-0.01}}{1 - e^{-0.01} + 0.01} = 0.496$$

- Slotted 1-persistent

$$S = \frac{e^{-1.01}(1 + 0.01 - e^{-0.01})}{(1 + 0.01)(1 - e^{-0.01}) + 0.01e^{-1.01}} = 0.531$$

14.7 Error Control Schemes

Channel coding and Automatic Repeat Request (ARQ) schemes are used to increase the performance of mobile communication systems. In the physical layer of the DS-CDMA system, error detection and correction techniques such as Forward Error Correction (FEC) schemes are used. For some of the data services, higher-layer protocols use ARQ schemes to enable retransmission of any data frames in which an error is detected. The ARQ schemes are classified as follows [21,22]:

- **Stop and Wait.** The sender transmits the first packet numbered 0 after storing a copy of that packet. The sender then waits for an ACK numbered 0 (ACK0) of that packet. If the ACK0 does not arrive before a time-out, the sender makes another copy of the first packet, also numbered 0, and transmits it. If the ACK0 arrives before a time-out, the sender discards the copy of the first packet and is ready to transmit the next packet, which it numbers 1. The sender repeats the previous steps, with numbers 0 and 1 interchanged. The advantages of the Stop and Wait protocol are its simplicity and its small buffer requirements. The sender needs to keep only a copy of the last packet transmitted, and the receiver does not need to buffer packets at the data link layer. The main disadvantage of the Stop and Wait protocol is that it does not use the communication link very efficiently and it is slow.

  The total time taken to transmit a packet and to prepare for transmitting the next one is

$$T = T_p + 2T_{\text{prop}} + 2T_{\text{proc}} + T_a \tag{14.8}$$

The protocol efficiency without error is

$$\eta(0) = \frac{T_p}{T} \tag{14.9}$$

where T = total time for transmitting a packet,
T_p = transmission time for a packet,
T_{prop} = propagation time of a packet or an ACK,
T_{proc} = processing time for a packet or an ACK, and
T_a = transmission time for an ACK.

If p is the probability that a packet or its ACK is corrupted by transmission errors, and a successful transmission of a packet and its ACK takes T seconds and occurs with probability $1 - p$, the protocol efficiencies for full-duplex (FD) and half-duplex (HD) operation are given as

$$\eta_{\text{FD}} = \frac{(1-p)T_p}{(1-p)T + pT_p} \tag{14.10}$$

$$\eta_{HD} = \frac{(1-p)T_p}{T} \tag{14.11}$$

- **Selective Repeat Protocol (SRP).** The data link layer in the receiver delivers exactly one copy of every packet in the correct order. The data link layer in the receiver may get the packets in the wrong order from the physical layer. This occurs, for example, when transmission errors corrupt the first packet and not the second one. The second packet arrives correctly at the receiver before the first packet gets there. The data link layer in the receiver uses a buffer to store the packets that arrive out of order. Once the data link layer in the receiver has a consecutive group of packets in its buffer, it can deliver them to the network layer. The sender also uses a buffer to store copies of the unacknowledged packets. The number of the packets that can be held in the sender/receiver buffer is a design parameter. Let W = the number of packets that the sender and receiver buffers can each hold and SRP = number of packets modulo-2 W. The protocol efficiency without any error and with an error probability of p is given as

$$\eta(0) = \min\left\{\frac{WT_p}{T}, 1\right\} \tag{14.12}$$

For very large W, the protocol efficiency is

$$\eta(p) = 1 - p \tag{14.13}$$

where T = time-out = WT_p

$$\eta(p) = \frac{2 + p(W-1)}{2 + p(3W-1)} \tag{14.14}$$

SRP is very efficient, but it requires buffering packets at both the sender and the receiver.

- **Go-Back-N (GBN).** The Go-Back-N protocol allows the sender to have multiple unacknowledged packets without the receiver having to store packets. This is done by not allowing the receiver to accept packets that are out of order. When a time-out timer expires for a packet, the transmitter resends that packet and all subsequent packets. The Go-Back-N protocol improves on the efficiency of the Stop and Wait protocol, but is less efficient than SRP. The protocol efficiency is given as

$$\eta_{FD} = \frac{1}{1 + \left(\dfrac{p}{1-p}\right)W} \tag{14.15}$$

- **Window-Control Operation Based on Reception Memory (WORM) ARQ.** In digital cellular systems, bursty errors occur due to multipath fading, shadowing, and handoffs. The BER fluctuates from 10^{-1} to 10^{-6}. Therefore, the conventional ARQ schemes do not operate well in a digital cellular system. WORM ARQ has been suggested for control of dynamic error characteristics. It is a hybrid scheme that combines SRP with GBN. GBN protocol is chosen in severe error conditions, whereas SRP is selected in normal error conditions.

- **Variable Window and Frame Size GBN and SRP.** Since CDMA systems have bursty error characteristics, the error control schemes should have a dynamic adaptation to a bursty channel environment. SRP and GBN with variable window and frame size have been proposed in [22] to improve error control in CDMA systems. Table 14-3 provides the window and frame sizes for different bit error rates. If the error rate increases, the window and frame size are decreased. If the error rate is small, the window and frame size are increased. The optimum threshold values of BER and window and frame size were obtained through computer simulation.

In CDMA systems, the forward link consists of pilot, sync, paging, and traffic channels. System information sent on the sync and paging channels allows each mobile station to evaluate the BER easily by measuring the ratio of the number of retransmitted frames to the number of transmitted frames over a 2-second period. Thus, the mobile station can change the window and frame sizes according to the BER.

Table 14-3   Bit Error Rate vs. Window and Frame Size

BER	Window Size (W)	Frame Size (bits)
$BER \leq 10^{-4}$	32	172
$10^{-4} < BER < 10^{-3}$	8	80
$10^{-3} < BER < 10^{-2}$	4	40
$10^{-2} < BER$	2	16

EXAMPLE   14.4

Consider a WLAN in which the maximum propagation delay is 4 μsec. The WLAN operates at a data rate of 10 Mbps. The data and ACK packet are of 400 and 20 bits, respectively. The processing time for a data or ACK packet is 1 μsec. If the probability $p = 0.01$ that a data packet or its ACK can be corrupted during transmission, find the data link protocol efficiency with (1) Stop and Wait protocol, full duplex, (2) SRP with window size $W = 8$, and (3) Go-Back-N protocol with window size $W = 8$.

$$T_p = \frac{400}{10} = 40 \, \mu sec$$

$$T_a = \frac{20}{10} = 2 \, \mu sec$$

$$T_{prop} = 4 \, \mu sec$$

$$T_{proc} = 1 \, \mu sec$$

Stop and Wait:

$$T = 40 + 2 \times 4 + 2 \times 1 + 2 = 52 \, \mu sec$$

$$\eta = \frac{(1 - 0.01) \times 40}{52} = 0.762$$

SRP:

$$\eta = \frac{2 + 0.01(8 - 1)}{2 + 0.01(24 - 1)} = 0.954$$

Go-Back-N:

$$\eta = \frac{1}{1 + 8\left(\frac{0.01}{1 - 0.01}\right)} = 0.925$$

14.8 Data Services in IS-95

The data services in IS-95 include circuit-switched async data, circuit-switched digital fax, packet data, and analog fax. The current data standards are the following:

- **IS-658**

    - Defines the interface between the MSC and an external data interworking function (IWF)
    - Applies to circuit-switched async data, digital fax, and packet data services

- **Medium Data Rates (MDR)**

    - Included as part of IS-95B standardization

- **IS-707**

    - *RS1 Service Options (SOs)*
        SO 4100: Async Data
        SO 4101: Digital fax
        SO 4103: IP/Mobile IP
        SO 4104: CDPD

    - *RS2 SOs*
        SO 12: Async Data
        SO 13: Digital fax
        SO 15: IP/Mobile IP
        SO 16: CDPD

The cdmaOne[*] (family of IS-95 standards) packet data technology uses a TCP/IP-compliant CDPD protocol stack to provide seamless connectivity with enterprise networks and expedite third-party application development. Adding data to the cdmaOne network allows an operator to

---

[*] Trade name used by CDMA Users Group.

continue using its existing radios, backhaul facilities, infrastructure, and handsets while merely implementing a software upgrade with an IWF.

TIA IS-95B allows for code or channel aggregation to provide data rates of 64–115 kbps, as well as offering improvements in soft handoffs and interfrequency hard handoffs. To achieve a 114-kbps rate, up to eight CDMA traffic channels, each offering 14.4 kbps, are needed to be aggregated.

14.9 Asynchronous Data and Group-3 Facsimile

The general approach taken in TIA IS-95A [14] for data services reuses the previously specified physical layer of the IS-95A protocol stack as the physical layer. Fig. 14-3 shows the air interface (U_m) protocol stack.

IS-95A asynchronous data has been structured as a circuit-switched service, in which a dedicated path is established between the data devices for the duration of the call. It is used for connectivity through the PSTN when point-to-point communication to a PC or fax user is required. For example, for a file transfer involving PC-to-PC communications, asynchronous data service is the preferred cellular service mode.

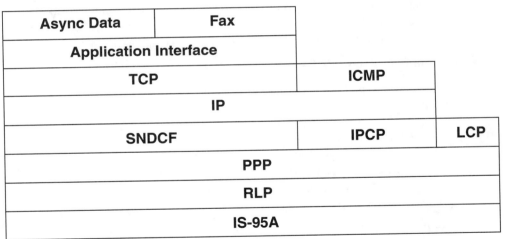

TCP:	Transmission Control Protocol
ICMP:	Internet Control Message Protocol
IP:	Internet Protocol
SNDCF:	Subnetwork Dependent Convergence Function
IPCP:	Internet Protocol Control Protocol
LCP:	Link Control Protocol
PPP:	Point-to-Point Protocol
RLP:	Radio Link Protocol

Figure 14-3 The U_m Protocol Stack

The Radio Link Protocol (RLP) employs ARQ, Forward Error Correction (FEC), and flow control. Flow control and retransmission of data blocks with errors are used to provide improved performance in the mobile segment of the data connection at the expense of variations in throughput and delay. Typical raw channel data error rates for cellular transmission are approximately 10^{-2}. However, an acceptable data transmission usually requires a BER of about 10^{-6}. In order to achieve this, it requires the design of efficient ARQ and error correction codes to deal with error characteristics in the mobile environment.

The CDMA protocol stack for data and facsimile (Figure 14-3) has the following layers:

- **Application Interface Layer.** This layer includes an application interface between the data source/destination in the mobile terminal (MT0) or terminal equipment (TE2) and the transport protocol layer. In the base station, the application interface resides between the data source/destination on the network (A_i interface) side and the transport protocol layer. The application interface provides: modem control, AT command processing,[*] negotiation of air interface data compression, and data compression over the air interface (optional).

- **Transport Layer.** The transport layer for CDMA asynchronous data and fax services is based on Internet transport layer protocol known as Transmission Control Protocol (TCP) [7]. The implementation complies with the requirements for TCP with modifications as described in IS-95 [14]. If the modified procedure is disabled, there is no maximum number of retransmission attempts during synchronization, and an established TCP connection remains open until explicitly closed by the mobile station or base station. The application interface sets the value of R_2 in the protocol. The base station follows either the procedure of the Internet Control Message Protocol (ICMP) [6] or the procedure given above for TCP.

- **Network Layer.** The network layer for CDMA async data and fax services is based on Internet network layer protocol known as the Internet Protocol (IP) [5]. The network layer includes the ICMP [6]. The implementation complies with the requirements of the IP [5] and requirements for Internet hosts [8] with modifications as described in IS-95 [14]. The interface between the network and transport layer complies with the requirements of the ICMP.

- **Sub-Network Dependent Convergence Function (SNDCF).** The SNDCF performs header compression on the headers of the transport and network layers. This function is negotiated using Point-to-Point Protocol (PPP) and Internet Protocol Control Protocol (IPCP) [10]. Mobile stations support Van Jacobson TCP/IP header compression. A minimum of one compression slot is negotiated. Base stations support TCP/IP header compression compatible with that required for mobile stations. Negotiation of the

---

* The AT commands were originally defined by Hayes Microcomputer Company for its wireline modems. The command set has now been adopted by most wireline and wireless modems. The name AT is derived from the use of an "AT" to preface all commands to the modem.

parameters of header compression is carried out using IPCP. The SNDCF sublayer accepts the network-layer datagram from the network layer, performs header compression as required, and passes the datagram to the PPP layer, indicating the appropriate PPP protocol identifier. The SNDCF sublayer receives network-layer datagrams with compressed or uncompressed headers from the PPP layer, decompresses the datagram header as necessary, and passes the datagram to the network layer.

- **Data Link Layer.** This layer uses PPP [11]. The PPP Link Control Protocol (LCP) is used for initial link establishment and for the negotiation of optional link capabilities. The data link layer uses the PPP IPCP to negotiate IP addresses and TCP/IP header compression. The data link layer accepts network-layer datagrams from the SNDCF and encapsulates them in the PPP information field. The packet is framed using the octet synchronous framing protocol, except that there is no interframe fill. No flag octets are sent between a flag octet that ends one PPP frame and the flag octet that begins the subsequent PPP frame. The framed PPP packets are passed to the RLP layer for transmission. The data link layer accepts received octets from the RLP layer and reassembles the original PPP packets. The PPP process discards any PPP packet for which the received Frame Check Sequence (FCS) is not equal to the computed value.

- **Internet Protocol Control Protocol Sublayer.** This sublayer supports negotiation of the IP-address (type = 3) and IP-compression (type = 2) protocol parameters. IPCP negotiates a temporary IP address for the mobile station whenever a transport layer connection is actively opened. Mobile stations maintain the temporary IP address only while a transport layer connection is open or being opened, and discards the temporary IP address when the transport layer connection is closed.

- **Link Control Protocol.** If the protocol identifier is 0xC021, the PPP layer processes the packet according to PPP LCP. For other supported protocol identifiers, the PPP layer removes the PPP encapsulation and passes the datagram and protocol identifier to the SNDCF. For unsupported protocol identifiers, the LCP Protocol-Reject is passed to the RLP layer for transmission. The mobile station supports the PPP LCP Configure-Request, Configure-ACK, Configure-Negative Acknowledgment (NAK), Configure-Reject, Terminate-Request, Terminate-ACK, Code-Reject, and Protocol-Reject. Other LCP packet types may also be supported. The PPP LCP negotiates the following configuration options:

    1. Async control character map: The mobile station does not require any mapping of control characters. The base station may negotiate mapping of control characters.

    2. Protocol field compression: this applies when the protocol number is less than 0xFF.

    3. Address and control field compression: this applies when the protocol number is not 0xC021.

The mobile station may also support other configuration options (such as maximum receive unit, authentication protocol, link quality protocol, or magic number). When an option is received that is not supported, Configure-Reject is sent as an indication to the peer.

- **Radio Link Protocol Layer.** The RLP layer provides an octet stream service over the forward and reverse traffic channels and substantially reduces the error rate typically exhibited by these channels. This service is used to carry the variable-length data packets on the PPP layer. The RLP divides the PPP packets into TIA IS-95A traffic channel frames for transmission. There is no direct relationship between PPP packets and traffic channel frames. A large packet may span multiple traffic channel frames, or a single traffic channel frame may contain all or part of several small PPP packets. The RLP is unaware of higher-layer framing; it operates on a featureless octet stream, delivering the octets in the order received from the PPP layer. For service options supporting an interface with multiplex option 1, RLP frames may be transported as primary or secondary traffic or as signaling traffic via data burst messages. For the primary or secondary traffic, the RLP generates and supplies exactly one frame to the multiplex sublayer every 20 ms. The frame contains the service option information bits. The multiplex sublayer in the mobile station categorizes every received traffic frame and supplies the frame type and accompanying bits, if any, to the RLP layer. The frame type and frame category for primary and secondary traffic are given in Tables 14-4 and 14-5. A blank frame is used for blank-and-burst transmission of signaling traffic.

  The signaling subchannel may carry frames from multiple RLPs, with each RLP having a distinct BURST_TYPE. Each service option defines a unique BURST_TYPE used for RLP. The primary and secondary multiplex subchannels each carry at most a single RLP layer. RLP data frames sent on one multiplex subchannel are not to be transmitted on another subchannel. RLP frames are not sent on the access and paging channels.

- **Radio Interface.** The mobile station and base station support the physical layer, multiplex sublayer, radio link management, and call control as defined in TIA IS-95A. They use service option 4 for async data services and service option 5 for Group-3 fax

Table 14-4   RLP Frame with Primary Traffic

RLP Frame Type	Bits/Frame	Multiplex Option 1 Frame Categories
Full rate	171	1
Half rate	80	2, 6, 11
Erasure	0	all others
1/8 rate	16	4, 8, 13
Blank	0	5, 14
Erasure	0	all others

Table 14-5   RLP Frame with Secondary Traffic

RLP Frame Type	Bits/Frame	Multiplex Option 1 Frame Categories
Rate 1	168	14
Rate 7/8	152	13
Rate 3/4	128	12
Rate 1/2	88	11
Blank	0	1–8
Erasure	0	9, 10

services, but they do not transmit 1/4-rate frames when service option 4 or service option 5 is active. Service options 4 and 5 support an interface with multiplex option 1. RLP frames for service options 4 and 5 are transported only as primary traffic or signaling traffic. The mobile station and the base station perform service option negotiation for service options 4 and 5 as described in TIA IS-95A (sections 6.6.4.1.2 and 7.6.4.1.2). Initialization and connection in the mobile station and the base station to accept service option 4 or service option 5 in response to a Service Option Request order is performed according to the specifications in TIA IS-99 (sections 3.8.4.1 and 3.8.4.2).

14.10 Short Message Service

The Short Message Service (SMS) [17] allows the exchange of short alphanumeric messages between a mobile station and the cellular system and between the cellular system and an external device capable of transmitting and optionally receiving short messages. The external device may be a voice telephone, a data terminal, or a short message entry system. The SMS consists of message entry features, administration features, and message transmission capabilities. These features are distributed between a cellular system and the SMS Message Center (MC) which together make up the SMS system. The MC may be either separate from or physically integrated into the cellular system.

Short message entry features are provided through interfaces to the MC and the mobile station. Senders use these interfaces to enter short messages, intended destination addresses, and various delivery options. MC interfaces may include features such as audio response prompts and Dual-Tone Multifrequency (DTMF) reception for dial-in access from voice telephones, as well as appropriate menus and message entry protocols for dial-in or dedicated data terminal access. Mobile station interfaces may include keyboard and display features to support message entry. Also, a cellular voice service subscriber can use normal voice or data features of the mobile station to call an SMS system to enter a message.

An SMS teleservice can provide the option of specifying priority level, future delivery time, message expiration interval, or one or more of a series of short, predefined messages. If supported by the teleservice, the sender can request acknowledgment that the message was received by the mobile station. An SMS recipient, after receiving a short message, can manually acknowledge the message. Optionally, the recipient can specify one of a number of predefined messages to be returned with acknowledgment to the sender.

SMS administration features include message storage, profile editing, verification of receipt, and status inquiry capabilities. The SMS transmission capabilities provide for the transmission of short messages to or from an intended mobile station as well as return of acknowledgments and error messages. These messages and acknowledgments are transmitted to or from the mobile station whether it is idle or engaged in a voice or data call. The cellular service provider may offer SMS transmission to its cellular voice and data customers only or may provide an SMS-only service without additional voice or data transmission capability. All available mobile stations on a CDMA paging channel can receive a broadcast message. A broadcast message is not acknowledged by the mobile station. Broadcast messaging services may be made available to mobile stations on CDMA paging channels as well as to mobile stations in a call on a CDMA traffic channel.

Fig. 14-4 shows the network reference model for SMS. The base station (BS) contains the transceiver equipment, the MSC, and any IWF required for network connection. These elements are grouped together because there is no need to distinguish them. The MC element in the model represents a generic SMS MC function. The N reference point represents one or more standardized interfaces between an SMS MC and a BS. The TE is voice or data equipment connected either directly or indirectly to the MC. It is possible for the MC to be included in or co-located with a BS. In this case, the N interface is internal to the BS.

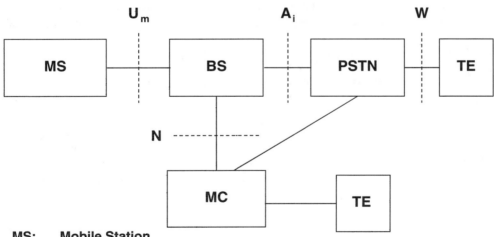

MS: **Mobile Station**
BS: **Base Station**
MC: **Message Center**
PSTN: **Public Switching Telephone Network**
TE: **Terminal Equipment**

Figure 14-4 Simplified SMS Reference Model

The SMS protocol stack for the CDMA mode of operation is shown in Fig. 14-5. The SMS bearer service is the portion of the SMS system responsible for delivery of messages between the MC and mobile user equipment. The bearer service is provided by the SMS transport and relay layers.

The SMS transport layer is the highest layer of the bearer service protocol. The transport layer manages the end-to-end delivery of messages. In an entity serving as a relay point, the transport layer is responsible for receiving SMS transport layer messages from an underlying SMS relay layer, interpreting the destination address and other routing information, and forwarding the message via an underlying SMS relay layer. In entities serving as end points, the transport layer provides the interface between the SMS bearer service and SMS teleservice.

SMS uses the following layers:

- **SMS Relay Layer.** The SMS relay layer provides the interface between the transport layer and the link layer used to carry short message traffic. On the U_m interface, the SMS relay layer supports the SMS transport layer by providing the interface to the TIA IS-95A transmission protocols required to carry SMS data between CDMA mobile stations and the base stations. On the N interface, the SMS relay layer supports the SMS transport layer by providing the interface to the network protocols required to carry SMS data between the MC and TIA IS-95A base stations. The N reference point is assumed to be an intersystem network link with connectivity to the MC. Intersystem links can use a variety of public and private protocols. SMS protocols and message formats on intersystem links may differ from those used on the CDMA air interface. The N interface relay layer is responsible for formatting and parsing SMS

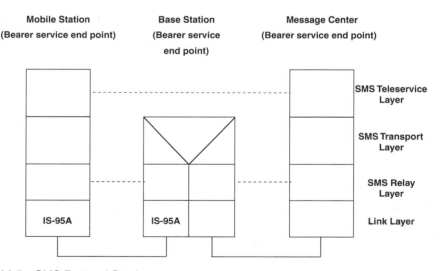

Figure 14-5 SMS Protocol Stack

messages as necessary when transmitting and receiving messages on the intersystem links. The SMS relay layer performs the following functions:

1. Accepting transport layer messages and delivering them to the next indicated relay point or end point
2. Providing error indications to the transport layer when messages cannot be delivered to the next relay point or end point
3. Receiving messages and forwarding them to the transport layer
4. Interfacing to and controlling the link layer used for message relay
5. Formatting messages according to SMS standards and/or other message standards, as required by the link layer and/or peer SMS layer

- **SMS Transport Layer.** This layer resides in SMS bearer service end points and relay points. In a bearer service end point, the SMS transport layer provides the means of access to the SMS system for teleservices that generate or receive SMS messages. In a bearer service relay point, the transport layer provides an interface between relay layers. The SMS transport layer uses relay layer services to originate, forward, and terminate SMS messages sent between mobile stations and MCs. It is assumed that the link layers used by the relay layers support message addressing, so that certain address parameters can be inferred by the relay layer from link layer headers and are therefore not required in transport layer messages. It is assumed that an SMS point-to-point message does not require certain address parameters because the link layers will provide this address. On the CDMA paging channel, for example, it can be assumed that the relay layer can extract the address from the ADDRESS field of the TIA-IS-95A data burst message. SMS transport layers have different functions in SMS bearer service end points and relay points. In an SMS bearer service end point, the transport layer provides the following functions:

1. Receiving message parameters from SMS teleservices, formatting SMS transport layer messages, and passing the message to the relay layer using the appropriate relay layer service primitives
2. Informing relay layer when all expected acknowledgments of submitted messages have been received
3. Informing the teleservices when relay layer errors are reported
4. Receiving SMS messages from the relay layer and passing the messages to the SMS teleservice
5. In mobile stations, performing authentication calculations

In an SMS bearer service relay point, the transport layer provides the following functions:

1. Receiving SMS messages from a relay layer, reformatting the SMS transport layer message if necessary, and passing the message to another relay layer using the appropriate relay layer service primitives

2. Passing confirmations or error reports between the relay layers if requested

3. In the TIA IS-95A base stations, performing authentication calculations or interfacing with the entities performing authentication calculations

The transport layer requires the following services from the relay layer:

1. Accepting transport layer messages and delivering them to the next indicated relay point or end point

2. Returning confirmations or error reports for messages sent

3. Receiving messages and forwarding them to the transport layer with the appropriate parameters

- **SMS Teleservice Layer.** The teleservice layer resides in a bearer service end point and supports basic SMS functions through a standard set of subparameters of the transport layer's bearer data parameter. When a mobile station sends an SMS User Acknowledgment message, the teleservice layer performs the following:

    1. The teleservice layer supplies the destination address parameter to the transport layer and sets the destination address parameter equal to the address contained in the originating address field of the SMS message being acknowledged.

    2. The MESSAGE_ID field of the message identifier subparameters is set to the value of MESSAGE_ID field in the SMS message being acknowledged.

Broadcast Messaging Service Teleservice messages are sent using the SMS Deliver message. For more details refer to TIA IS-637.

14.11 Packet Data Services for CDMA Cellular/PCS Systems [18]

14.11.1 Network Reference Model

The network reference model for packet data services and protocol options specified in TIA IS-657 is shown in Fig. 14-6.

14.11.2 Network Element

The reference model elements are

- **Terminal Equipment 2 (TE2).** A TE2 is a data terminal device that has a non-ISDN user-network interface.

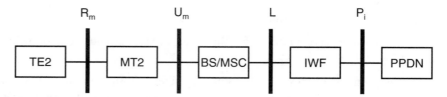

Figure 14-6 Network Reference Model

- **Mobile Terminal 0 (MT0).** An MT0 is a self-contained mobile terminal that does not support an external interface.
- **Mobile Terminal 2 (MT2).** An MT2 provides a non-ISDN (R_m) user interface.
- **Base Station (BS).** A base station represents the equipment on the land side of the U_m interface, including radio processing and management and protocol processing and management.
- **Mobile Switching Center (MSC).** The MSC represents the functions provided by the cellular switch including circuit-switched call management, mobile location, and mobile management.
- **Interworking Function (IWF).** An IWF provides functions needed for terminal equipment connected to a mobile termination to interface with other networks such as PSTN or CDPD network. A CDPD Mobile Data Intermediate System (MD-IS) is an example of an IWF.
- **Public Packet Data Network (PPDN).** A public packet-switched data network (such as the Internet) provides a transport mechanism for packet data between processing elements capable of using such service.

14.11.3 Network Reference Points

The reference points are

- **Reference Point R_m.** A physical interface connecting a TE2 to an MT2.
- **Reference Point U_m.** A physical interface connecting an MT0 or MT2 to a BS/MSC. This is an air interface.
- **Reference Point L.** A physical interface connecting a BS/MSC to an IWF.
- **Reference Point P_i.** A physical interface connecting an IWF to a PPDN.

14.11.4 Protocol Options

TIA IS-657 defines the requirements for communication protocols on the links between a mobile and an IWF, including requirements for R_m, U_m, and L interfaces. The relay layer provides lower-layer communication and packet framing between the entities of the packet data service reference model. Over the R_m interface between the TE2 and the MT2, the relay layer is a simple EIA/TIA 232E interface. Over the U_m interface, the relay layer is a combination of RLP (defined in TIA IS-99 [15]) and TIA IS-95A [14] protocols. On the L interface, the relay layer uses the protocols defined in TIA IS-687 [19]. The two protocol stack options are discussed in the following sections.

14.11.4.1   Relay Layer R_m Interface Protocol Option

The relay layer R_m interface protocol option supports TE2 applications in which the TE2 is responsible for all aspects of packet data service mobility management and network address management (e.g., IPCP and CDPD registration and authentication protocols). The link layer is implemented using PPP. When using the relay layer R_m interface protocol option, the link layer connection is between the TE2 and the IWF. The network layer includes protocols such as IP and CLNP, and packet data network registration and authentication protocols such as MNRP (see Fig. 14-7).

14.11.4.2   Network Layer R_m Interface Option

The network layer R_m interface protocol option supports TE2 applications in which the MT2 is responsible for all aspects of packet mobility management and network address management (e.g., IPCP and CDPD registration and authentication protocols). In this option, there are independent link layer connections between the TE2 and the MT2, as well as between the TE2 and IWF. The IWF link layer (between the MT2 and IWF) is implemented using Interned PPP. The R_m link layer (between the MT2 and TE2) can be implemented using the Interned PPP protocol to support the IP network layer protocol. For this option, the network layer also provides independent service between the TE2 and MT2 and between the MT2 and IWF. The TE2 includes routing protocols and operates as if locally connected to a network routing server. The MT2 includes both routing and packet data network registration and authentication protocols (see Fig. 14-8).

14.11.4.3   Radio Link Protocol (RLP)

The RLP employs the *link layer retransmission* approach to improve TCP performance. Each TCP packet is segmented into several 20-msec radio link frames (192 bits at a 9.6-kbps transmission rate). The RLP improves the FER seen by the TCP layer. The RLP is a NAK-based selective repeat ARQ scheme and performs partial link recovery through a limited number of retransmissions, n, in case of frame error. The RLP frame size is 20 bytes. About 30 RLP frames per TCP segment are used.

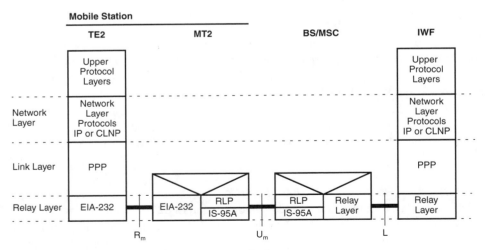

Figure 14-7   Relay Layer R_m Interface Protocol Option

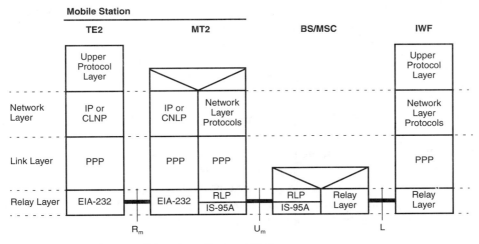

Figure 14-8 Network Layer R Interface Protocol Option

14.11.5 Applicable Mobile Type

The CDMA-CDPD feature applies to mobiles that comply with IS-657 and support service negotiation and configuration. The mobiles must support the optional RLP data frame encryption as specified in IS-657.

Two mobile configurations are defined in IS-657: the relay layer R_m protocol option and the network layer R_m protocol option. From the system's perspective, both mobile configurations are supported equally and transparently. However, the network layer option has an advantage over the relay layer option—since its mobility management protocols reside with the mobile unit (MT2), only the standard Internet protocol stacks run on laptop (TE2).

14.11.6 Packet Data Protocol States

The IWF and the mobile use a link layer connection to transmit and receive packets if the link layer connection is opened when a packet data service option is first connected. Once an IWF link layer connection is opened, bandwidth (in the form of traffic channel assignment) is allocated to the connection on an as-needed basis. The IWF link layer connection can be in any of the following states:

- **Closed.** The IWF link layer connection is closed when the IWF has no link layer connection state information for the mobile. In this state the mobile does not provide packet data service.
- **Opened.** The IWF layer connection is opened when the IWF has link layer connection state information for the mobile. The opened state has two substates:

    1. *Active:* An open IWF link layer connection is active when there is an L interface

virtual circuit for the mobile and the mobile is on a traffic channel with the packet data service option connected.

2. *Dormant:* An open IWF link connection is dormant when there is no L interface virtual circuit for the mobile and the mobile is not on a traffic channel with the packet data service option connected.

The BS/MSC and IWF maintain the state of link layer connection. The mobile maintains the state of PPP Link Control Protocol (LCP) and manages the IWF link layer connection using LCP opening and closing procedures.

With IWF link layer connection in a dormant state, if either the mobile or BS/MSC has data to send, it is not necessary to reopen the link layer connection or to reinitialize any upper layer protocols, provided the packet data service type has not changed since the link layer last entered the dormant state. The mobile and BS/MSC can freely mix packet data service requests using any supported rate set within a service type.

14.11.6.1 MS Packet Data Service States

MS packet data service states are

- **Inactive State.** The mobile does not provide packet data service.
- **Active State.** The mobile provides packet data service.

14.11.6.2 MS Packet Data Service Call Control Functions

The mobile performs the packet data service call control function using the following states (see Fig. 14-9):

- **Null State.** The packet data service call control function is in this state when packet data service has been activated.
- **Initialization/Idle State.** The mobile attempts to establish a traffic channel for the purpose of initiating packet data service.
- **Initialize/Traffic State.** The mobile communicates with the BS/MSC on a traffic channel and attempts to connect a packet data service option for the purpose of initiating packet data service.
- **Connected State.** A packet data service option is connected. The mobile can transfer packet data.
- **Dormant/Idle State.** The mobile is not on a traffic channel. The mobile cannot transfer packet data.
- **Dormant/Traffic State.** The mobile is communicating with the BS/MSC on a traffic channel, but the packet data service option has been disconnected. The mobile cannot transfer packet data.
- **Reconnect/Idle State.** The mobile attempts to establish a traffic channel.
- **Reconnect/Traffic State.** The mobile communicates with the BS/MSC on a traffic channel and attempts to connect a packet data service option.

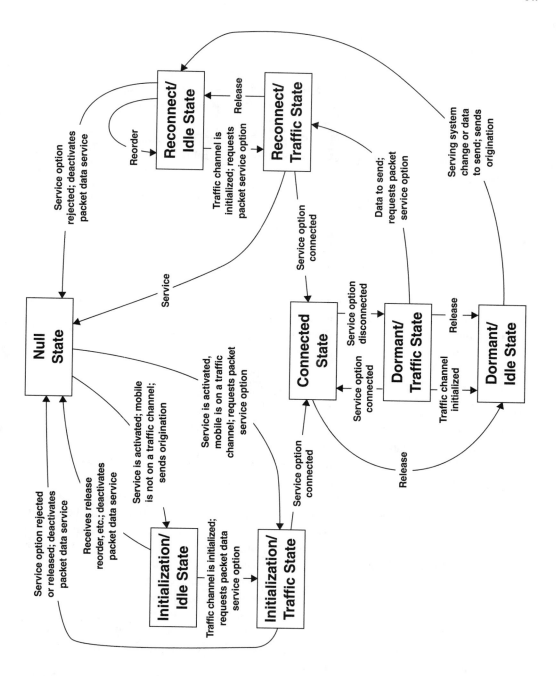

Figure 14-9 MS Packet Data Service Call Control Functions

14.11.6.3 BS/MSC Packet Data Service States

Packet data service processing in the BS/MSC consists of the following states (see Fig. 14-10):

- **Inactive State.** The BS/MSC does not provide packet data service to the mobile.
- **Active State.** The BS/MSC provides packet data service to the mobile.

14.11.6.4 BS/MSC Packet Data Service Call Control Functions

The BS/MSC performs the packet data service call control functions which consist of the following states (see Fig. 14-10):

- **Null State.** The BS/MSC has no connection of a packet data service option to the mobile.
- **Paging State.** The IWF has requested that the BS/MSC connect a packet data service option to the mobile for delivery of packet data, and the BS/MSC pages the mobile.
- **Initialization/Idle State.** The BS/MSC is awaiting initialization of a traffic channel with the mobile.
- **Initialization/Traffic State.** The mobile is on a traffic channel. The BS/MSC awaits connection of a packet data service option.
- **Connected State.** A packet data service option has been connected. Packet data is exchanged with the mobile.

14.11.7 Packet Mode Data Service Features

The following are packet mode data service features.

- **Dual 9.6- and 14.4-kbps Speed.** The CDMA packet data services support both 9.6- and 14.4-kbps data rates.
- **Bandwidth Management.** The dormant mode is supported to insure that bandwidth is not wasted if an end point enters an inactive state.
- **Mobility.** Mobility support for packet mode data services allows the user's data application to continue correct operation during movement of the mobile. Subscriber mobility is supported for intra-MSC and inter-MSC. Inter-MSC support requires that a unique System Identification/Network Identification (SID/NID) be assigned for each MSC. This is because a new IWF will provide the packet data service when a mobile moves into a new MSC serving area, so an IWF link layer connection transfer must take place to maintain the user's packet data connectivity. According to IS-657, the mobile is required to reopen the link layer connection when it detects a change in the SID or NID of the serving system. Inter-MSC packet data mobility is easily supported in a network where each MSC has a unique SID/NID. Intersystem mobility is supported if both the CDMA and CDPD have roaming agreements and the systems share Shared Secret Data (SSD).

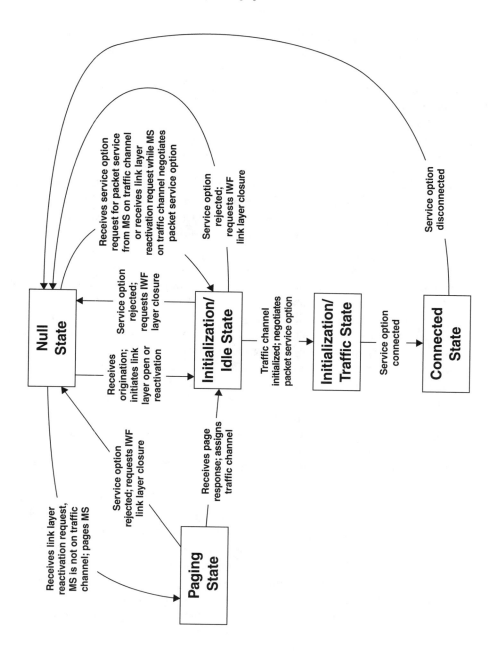

Figure 14-10 Packet Data Service Call Control States in BS/MSC

- **Fraud Containment.** The CDPD authentication is performed as a part of CDPD registration procedures when a mobile requests the CDPD service. The mobile user is validated with the home system based on a triplet formed by the Network Entity Identifier (NEI), the Authentication Sequence Number (ASN), and the Authentication Random Number (ARN) assigned to the mobile. The NEI authentication procedure applies to both intra- and intersystem service requests. The user data sent over the air interface are encrypted.

- **Accounting/Billing Interface.** CDPD accounting records consist of information in terms of data packet count, data octet count, control packet count, control octet count, discarded packet count, and packet data service connection time. The packet data service connection time starts when the mobile first registers with the CDPD network and ends when the mobile deregisters from the CDPD network. The accounting information is collected at the serving IWF and distributed to the home IWF in real time. The CDMA network will continue to collect the accounting records in the same way as the voice and circuit-mode data calls. It collects only air link connection time, not the data volume. Additional information may be added to AMA records for packet data calls to keep track of active and dormant links.

- **CDMA Data Encryption.** CDMA data encryption is performed on mobile RLP data frames carried on the digital traffic channel for mobiles conforming to IS-657 or IS-99 according to "Common Crytographic Algorithms, Revision A.1," defined in TIA IS95.0.A. The data encryption feature increases the security of user data transmitted over the air interface. It is mandatory for CDMA-CDPD services (i.e., service options 7 and 15). The feature requires activation of the CDMA authentication feature with global authentication because the DataKey and L table used in the encryption process is calculated for each data service connection only after a successful CDMA authentication. CDMA authentication is necessary to insure that the authentication center or Visited Location Register (VLR) has the same SSD and, therefore, the same DataKey and L table as the mobile.

- **Voice or Circuit-Mode Data Service while Packet Data Dormant.** The feature allows subscribers with a multifunction packet data mobile device with voice and/or circuit data capabilities to use voice or circuit data services while the packet data is in the dormant mode.The mobile maintains the packet data service call control state information while handling the voice call and restarts the packet data protocol stack when the dormant link layer connection is reactivated. It is assumed that the mobile is set up either manually or automatically in the proper mode for the requested service. The system is not involved in setting up the service mode on the mobile. The dormant link layer connection is supported with some limitations during a voice call. The limitation is that a request to reactivate the dormant link layer connection will be rejected if there is a voice call. When a packet data call is rejected, the IWF will gracefully discard the data received. This may eventually cause the user data application to be terminated if the voice call continues longer than is allowed by the user application. When a link layer connection enters the dormant state, subscriber call features that were deactivated for packet data calls will be reactivated. Tones and announcements will be played for the voice call. When a packet data call is established, subscriber call features that are not supported for packet data calls will be deactivated, and tones and announcements will be blocked.

- **Simultaneous Voice and Data Support.** This feature allows voice and packet data traffic to be carried simultaneously within a single traffic channel. Subscribers must have a mobile device that is capable of simultaneous voice and data service. Normal voice calls are allowed and carried as primary traffic at any time, whether the mobile is connected to a packet data call or not. If a new voice call is requested while the packet data call is connected, the packet data call will be dropped first to start the voice call. The packet data call will then be added using service negotiation procedures to provide a service configuration containing voice as primary traffic and packet data as secondary traffic. The transition of packet data from primary to secondary traffic may cause loss of user data, but this will be detected and retransmitted as necessary by the upper layer protocols without disruption to the user application. Thus, during a voice call, a link layer connection may be opened and a packet data call established and data carried as secondary traffic without disrupting the voice call. Both 9.6- and 14.4-kbps data rates are supported for this feature.

- **Service Options 4103 and 15.** Service option 4103 provides the standard Internet protocol packet-mode data services instead of the CDPD data service. Service option 4103 is for a data rate of 9.6 kbps, whereas service option 15 is for the same service at 14.4 kbps. Service options 4103 and 15 may be provided with or without mobility support. Mobility management for service options 4103 and 15 will be based on mobile IP protocol. Service options 4103 and 15 without mobility is suitable for users requiring only portable services (i.e., users not moving while a data application is running).

14.12 Summary

Since wireless data networks do not operate without interconnection to other networks, this chapter covered a variety of wireless data systems including the wide-area wireless data systems, high-speed Wireless Local Area Networks (WLANs), and the specific systems supported by CDMA. We examined the various standards being adopted by the IEEE, WINForum, ETSI (in Europe), and ARPA for wireless LANs. Since packet networks are an important part of wireless networks, the chapter briefly listed the characteristics of the access methods in common use and defined their throughput equations. The common packet protocols are ALOHA, slotted ALOHA, and Carrier-Sense-Multiple Access (CSMA) ALOHA.

We then presented the methods used to control errors for wireless data systems and concluded with the highlights of the TIA IS-99, TIA IS-637, and TIA IS-657 standards for CDMA cellular systems. CDMA supports asynchronous data, facsimile, packet data, and short message service (SMS) to end points in another wireless network or to the wireline network. We examined the reference models and protocol stacks for each of these data services.

14.13 References

1. Bates, R. J., *Wireless Networked Communication*, McGraw-Hill, Inc., New York, 1994.

2. Habab, I. M., Kavehrad, M., and Sundberg, C. E. W., "ALOHA with Capture over Slow and Fast Fading Radio Channels with Coding and Diversity," *IEEE Journal of Selected Areas of Communications* 6, 1988, pp. 79–88.

3. Hammond, J. L., and O'Reilly, J. P., *Performance Analysis of Local Computer Networks*, Addison-Wesley Publishing Company, Reading, MA, 1986.

4. Pahlavan, K., and Levesque, A. H., *Wireless Information Networks,* John Wiley and Sons, Inc., New York, 1995.

5. RFC 791, "Internet Protocol."

6. RFC 792, "Internet Control Message Protocol."

7. RFC 793, "Transmission Control Protocol."

8. RFC 1122, "Requirements for Internet Hosts—Communication Layers."

9. RFC 1144, "Compressing TCP/IP Headers for Low-Speed Serial Links."

10. RFC 1332, "The PPP Internet Protocol Control Protocol (IPCP)."

11. RFC 1661, "The Point-to-Point Protocol (PPP)."

12. RFC 1700, "Assigned Numbers."

13. Skalar, B., *Digital Communications—Fundamentals and Applications*, Prentice Hall, Englewood Cliffs, NJ, 1988.

14. TIA IS-95A, "Mobile Station–Base Station Compatibility Standard for Dual-Mode Wideband Spread Spectrum Cellular System."

15. TIA IS-99, "Data Services Option Standard for Wideband Spread Spectrum Digital Cellular System."

16. TIA 232E, "Interface between DTE and DCE Employing Serial Binary Data Interchange."

17. TIA IS-637, "Short Message Services for Wideband Spread Spectrum Cellular Systems."

18. TIA IS-657, "Packet Data Services Option for Wideband Spread Spectrum Cellular System."

19. TIA IS-687, "Data Services Inter-Working Function Interface Standard for Wideband Spread Spectrum Digital Cellular System."

20. Viterbi, A. J., and Padovani, Roberto, "Implications of Mobile Cellular CDMA," *IEEE Communication Magazine* 30 (12), 1992, pp. 38–41.

21. Walrand, J., *Communications Networks: A First Course*, Irwin, Homewood, IL, 1991.

22. Woo, Ill, and Cho, Dong-Ho, "A Study on the Performance Improvements of Error Control Schemes in Digital Cellular DS/CDMA Systems," *IEICE Trans. Communications* E77-B (7), July 1994.

cdma2000 System

15.1 Introduction

The International Telecommunications Union-Radio Communication (ITU-R) standardization sector developed specifications for International Mobile Telecommunications—2000 (IMT-2000). IMT-2000 has greatly expanded the range of service capabilities and covers a wide range of environments. IMT-2000 [5] specifications are aimed to facilitate the introduction of new capabilities and to provide a seamless evolution from the substantially installed second-generation (2G) telecommunications base by the year 2000+. The third-generation (3G) telecommunications systems based on IMT-2000 specifications will be introduced in service in the years 2000–2002. The 3G systems will offer a plethora of telecommunications services including voice, low- and high-bit-rate data, multimedia, and video to mobile users via a range of mobile terminals, operating in both public and private environments (office areas, residential areas, transportation media, etc.).

The ITU World Administration Radio Conference in 1992 (WARC-92) identified 230 MHz in a 2-GHz band for use on a worldwide basis for the satellite and terrestrial components of IMT-2000. The WARC-95 revised 2-GHz frequency allocations for mobile satellite services (MSS) to provide a satellite component of IMT-2000 (see Fig. 15-1).

Table 15-1 provides a summary of the current IMT-2000 3G air interface proposals to ITU and their network interfaces.

The cdma2000 Radio Transmission Technology (RTT) is a wideband, spread spectrum radio interface that uses CDMA technology to satisfy the needs of 3G wireless communication systems. This RTT meets all requirements specified in the ITU circular letter and the corresponding documents of the IMT-2000. The service requirements are satisfied for indoor office, indoor-to-outdoor/pedestrian, and vehicular environments. The cdma2000 system will also be backward compatible with the current cdmaOne (IS-95) family of standards.

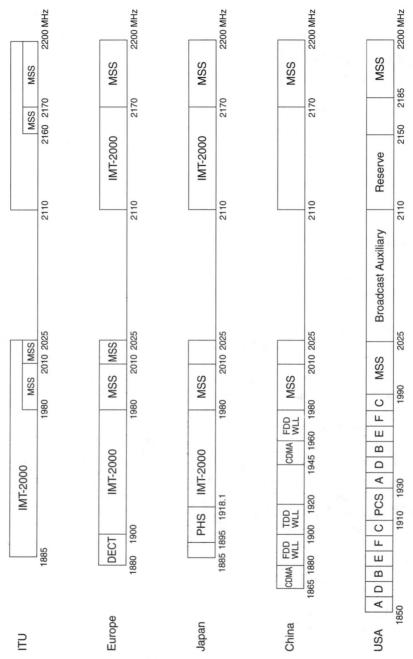

Figure 15-1 ITU Frequency Allocation Worldwide for 3G

Table 15-1 Current IMT-2000 Proposals

	cdma2000	ARIB/DOCOMO	UMTS	UWC-136
2G System	IS-95/cdmaOne	PDC	GSM	IS-136
3G Air Interface	cdma2000	W-CDMA	UTRA (W-CDMA/ TD-CDMA)	IS-136+/ 136-HS/ 136+HS
3G Network Interface	Evolved ANSI-41	Evolved GSM MAP	Evolved GSM MAP	Evolved ANSI-41
Standards Bodies	TIA TR-45 (supported by CDG)	ARIB	ETSI	TIA TR-45 (supported by UWCC)

UMTS = Unversal mobile telecommunications system
UWC = Universal wireless communication
PDC = Pacific digital cellular
HS = High speed
UTRA = UMTS terrestrial radio access
UWCC = UWC Consortium
ARIB = Association of Radio Industries Business (Japan)

The cdma2000 system provides a wide range of implementation options to support data rates (both circuit switched and packet switched) starting from the TIA IS-95B-compatible rate of 9.6 kbps up to greater than 2 Mbps. The cdma2000 system provides maximum flexibility to carriers in making engineering trade-offs between

- Channel sizes of 1, 3, 6, 9, and 12×1.25 MHz
- Support for advanced antenna technologies
- Cell sizes (e.g., the cdma2000 system's increased performance can be realized in terms of increased range to permit carriers to reduce the total number of cell sites)
- Higher data rates that can be supported in all channel sizes
- Support for advanced services possible or practical in other systems (e.g., high-speed circuit data, B-ISDN, or H.224/223 teleservices)

The cdma2000 system can be operated economically in a wide range of environments, including

- Outdoor megacells (cell >35 km radius)
- Outdoor macrocells (cell 1–35 km radius)
- Indoor/outdoor microcells (up to 1 km radius)
- Indoor/outdoor picocells (<50 m radius)

The cdma2000 system can be deployed in

- Indoor/outdoor environment
- Wireless local loop (WLL)

- Vehicular environment
- Mixed vehicular and indoor/outdoor environment

The cdma2000 system's mobility is variable, ranging from fixed wireless to high speeds of up to 300 mph. cdma2000 provides a layered structure to support the integration of the bottom two layers of the RTT into systems that implement any network standards (e.g., ITU-T-defined signaling services). It also provides backward compatibility to TIA IS-95B signaling and call control models. An extended cdma2000 upper layer signaling structure is capable of supporting a wide range of advanced services (e.g., multimedia) in an optimized and efficient manner.

cdma2000 supports the 3G wireless intelligent networking (WIN) services and services defined by the ITU or other international standards organizations and provides a graceful evolution from existing 2G TIA IS-95B technology. It includes the following features:

- Support for overlay configurations
- Support for backward compatibility to TIA IS-95B signaling and network
- Support for graceful and gradual upgrade from 2G system to 3G system
- Sharing of common channels with an underlay TIA IS-95B system during transition periods

cdma2000 provides an evolutionary path by reusing existing TIA IS-95B standards, including

- TIA IS-95B: Mobile Station and Radio Interface Specifications
- IS-707: Data Services (Packet, Async, and Fax)
- IS-127: Enhanced Variable-Rate Codec (EVRC) 8.5-kbps speech coder
- IS-733: 13-kbps speech coder
- IS 637: Short Message Service (SMS)
- IS 638: Over-the-Air Activation and Parameter Administration (supporting the configuration and service activation of mobile stations over the radio interface)
- IS-97 and IS-98 (Minimum Performance Specifications)
- The basic TIA IS-95B channel structure
- Extensions to TIA IS-95B Fundamental/supplemental channel structure, multiplex layer, and signaling to support higher rate operation, common broadcast channels (pilot, paging, and sync)
- IS-634A: no significant changes expected for cdma2000; the layered structure of cdma2000 integrates smoothly with the component structure of IS-634A
- TIA IS-41D: No significant changes needed for cdma2000; the layered structure of cdma2000 offers the potential for easy integration with enhanced network services (WIN)

15.2 cdma2000 Layering Structure

15.2.1 Upper Layer

Fig. 15-2 shows the layer structure of cdma2000. The *upper layers* contain three basic services:

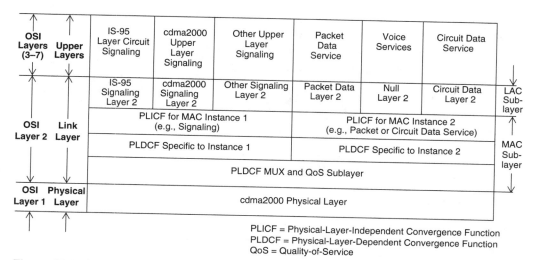

Figure 15-2 cdma2000 Layering Structure

- **Voice services.** Voice telephony services, including PSTN access, mobile-to-mobile voice services, and Internet telephony.
- **End-user data-bearing services.** Services that deliver any form of data on behalf of mobile end user, including packet data (e.g., IP service), circuit data services (e.g., B-ISDN emulation services), and SMS. Packet data services conform to industry standard connection-oriented and connectionless packet data including IP-based protocols (e.g., TCP and UDP) and ISO/OSI Connectionless Interworking Protocol (CLIP). Circuit data services that emulate international-standards-defined, connection-oriented services such as asynchronous (async) dial-up access, fax, V.120 rate-adapted ISDN, and B-ISDN services.
- **Signaling.** Services that control all aspects of operation of the mobile.

15.2.2 Link Layer

The *link layer* provides varying levels of reliability and QoS characteristics according to the needs of the specific upper layer service. It gives protocol support and control mechanisms for data transport services and performs all the functions necessary to map the data transport needs of the upper layers into specific capabilities and characteristics of the physical layer. The link layer is subdivided into sublayers.

- Link Access Control (LAC) sublayer (see Fig. 15-3)
- Media Access Control (MAC) sublayer (see Fig. 15-4)

The LAC sublayer manages point-to-point communication channels between peer upper layer entities and provides framework to support a wide range of different end-to-end reliable link layer protocols.

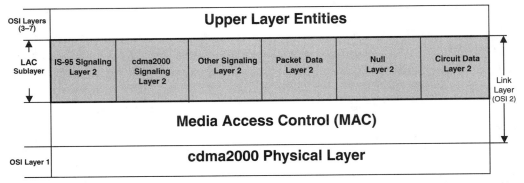

Figure 15-3 Link Layer of cdma2000

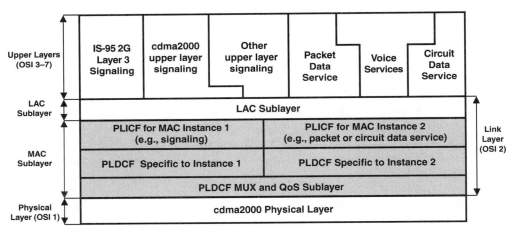

Figure 15-4 cdma2000 MAC Sublayer

The cdma2000 system includes a flexible and efficient MAC sublayer that supports multiple instances of an advanced-state machine, one for each active packet or circuit data instance. Together with the QoS control entity, the MAC sublayer realizes the complex multimedia, multiservice capabilities of 3G wireless systems with QoS management capabilities for each active service. The MAC sublayer provides three important functions:

- **MAC control state.** Procedures for controlling the access of data services (packet and circuit) to the physical layer (including contention control between multiple services from a single user as well as between competing users).
- **Best-effort delivery.** Reasonably reliable transmission over the radio link with radio link protocol (RLP) providing a *best-effort* level of reliability.

- **Multiplexing and QoS control.** Enforcement of negotiated QoS levels by mediating conflicting requests from competing services and appropriately prioritizing access requests.

The MAC sublayer provides differing QoS to the LAC sublayer (e.g., different modes of operation). It may be constrained by backward compatibility (e.g., for IS-95B signaling layer 2), and it may have to be compatible with other link layer protocols (e.g., for compatibility with non-IS-95 air interfaces or for compatibility with future ITU-defined protocol stacks). The MAC sublayer is subdivided into

- Physical-Layer-Independent Convergence Function (PLICF)
- Physical-Layer-Dependent Convergence Function (PLDCF) which is further subdivided into

  - Instance-specific PLDCF
  - PLDCF MUX and QoS sublayer

PLICF provides service to the LAC sublayer and includes all MAC operational procedures and functions that are not unique to physical layer. Each instance of PLICF maintains service status for the corresponding service. PLICF uses services provided by PLDCF to implement actual communications activities in support of MAC layer service. Services used by PLICF are defined as a set of logical channels that carry different types of control or data information. The PLICF data service consists of the following states/substates (see Figs. 15-5 and 15-6):

- Null state
- Initialization state
- Control hold state

  - Normal substate
  - Slotted substate

- Active state
- Suspended state

  - Virtual traffic substate
  - Slotted substate

- Dormant state

  - Dormant/idle substate
  - Dormant/burst substate

The *null state* is considered to be the default state prior to activation of packet data service. After the packet service is invoked, a transition to the *initialization state* occurs during which an attempt is made to connect the packet service.

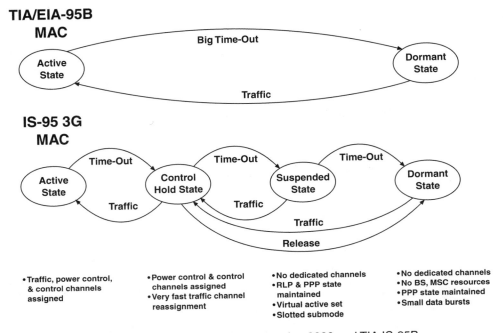

•Traffic, power control, & control channels assigned

•Power control & control channels assigned
•Very fast traffic channel reassignment

•No dedicated channels
•RLP & PPP state maintained
•Virtual active set
•Slotted submode

•No dedicated channels
•No BS, MSC resources
•PPP state maintained
•Small data bursts

Figure 15-5 Packet Data MAC Operation States in cdma2000 and TIA IS-95B

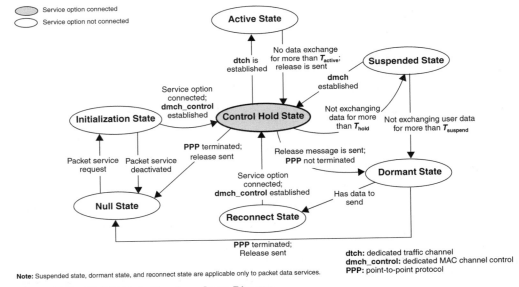

Note: Suspended state, dormant state, and reconnect state are applicable only to packet data services.

Figure 15-6 PLICF Data Services State Diagram

Traffic, power control, and control channels are assigned in the *active state*. In the *control hold state*, a dedicated control channel is maintained between the user and the base station on which any MAC control command (for example, the command to begin a high-speed data burst) can be transmitted with virtually no latency. Power control is also maintained so that high-speed burst operation can begin without any delay due to stabilization of power control.

In the *suspended hold state*, there are no dedicated channels maintained to or from the user. However, the state information for RLP is maintained, and the base station and user maintain a virtual active set that allows either one to know which base station can best be used (accessed by the user or paged by the base station) in the event that packet data traffic occurs for the user. This state also supports a slotted substate that permits the user's mobile device to preserve power in a highly efficient manner.

A short data burst mode is added to the cdma2000 *dormant state* to support the delivery of short messages without incurring the overhead of a transition from dormant to active state. Transitions between MAC states can be indicated by MAC control signaling or by the expiration of timers. By properly selecting the values for the timers, the cdma2000 MAC can be adapted to a wide variety of data services and operating environments.

The states are categorized as either connected or not connected depending on the status of data service option. The data service option is connected in the *control hold*, *active*, and *suspended hold states*. The data service option is not connected in the *null*, *initialization*, *dormant*, or *reconnect state*s. Fig. 15-6 shows the state diagram for the PLICF data service option.

PLDCF performs mapping of logical channels from PLICF to logical channels supported by the specific physical layer. PLDCF performs multiplexing, demultiplexing, and consolidation of control information with bearer data from the control and traffic channels from multiple PLICF instances in the same mobile. PLDCF implements QoS capabilities, including resolution of priorities between competing PLICF instances, and maps QoS requests from PLICF instances into the appropriate physical layer service requests to deliver the desired QoS. The major functions of this sublayer are to

- Perform any required mapping of the simpler logical channels from the PLICF into the logical channels supported by physical layer
- Perform any (optional) automatic repeat request (ARQ) protocol functions that are tightly integrated with physical layer
- Perform some of the physical-layer-specific low-level functions of IS-95B RLP

For cdma2000, four specific PLDCF ARQs are defined.

1. **Radio Link Protocol (RLP).** This protocol provides a highly efficient streaming service that makes a best effort to deliver data between peer PLICF entities. RLP provides both transparent and nontransparent modes of operation. In the nontransparent mode, RLP uses ARQ protocol to retransmit data segments that were not delivered properly by the physical layer. In the nontransparent mode, RLP can introduce some delay. In the transparent mode, RLP does not retransmit missing data segments. However, RLP

maintains byte synchronization between the sender and receiver and notifies the receiver of the missing parts of the data stream. Transparent RLP does not introduce any transmission delay and is useful for implementing voice services over RLP.

2. **Radio Burst Protocol (RBP).** This protocol provides a mechanism for delivering relatively short data segments with best-effort delivery over a shared access common traffic channel (ctch). This capability is useful for delivering a small amount of data without incurring the overhead of establishing a dedicated traffic channel (dtch).

3. **Signaling Radio Link Protocol (SRLP).** This protocol provides best-effort streaming service for signaling information analogous to RLP, but optimized for the dedicated signaling channel (dsch).

4. **Signaling Radio Burst Protocol (SRBP).** This protocol provides a mechanism to deliver signaling messages with best-effort delivery analogous to RBP, but optimized for signaling information and the common signaling channel (csch).

PLDCF includes a Radio Link Access Control (RLAC) function that abstracts the RLP and RBP from PLICF and coordinates the transmission of data (traffic or signaling) between RLP and RBP according to the current operational state of MAC (e.g., restricts the use of RBP to cases in which PLICF is in the packet data dormant state).

The PLDCF MUX and QoS sublayer coordinates multiplexing and demultiplexing of code channels from multiple PLICF instances. It implements and enforces QoS differences between instances and maps the data streams and control information on multiple logical channels from different PLICF instances into requests for logical channels, resources, and control information from the physical layer.

15.3 cdma2000 Channels

15.3.1 Channel-Naming Convention

A logical channel is denoted by three or four lowercase acronyms followed by "ch" for *channel*. The fourth letter applies to common channels used in dormant or suspended states. Table 15-2 lists the conventions for logical channels.

A physical channel (Table 15-3) is represented by uppercase abbreviations. The first letter in the name of the channels indicates the direction of the channel, except for the paging and access channels where the direction is implicitly specified.

15.4 Logical Channels Used by PLICF

The following sections describe the logical channels used by PLICF.

15.4.1 Dedicated Traffic Channel (f/r-dtch)

dtch is the forward or reverse logical channel that is used to carry user data traffic. This logical channel is a point-to-point channel and is allocated for use throughout the active state of data service. It carries data dedicated to a single PLICF instance.

Table 15-2 Logical Channel-Naming Convention

First Letter	Second Letter	Third Letter
f = forward (BS to MS) r = reverse (MS to BS)	d = dedicated c = common	t = traffic m = MAC s = signaling

Table 15-3 Physical Channel-Naming Convention

Channel Name	Physical Channel
F/R-FCH	Forward/Reverse Fundamental Channel
F/R-SCH	Forward/Reverse Supplementary Channel
F/R-DCCH	Forward/Reverse Dedicated Control Channel
F-PCH	Forward Paging Channel
R-ACH	Reverse Access Channel
F/R-CCCH	Forward/Reverse Common Control Channel
F-DAPICH	Forward Dedicated Auxiliary Pilot Channel
F-CAPICH	Forward Common Auxiliary Pilot Channel
F/R-PICH	Forward/Reverse Pilot Channel
F-SYNC	Forward Sync Channel

15.4.2 Common Traffic Channel (f/r-ctch)

ctch is the forward or reverse logical channel that is used to carry short data bursts associated with the data service in the dormant/burst substate of the dormant state. This logical channel is a point-to-point channel and is allocated for the duration of the short burst. It shares access among many mobiles and/or PLICF instances.

15.4.3 Dedicated MAC Channel (f/r-dmch_control)

dmch_control is the forward or reverse logical channel that is used to carry MAC messages. This logical channel is a point-to-point channel and is allocated throughout the active state and control hold state of data service. It carries control information dedicated to a single PLICF instance.

15.4.4 Reverse Common MAC Channel (r-cmch_control)

The r-cmch_control is the reverse logical channel used by the mobile while data service is in the dormant/idle substate of the dormant state or suspended state. This logical channel is used to carry MAC messages. It is shared by a group of mobiles in the sense that access to this channel is gained on a contention basis.

15.4.5 Forward Common MAC Channel (f-cmch_control)

The f-cmch_control is the forward logical channel used by the base station while data service is in the dormant/idle substate of the dormant state or suspended state. This logical channel is used to carry MAC messages. It is a point-to-multipoint channel.

15.4.6 Dedicated Signaling Channel (dsch)

dsch carries upper layer signaling data dedicated to a single PLICF instance.

15.4.7 Common Signaling Channel (csch)

csch carries upper layer signaling data with shared access among many mobiles and/or PLICF instances.

15.5 Physical Layer

The physical layer provides coding and modulation services for a set of logical channels used by the PLDCF MUX and QoS sublayer. The physical channels (see Fig. 15-7) are classified as

- **Forward/Reverse Dedicated Physical Channels (F/R-DPHCH).** The collection of all physical channels that carry information in a dedicated, point-to-point manner between the base station and a single mobile (see Fig. 15-8)
- **Forward/Reverse Common Physical Channels (F/R-CPHCH).** The collection of all physical channels that carry information in a shared access, point-to-multipoint manner between the base station and multiple mobile stations (see Fig. 15-9)

15.6 Forward Link Physical Channels

Forward dedicated channels carry information between the base station and a specific mobile; common channels carry information from the base station to a set of mobiles in a point-to-multipoint manner. Table 15-4 lists these channels.

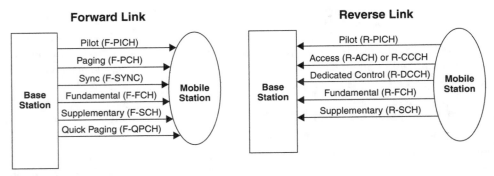

Figure 15-7   cdma2000 Physical Channels

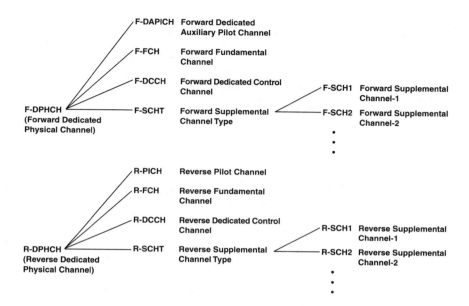

Figure 15-8   cdma2000 Overview of Dedicated Physical Channels

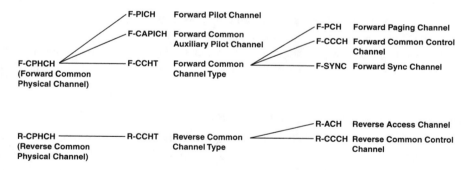

Figure 15-9   cdma2000 Overview of Common Physical Channels

15.6.1   Forward Pilot Channel (F-PICH)

This channel is continuously broadcast throughout the cell in order to provide timing and phase information. The common pilot is an all-0s sequence prior to Walsh spreading with Walsh 0. The F-PICH is shared by all traffic channels and is used for

- Estimating channel gain and phase

Table 15-4   Forward Link Channels

	Physical Channel	Channel Name
Forward Common Physical Channels (control and overhead channels)	Forward Pilot Channel	F-PICH
	Forward Paging Channel	F-PCH
	Forward Sync Channel	F-SYNC
	Forward Common Control Channel	F-CCCH
	Forward Common Auxiliary Pilot Channel	F-CAPICH
	Forward Quick Paging Channel	F-QPCH
	Forward Broadcast Common Channel	F-BCCH
Forward Dedicated Physical Channels	Forward Dedicated Auxiliary Pilot Channel	F-DAPICH
	Forward Dedicated Common Control Channel	F-DCCH
	Forward Traffic Channel —Fundamental —Supplementary	F-FCH F-SCH

- Detecting multipath rays so that RAKE fingers are efficiently assigned to the strongest multipath
- Cell acquisition and handoff

With a common pilot, it is possible to send the pilot signal without incurring significant overhead for each user. A system with a common pilot approach can achieve better performance than a system using a per-user pilot approach. For voice traffic, the common pilot can provide better channel estimation and lower overhead, resulting in improved receiver performance. It can also provide improved search and handoff performance.

15.6.2   Forward Sync Channel (F-SYNC)

The sync channel is used by mobiles operating within the coverage area of the base station to acquire initial time synchronization. There are two types of F-SYNC: *shared F-SYNC* and *wideband F-SYNC*. The shared F-SYNC provides service to both the IS-95B and cdma2000 when using the F-SYNC in an IS-95B underlay channel. This mode is applicable only in overlay configurations.

The wideband F-SYNC is modulated across the entire wideband channel. The wideband F-SYNC is modulated as a separate channel within the forward common physical channel (F-CPHCH). This mode is applicable to both overlay and nonoverlay configurations.

15.6.3   Forward Paging Channel (F-PCH)

A cdma2000 system can have multiple paging channels per base station. A paging channel is used to send control information and paging messages from the base station to mobiles and operates at a data rate of 9.6 or 4.8 kbps (same as IS-95). The F-PCH carries overhead messages, pages, acknowledgments, channel assignments, status requests, and SSD updates from the base station to the mobile.

There are two types of paging channels: *shared F-PCH* and *wideband F-PCH*. The shared F-PCH provides service to both the IS-95B and cdma2000 when using the F-PCH in an IS-95B underlay channel. This mode is applicable only in overlay configurations. The wideband F-PCH is modulated across the entire wideband channel. The wideband F-PCH is modulated as a separate channel within the F-CPHCH. This mode is applicable to both overlay and nonoverlay configurations.

Figs. 15-10 and 15-11 show F-CPHCH (F-PICH, F-SYNC, and F-PCH) for $N = 1$ and $N \geq 3$.

15.6.4 Forward Common Control Channel (F-CCCH)

The F-CCCH is a common channel used for communication of layer 3 and MAC messages from the base station to the mobile. Possible frame sizes for F-CCCH are 5 ms, 10 ms, and 20 ms, depending upon the operating environment. It is identical with the F-PCH for a 9.6-kbps rate (20-ms frame).

15.6.5 Forward Common Auxiliary Pilot Channel (F-CAPICH)

This channel is used with antenna beam-forming applications to generate spot beams. Spot beams can be used to increase coverage in a particular geographical area or to increase capacity toward hot spots. The F-CAPICH can be shared among multiple mobiles in the same spot beam.

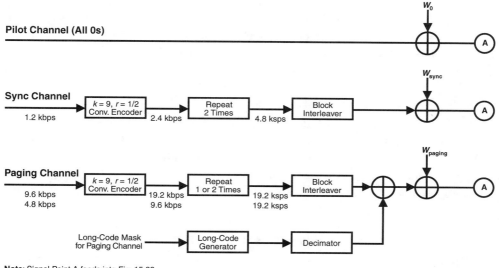

Note: Signal Point A feeds into Fig. 15.22.

Figure 15-10 cdma2000 F-CPHCH for $N = 1$

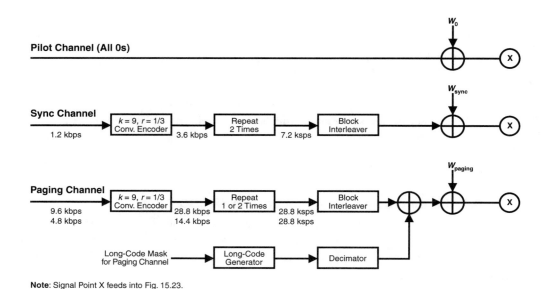

Note: Signal Point X feeds into Fig. 15.23.

Figure 15-11   cdma2000 F-CPHCH for $N \geq 3$

Auxiliary pilots are code multiplexed with other forward link channels, and they use orthogonal Walsh codes. Since a common pilot contains no data (all 0s), auxiliary pilots may use a longer Walsh sequence to lessen the reduction of orthogonal Walsh codes available for traffic channels. Auxiliary pilots can also be used for orthogonal diversity transmission in the direct spread forward link. Furthermore, if the CDMA system uses a separate antenna array to support directional or spot beams, it is necessary to provide a separate forward link pilot for channel estimation.

15.6.6   Forward Broadcast Common Channel (F-BCCH)

This is a paging channel dedicated to carrying only overhead messages and possible SMS broadcast messages. It removes the overhead messages from the paging channel to a separate broadcast channel. This improves the mobile initialization time and system access performance. At the same time, by reducing the number of messages on the F-PCH, the paging capacity is improved. The F-BCCH has a fixed Walsh code that is communicated to the mobile on the F-SYNC.

15.6.7   Forward Quick Paging Channel (F-QPCH)

The F-QPCH is a new type of paging channel used by a base station when it needs to contact the mobile in the slotted mode. Its use reduces the time the mobile needs to be "awake" resulting in increased battery life for the mobile.

The F-QPCH will contain a single bit message, the Quick Page message, to direct a slotted-mode mobile to monitor its assigned slot on the paging channel that immediately follows. The Quick Page message is sent up to 80 ms before the page message to alert the mobile to listen to the paging channel. The F-QPCH uses a different modulation, so it will appear as a different physical channel.

15.6.8 Forward Dedicated Auxiliary Pilot Channel (F-DAPICH)

An optional auxiliary pilot can be generated for a particular mobile. The F-DAPICH is used with beam-forming applications and beam-steering techniques to increase the coverage or data rate toward a particular mobile.

15.6.9 Forward Fundamental Channel (F-FCH)

This channel is transmitted at a variable rate as in IS-95B and consequently requires rate detection at the receiver. Each F-FCH is transmitted on a different orthogonal code channel and uses frame sizes corresponding to 20 ms and 5 ms. The 20-ms frame structure supports the data rate corresponding to Rate Set 1 (RS1) and RS2, where the rates are 9.6, 4.8, 2.7, and 1.5 kbps for RS1 and 14.4, 7.2, 3.6, and 1.8 kbps for RS2. The $N = 1$ RS1 F-FCH is shown in Fig. 15-12 and $N = 1$ RS2 F-FCH in Fig. 15-13. For $N = 1$ and RS1, a rate 1/2 convolutional encoder is used. For $N = 1$ and RS2, a rate 1/3 convolutional code followed by puncturing every ninth bit effectively provides a 3/8 code rate.

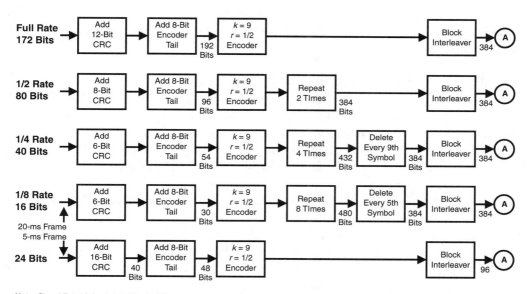

Note: Signal Point A feeds into Fig. 15.22.

Figure 15-12   F-FCH for $N = 1$ RS1

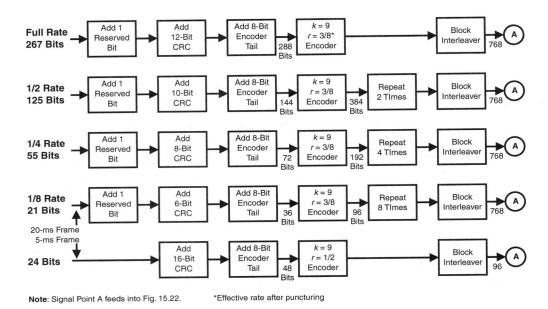

Note: Signal Point A feeds into Fig. 15.22.   *Effective rate after puncturing

Figure 15-13   F-FCH for $N = 1$ RS2

For $N \geq 3$, the RS1 and RS2 F-FCHs are shown in Figs. 15-14 and 15-15, respectively. For $N = 3$, 6, 9, and 12 and RS1, a 1/3 code rate is used. For $N = 3$, 6, 9, and 12 and RS2, code rates of 1/2 and 1/4 are supported.

15.6.10   Forward Supplemental Channel (F-SCH)

The F-SCH can be operated in two distinct modes. The first mode is used for a data rate not exceeding 14.4 kbps and uses blind rate detection (no scheduling or rate information provided). In the second mode the rate information is explicitly provided to the base station. In the first mode, the variable rates provided are those derived from IS-95B RS1 and RS2. The structures for the variable-rate modes are identical to the 20-ms F-FCH. In the second mode, the high data rate modes can have $k = 9$ convolutional coding or turbo coding with $k = 4$ component encoders. For the case of convolutional codes, there are 8 tail bits. For the case of turbo codes, 6 tail bits and 2 reserve bits are used.

There may be more than one F-SCH in use at a given time. The individual F-SCH target FERs may be set independently with respect to the F-FCH and other F-SCHs, since optimal FER for data is different than for voice. For classes of data services that have less stringent delay requirements, the FER may also be managed by retransmissions.

The F-SCH supports 20-ms frames. For data rates derived from RS1, the F-SCH supports data rates from 9.6 to 307.2 kbps (see Figs. 15-16 and 15-17).

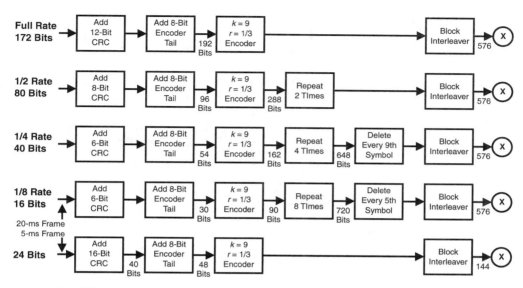

Note: Signal Point X feeds into Fig. 15.23.

Figure 15-14   F-FCH for $N \geq 3$ and RS1

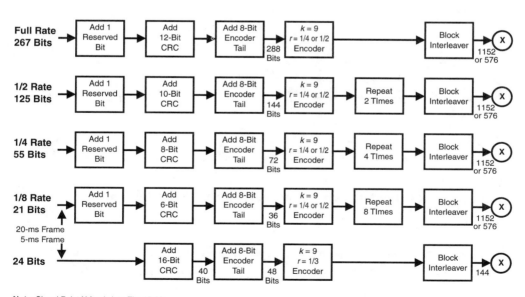

Note: Signal Point X feeds into Fig. 15.23.

Figure 15-15   F-FCH for $N \geq 3$ and RS2

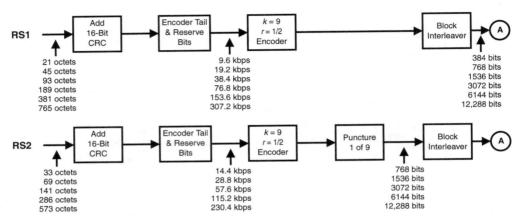

Note: Signal Point A feeds into Fig. 15.22.

Figure 15-16   F-SCH for *N* = 1 and Channel Bandwidth of 1.25 MHz

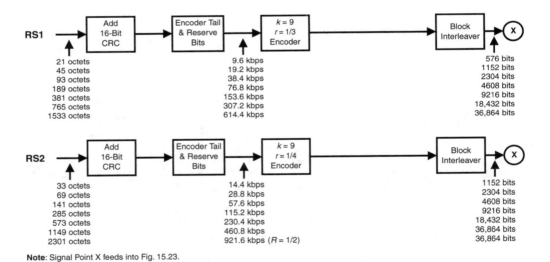

Note: Signal Point X feeds into Fig. 15.23.

Figure 15-17   F-SCH for *N* ≥ 3 and Channel Bandwidth of *N* × 1.25 MHz, *N* ≥ 3

15.6.11   Forward Dedicated Control Channel (F-DCCH)

The F-DCCH supports 5-ms and 20-ms frames at the 9.6-kbps encoder input rate. Sixteen CRC bits are added to the information bits for 5-ms frames or 12 CRC bits are added for 20-ms frames, followed by the addition of 8 tail bits, convolutional encoding, interleaving, and scrambling (see Figs. 15-18 and 15-19).

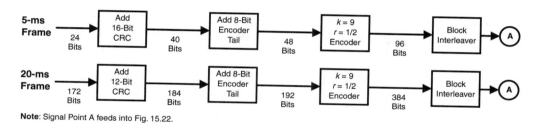

Note: Signal Point A feeds into Fig. 15.22.

Figure 15-18 Forward Dedicated Control Channel (F-DCCH) for $N = 1$

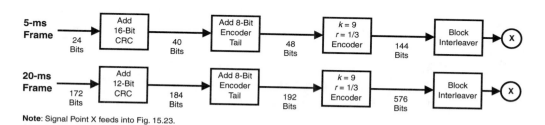

Note: Signal Point X feeds into Fig. 15.23.

Figure 15-19 Forward Dedicated Control Channel (F-DCCH) for $N \geq 3$

15.7 Forward Link Features

The forward link supports chip rates of $N \times 1.2288$ Mcps (where $N = 1, 3, 6, 9, 12$). For $N = 1$, the spreading is similar to IS-95B (see chapter 7); however QPSK modulation and fast closed-loop power control are used. There are two options for chip rates for $N > 1$: multicarrier and direct spread (see Fig. 15-20). The multicarrier approach demultiplexes modulation symbols onto N separate 1.25-MHz carriers ($N = 3, 6, 9, 12$). Each carrier is spread with a 1.2288-Mcps rate. The $N > 1$ direct-spread approach transmits modulation symbols on a single carrier which is spread with a chip rate of $N \times 1.2288$ Mcps ($N = 3, 6, 9, 12$).

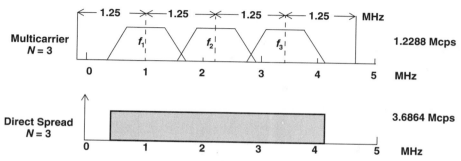

Figure 15-20 Multicarrier and Direct Spread Approach on Forward Link in cdma2000

15.7.1 Transmit Diversity

Transmit diversity can reduce the required E_b/I_t or (required transmit power per channel) and thus enhance system capacity. Transmit diversity can be implemented in the following ways:

- **Multicarrier Transmit Diversity.** Antenna diversity can be implemented in a multicarrier forward link with no impact on the subscriber terminal, where a subset of carriers is transmitted on each antenna. The main characteristics of the multicarrier approach are

  - Coded information symbols are demultiplexed among multiple 1.25-MHz carriers.
  - Frequency diversity is equivalent to spreading the signal over the entire bandwidth.
  - Both time and frequency diversity are captured by convolutional coder/symbol repetition and interleaver.
  - RAKE receiver captures signal energy from all bands.
  - Each forward link channel may be allocated an identical Walsh code on all carriers.
  - Fast power control.

In 3×1.25-MHz multicarrier transmitter, the serial coded information symbols are divided into three parallel data streams, and each data stream is spread with a Walsh code and a long PN sequence at a rate of 1.2288 Mcps. At the output of the transmitter, there are three carriers—A, B, and C (see Fig. 15-21).

After processing the serial coded information symbols with parallel carriers, the multicarrier will be transmitted by multiantenna, which is called Multicarrier Transmit Diversity (MCTD). In the MCTD, the total carriers are divided into subsets; then each subset of the carriers is transmitted on each antenna, where frequency filtering provides near-perfect orthogonality between antennas. This provides improved frequency diversity and hence increases forward link capacity.

- **Direct-Spread Transmit Diversity.** Orthogonal Transmit Diversity (OTD) may be used to provide transmit diversity for direct spread. Coded bits are split into two data streams and are transmitted via separate antennas. A different orthogonal code is used per antenna for spreading. This maintains the orthogonality between the two output

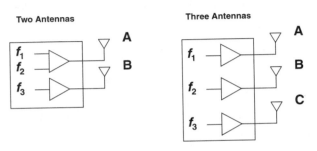

Figure 15-21 3×1.25-MHz Multicarrier Transmitter

streams, and hence self-interference is eliminated in flat fading. Note that, by splitting the coded bits into two separate streams, the effective number of spreading codes per user is the same as the case without OTD. An auxiliary pilot is introduced for the additional antenna.

15.7.2   Orthogonal Modulation

To reduce or eliminate intracell interference, each forward link physical channel is modulated by a Walsh code. To increase the number of usable Walsh codes, QPSK modulation is used before spreading. Every 2 information bits are mapped to a QPSK symbol. As a result, the available number of Walsh codes is increased by a factor of 2 relative to BPSK (prespreading) symbols. Walsh code length varies to achieve different information bit rates. The forward link may be interference limited or Walsh code limited depending on the specific deployment and operating environment. When a Walsh code limit occurs, additional codes may be generated by multiplying Walsh codes by the masking functions. The codes generated in this way are called *quasiorthogonal* functions.

15.7.3   Power Control

A new Fast-Forward Power Control (FFPC) algorithm for the forward link and a power control for the F-FCH and F-SCH is used in cdma2000. The standards specify a fast closed-loop power control at 800 Hz. Two schemes of power control have been proposed for the F-FCH and F-SCH.

* **Single-Channel Power Control.** This is based on the performance of the higher-rate channel between the F-FCH and F-SCH. The gain setting for the lower-rate channel is determined based on its relationship to the higher-rate channel.
* **Independent Power Control.** In this case, gains for the F-FCH and F-SCH are determined separately. The mobile runs two separate outer loop algorithms (with different E_b/N_t targets) and sends two forward E_b/I_t error bits to the base station.

15.7.4   Walsh Code Administration

IS-95A/B uses fixed-length 64-chip Walsh codes. The new rate sets in cdma2000 require variable length Walsh codes for traffic channels. The Walsh codes used are from 128 chips to 4 chips in length. The F-FCH Walsh code is fixed (128 chips for RS3 and RS5, and 64 chips for RS4 and RS6), whereas the length of the Walsh codes for F-SCH decreases as the information rate increases to maintain a constant bandwidth of the modulated signal. In addition to different Walsh code lengths, the coordination of the allocation of Walsh codes across the 2G and the 3G systems is necessary for the overlay systems.

The algorithm must ensure that Walsh codes assigned for different-rate supplemental channels are always orthogonal to each other as well as to the fundamental traffic channels, paging channels, sync channel, and pilot channel. For example, if an all-0s 4-chip Walsh code (0 0 0 0) is assigned, then there are two 8-chip Walsh codes that are not to be assigned at the same time

(0 0 0 0 0 0 0 0, 0 0 0 0 1 1 1 1); the remaining six 8-chip codes can be used since they are all orthogonal to it. By induction, four 16-chip, eight 32-chip, sixteen 64-chip, and thirty-two 128-chip codes must also be set aside to maintain orthogonality.

The 3G and 2G Walsh code assignments must be coordinated to insure that assigning the longer-length codes does not block out all of the shorter codes.

15.7.5 Modulation and Spreading

The $N = 1$ system can be deployed in a new spectrum or as a backward-compatible upgrade anywhere an IS-95B forward link is deployed in the same RF channel. The new cdma2000 channels can coexist in an orthogonal manner with the code channels of existing IS-95B system. The $N = 1$ spreading is shown in Fig. 15-22. First, the user data is scrambled by the user long PN code followed by I and Q mapping, channel gain, power control puncturing, and Walsh spreading. The power control bits may or may not be punctured onto the forward link depending on the specific logical-to-physical channel mapping. Next, as shown in Fig. 15-22, the signal is complex PN spread, followed by baseband filtering and frequency modulation.

The multicarrier system can be deployed in new spectrum or as a backward-compatible upgrade anywhere an IS-95B forward link is deployed in the same N RF channels. The new cdma2000 channels can coexist in an orthogonal manner with the code channels of existing IS-95B system.

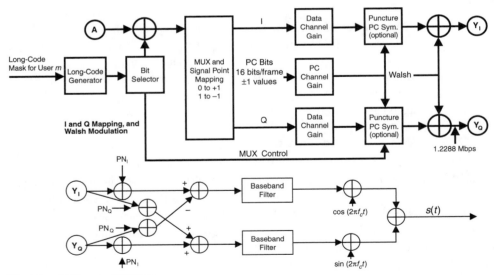

PN$_I$ = I-Channel PN Sequence 1.2288 Mcps
PN$_Q$ = Q-Channel PN Sequence 1.2288 Mcps
PC = Power Control

Figure 15-22 I and Q Mapping, Walsh Modulation, PN Spreading, and Frequency Modulation for $N = 1$

The overall structure of the multicarrier CDMA channel is shown in Fig. 15-23. After scrambling with the long PN code corresponding to user m, the user data is demultiplexed into N carriers, where $N = 3, 6, 9$, or 12. On each carrier, the demultiplexed bits are mapped onto I and Q followed by Walsh spreading. When applicable, power control bits, for reverse closed-loop power control, may be punctured onto the forward link channel at a rate of 800 Hz. The signal on each carrier is orthogonally spread by the appropriate Walsh code function in such a manner as to maintain a fixed chip rate of 1.2288 Mcps per carrier, where the Walsh code may differ on each carrier. The signal on each carrier is then complex PN spread, as shown in Fig. 15-24, followed by baseband filtering and frequency modulation.

$N = 1, 3, 6, 9$, and 12 direct spreading is shown in Fig. 15-25. The user data is first scrambled by the user long PN code followed by I and Q mapping, channel gain, power control

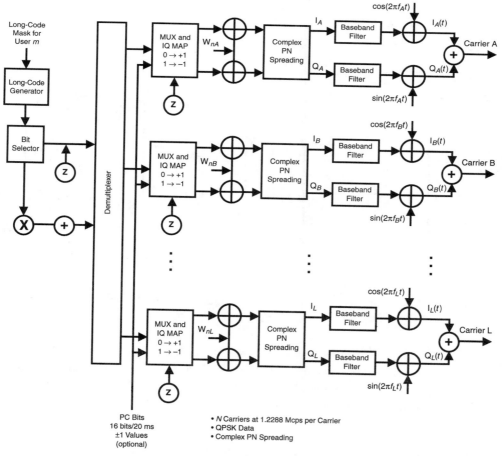

Figure 15-23 Multicarrier CDMA Forward Link Structure

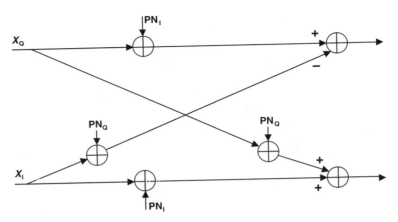

Figure 15-24 Complex PN Spreading

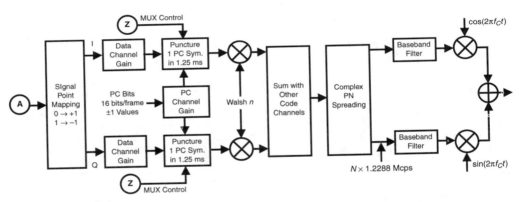

- One $N \times 1.2288$-Mcps Carrier
- QPSK Data
- Complex PN Spreading

Figure 15-25 $N = 1, 3, 6, 9,$ and 12, I and Q Mapping, and Walsh Modulation

puncturing, and Walsh spreading. The power control bits may not be punctured onto the forward link channel depending on the specific logical-to-physical channel mapping. Next, the signal is complex PN spread, followed by baseband filtering and frequency modulation.

Fig. 15-26 provides a comparison between the forward physical channels used in IS-95A/B and those used in cdma2000.

15.7.6 Key Characteristics of Forward Link

The key characteristics of the forward link are

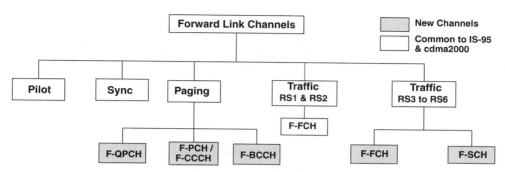

Figure 15-26 A Comparison between Forward Physical Channels for IS-95 and cdma2000

- Channels are orthogonal and use Walsh codes. Different-length Walsh codes are used to achieve the same chip rate for different information bit rates.
- QPSK modulation is used before spreading to increase the number of usable Walsh codes.
- Forward Error Correction (FEC) is used.

    ◆ Convolutional codes ($k = 9$) are used for voice and data.
    ◆ Turbo codes ($k = 4$) are used for high data rates on SCHs.

- Supports nonorthogonal forward link channelization.

    ◆ These are used when running out of orthogonal space (insufficient number of Walsh codes).
    ◆ Quasiorthogonal functions are generated by masking existing Walsh functions.

- Synchronous forward link.
- Forward link transmit diversity.
- Fast-forward power control (closed loop) 800 times per second.
- Supplemental channel active set, subset of fundamental channel active set. The maximum data rate supported for RS3 and RS5 for supplemental channel is 153.6 kbps (raw data rate). RS4 and RS6 will be supported only for voice calls with the fundamental channel rates of up to 14.4 kbps (raw data rate).
- Frame lengths:

    ◆ 20-ms frames are used for signaling and user information
    ◆ 5-ms frames are used for control information.

15.8 Reverse Physical Channels

Reverse physical channels (see Fig. 15-7) include dedicated channels to carry information from a single mobile to the base station and common channels to carry information from multiple mobiles to the base station. Table 15-5 lists the reverse physical channels.

Table 15-5 Reverse Physical Channels

	Physical Channels	Channel Name
Reverse Common Physical Channel	Reverse Access Channel	R-ACH
	Reverse Common Control Channel (9.6 kbps only)	R-CCCH
Reverse Dedicated Physical Channel	Reverse Pilot Channel	R-PICH
	Reverse Dedicated Control Channel	R-DCCH
	Reverse Traffic Channel: —Fundamental —Supplemental	R-FCH R-SCH

15.8.1   The Reverse Access Channel (R-ACH) and the Reverse Common Control Channel (R-CCCH)

These are common channels used for communication of layer 3 and MAC messages from the mobile to the base station. The R-CCCH differs from the R-ACH in that the R-CCCH offers extended capabilities beyond the R-ACH. For example, the R-CCCH supports lower latency access procedures required for efficient operation of the packet data in suspended state.

The R-ACH and R-CCCH are multiple access channels as mobile stations transmit without explicit authorization by the base station. The R-ACH and R-CCCH use a slotted ALOHA type of mechanism with higher capture probabilities due to the CDMA properties of the channel (simultaneous transmission of multiple users). There can be one or more access channels per frequency assignment. Different access channels are distinguished by different long PN codes. The R-CCCH is identical to the R-ACH for 9.6-kbps, 20-ms frames. Additional rates of 19.2 and 38.4 kbps and frames of 5 ms and 10 ms will be supported. Fig. 15-27 shows the structure of R-ACH and R-CCCH.

15.8.2   Reverse Pilot Channel (R-PICH)

The pilot channel for the reverse dedicated channels consists of a fixed reference value and multiplexed forward power control information. The time-multiplexed forward power control information is referred to as the power control subchannel. This subchannel provides information on the quality of the forward link at the rate of 1 bit per 1.25-ms Power Control Group (PCG) and is used by the forward link channels to adjust their power. The power control symbol repetition means that the 1-bit value is constant for that repeated symbol's duration. The power control bit uses the last portion of each PCG. The +1 pilot symbols and multiplexed power-control symbols are all sent with the same power level. The binary power control symbols are represented with ±1 values.

The R-PICH is used for initial acquisition, time tracking, RAKE-receiver coherent reference recovery, and power control measurements (see Fig. 15-28 for R-PICH structure).

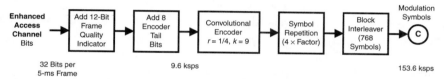

32 Bits per
5-ms Frame
9.6 ksps
153.6 ksps

• Channel Structure for the Header on the Enhanced Access Channel for Spreading Rate 1

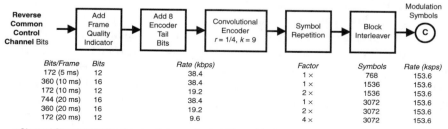

Bits/Frame	Bits	Rate (kbps)	Factor	Symbols	Rate (ksps)
172 (5 ms)	12	38.4	1 ×	768	153.6
360 (10 ms)	16	38.4	1 ×	1536	153.6
172 (10 ms)	12	19.2	2 ×	1536	153.6
744 (20 ms)	16	38.4	1 ×	3072	153.6
360 (20 ms)	16	19.2	2 ×	3072	153.6
172 (20 ms)	12	9.6	4 ×	3072	153.6

• Channel Structure for the Reverse Common Control Channel for Spreading Rate 1

Note: Signal Point C feeds into Fig. 15.30

Figure 15-27   Reverse Access and Common Control Channel Structure

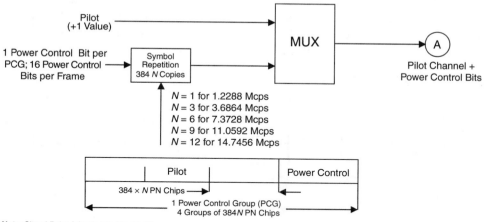

Note: Signal Point A feeds into Fig. 15.30

Figure 15-28   Reverse Pilot Channel (R-PICH) Structure for Reverse Dedicated Channels

15.8.3   Reverse Dedicated Control Channel (R-DCCH)

The R-DCCH, R-FCH, and R-SCH may or may not be used depending on the service scenario. Each physical channel is spread with a Walsh code sequence to provide orthogonal channelization among these physical channels. The spread pilot and R-DCCH are mapped to the in-phase (I) data channel. The spread R-FCH and R-SCH are mapped to the quadrature (Q) data channel. Then, the I and Q data channels are spread using a complex-multiply-type PN spreading approach (see Fig. 15-29 for reverse dedicated channel structure and Fig. 15-30 for reverse link I and Q mapping for 1× and 3×).

15.8.4   Reverse Fundamental Channel (R-FCH)

The R-FCH supports 5- and 20-ms frames. The 20-ms frame structures provide rates derived from the IS-95B RS1 or RS2. The 5-ms frames provide 24 information bits per frame with 16-bit CRC. Within each 20-ms frame interval, either one 20-ms R-FCH structure, up to four 5-ms R-FCH structure(s), or nothing can be transmitted. In addition, when a 5-ms R-FCH structure is used, it can be on or off in each of the four 5-ms segments of a 20-ms frame interval. The R-FCH is transmitted at different rates. The rates supported for the R-FCH are 1.5, 2.7, 4.8, and 9.6 kbps for RS3 and RS5, and 1.8, 3.6, 7.2, and 14.4 kbps for RS4 and RS6. (see Fig. 15-31 for R-FCH/R-SCH for 1×, Radio Configuration (RC) 3 and 3×, RC 5)

15.8.5   Reverse Supplementary Channel (R-SCH)

The R-SCH can be operated in two distinct modes. The first mode is used for data rates not exceeding 14.4 kbps and uses blind rate detection (no scheduling or rate information). In the second mode, the rate information is explicitly known by the base station. The R-SCH is used for data calls and can operate at different prenegotiated rates. Only RS3 and RS5 are supported for the R-SCH. Since only RS3 is supported on the R-FCH for high-speed packet data calls, the rates supported for the R-SCH are 9.6, 19.2, 38.4, 76.8, and 153.6 kbps (see Fig. 15-32 for R-FCH/R-SCH for 1×, RC 4 and 3×, RC 6).

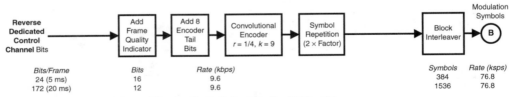

• **Reverse Dedicated Control Channel Structure for Radio Configuration (RC) 3 and 5**

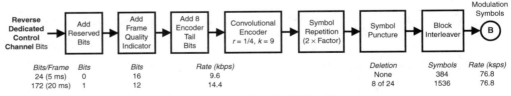

• **Reverse Dedicated Control Channel Structure for Radio Configuration (RC) 4 and 6**

Note: Signal Point B feeds into Fig. 15.30

Figure 15-29   Reverse Dedicated Channel Structure

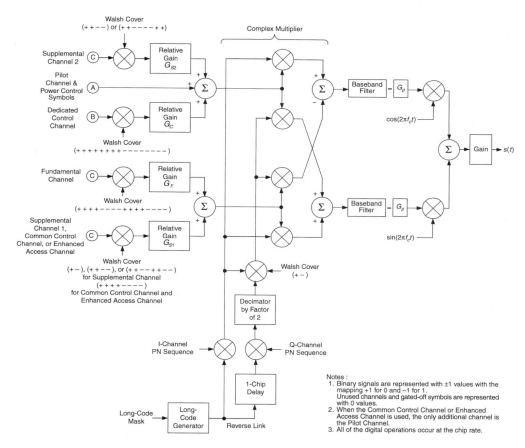

Figure 15-30   Reverse Link I and Q Mapping for 1× and 3×

Figure 15-33 provides a comparison between the reverse physical channels used in IS-95A/B and cdma2000.

15.8.6   FEC on Reverse Link

The reverse link uses a $k = 9$, $r = 1/4$ convolutional code for R-FCH. The distance properties of this code are better catered, providing performance gains vs. higher rate codes in fading and Additive White Gaussian Noise (AWGN) channel conditions. The constraint length $k = 9$, $r = 1/4$ convolutional code provides a gain of about 0.5 dB over a $k = 9$, $r = 1/2$ code in AWGN. The R-SCH uses convolutional codes for data rates of up to 14.4 kbps. Convolutional codes for higher data rates on the R-SCH are optional, and the use of turbo code is preferred. A common constituent code is used for the reverse link. Turbo codes of constraint length 4, rate 1/4, 1/3, and 1/2, are used for all R-SCH.

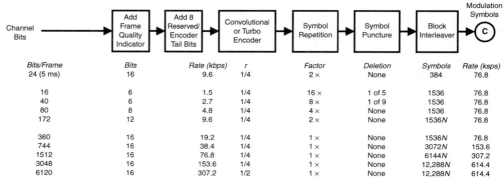

Bits/Frame	Bits	Rate (kbps)	r	Factor	Deletion	Symbols	Rate (ksps)
24 (5 ms)	16	9.6	1/4	2×	None	384	76.8
16	6	1.5	1/4	16×	1 of 5	1536	76.8
40	6	2.7	1/4	8×	1 of 9	1536	76.8
80	8	4.8	1/4	4×	None	1536	76.8
172	12	9.6	1/4	2×	None	1536N	76.8
360	16	19.2	1/4	1×	None	1536N	76.8
744	16	38.4	1/4	1×	None	3072N	153.6
1512	16	76.8	1/4	1×	None	6144N	307.2
3048	16	153.6	1/4	1×	None	12,288N	614.4
6120	16	307.2	1/2	1×	None	12,288N	614.4

Notes:
1. The 5-ms frame is used only for the fundamental channels, and only rates of 9.6 kbps or less are used for the fundamental channels.
2. Turbo coding may be used for the supplemental channels with rates of 19.2 kbps or more; otherwise $k = 9$ convolutional coding is used.
3. With convolutional coding, the reserved/encoder tail bits provide an encoder tail. With turbo coding, the first 2 of these bits are reserved bits that are encoded and the last 6 bits are replaced by an internally generated tail.
4. N is the number of consecutive 20-ms frames over which the interleaving is done (N = 1, 2 or 4).

Note: Signal Point C feeds into Fig. 15.30

Figure 15-31   R-FCH/R-SCH for 1×, RC3 and 3×, RC5

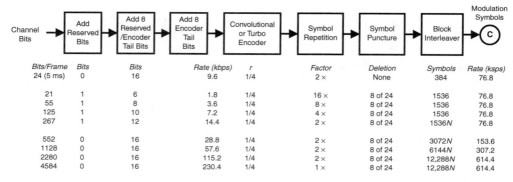

Bits/Frame	Bits	Bits	Rate (kbps)	r	Factor	Deletion	Symbols	Rate (ksps)
24 (5 ms)	0	16	9.6	1/4	2×	None	384	76.8
21	1	6	1.8	1/4	16×	8 of 24	1536	76.8
55	1	8	3.6	1/4	8×	8 of 24	1536	76.8
125	1	10	7.2	1/4	4×	8 of 24	1536	76.8
267	1	12	14.4	1/4	2×	8 of 24	1536N	76.8
552	0	16	28.8	1/4	2×	8 of 24	3072N	153.6
1128	0	16	57.6	1/4	2×	8 of 24	6144N	307.2
2280	0	16	115.2	1/4	2×	8 of 24	12,288N	614.4
4584	0	16	230.4	1/4	1×	8 of 24	12,288N	614.4

Notes:
1. The 5-ms frame is used only for the fundamental channels, and only rates of 14.4 kbps or less are used for the fundamental channels.
2. Turbo coding may be used for the supplemental channels with rates of 28.8 kbps or more; otherwise $k = 9$ convolutional coding is used.
3. With convolutional coding, the reserved/encoder tail bits provide an encoder tail. With turbo coding, the first 2 of these bits are reserved bits that are encoded and the last 6 bits are replaced by an internally generated tail.
4. N is the number of consecutive 20-ms frames over which the interleaving is done (N = 1, 2 or 4).

Note: Signal Point C feeds into Fig. 15.30

Figure 15-32   R-FCH/R-SCH for 1×, RC4 and 3×, RC6

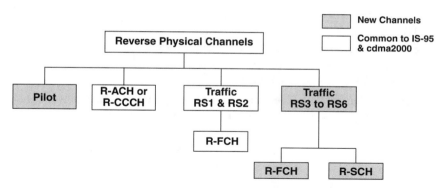

Figure 15-33   A Comparison of Reverse Physical Channels for IS-95 and cdma2000

15.8.7   Reverse Link Physical Layer Characteristics

- **Continuous waveform.** A continuous pilot and continuous data-channel waveform are used for all data rates. This continuous waveform minimizes interference to biomedical devices such as hearing aids and pacemakers and permits a range increase at lower transmission rates. The continuous waveform also enables the interleaving to be performed over the entire frame rather than just the portions that are not gated off. This enables the interleaving to achieve the full benefit of the frame time diversity. The base station uses the pilot for multipath searches, tracking, and coherent demodulation, as well as to measure the quality of the link for power control purposes. Separate orthogonal channels for the pilot and each of the data channels are used. Thus, the relative levels of the pilot and physical data channels can easily be adjusted without changing the frame structure or power levels of some symbols of a frame.
- **Orthogonal channels with different-length Walsh sequences.** The cdma2000 system uses orthogonal channels for the pilot and the other physical data channels. These orthogonal channels are provided with different-length Walsh sequences, with the higher-rate channels with shorter Walsh sequences. A short Walsh sequence allows high encoder output rates to be accommodated.
- **Rate matching.** Several approaches are needed to match the data rates to Walsh spreader input rates. These include adjusting the code rate using puncturing, symbol repetition, and sequence repetition. The design approach is to first try to use a low rate code, but not to reduce the rate below $r = 1/4$ since gains of smaller rates would be small and the decoder implementation complexity would increase significantly.
- **Low spectral side lobes.** The cdma2000 system achieves low spectral side lobes with nonideal mobile power amplifiers by splitting the physical channels into the I and Q channels and using a complex-multiply-type PN spreading approach.

- **Independent data channels.** Two types of physical data channels (R-FCH and R-SCH) are used on the reverse link that can be adapted to a particular type of service. The use of R-FCH and R-SCH enables the system to be optimized for multiple simultaneous services. These channels are separately coded and interleaved and may have different transmit power level and FER set points.

- **Reverse power control.** There are three components of reverse power control: *open loop*, *closed loop*, and *outer loop*. Open-loop power control sets the transmit power based upon the power that is received at the mobile. Open-loop power control compensates for the path loss from the mobile to the base station and handles very slow fading. Closed-loop power control consists of an 800-bps feedback loop from the base station to the mobile to set the transmit power of the mobile. Closed-loop power control compensates for medium to fast fading and inaccuracies in open-loop power control. Outer-loop power control is implementation specific but typically adjusts the closed-loop power control threshold in the base station to maintain a desired FER.

- **Separate dedicated control channel.** The reverse link consists of a separate low-rate, low-power, continuous, orthogonal, dedicated control channel. This allows for a flexible dedicated control channel structure that does not impact the other pilot and physical channel frame structures.

- **Frame length.** The cdma2000 system uses 5- and 20-ms frames for control information on fundamental and dedicated control channels and uses 20-ms frames for other types of data (including voice). Interleaving and sequence repetition are over the entire frame interval. This provides improved time diversity over systems that use shorter frames. The 20-ms frames are used for voice. A shorter frame would reduce the total voice delay, but it would degrade the demodulation performance due to the shorter interleaving span.

15.8.8   Reverse Link Modulation and Coding

The reverse link uses direct sequence spreading with an IS-95B chip rate of 1.2288 Mcps (denoted as 1× chip rate) or chip rates that are 3, 6, 9, or 12 times IS-95B chip rate. Higher-chip-rate systems are denoted as 3×, 6×, 9×, and 12× and they are respectively operated at 3.6864, 7.3728, 11.0592, and 14.7456 Mcps.

The 1× system can be used anywhere that an IS-95B reverse link is used. An IS-95B reverse link carrier frequency can also be shared with mobiles transmitting the IS-95B waveform and those transmitting the 1× cdma2000 waveform. The higher-chip-rate reverse links can be used in applications where larger bandwidth allocations are available. Mobiles that support a higher chip rate would typically also support the 1× chip rate. This will allow these mobiles to access base stations that support only the 1× and higher chip rate systems.

Within an operator's allocated band, the 1× cdma2000 reverse links would typically occupy the same bandwidth as the IS-95B reverse link system (i.e., 1.25 MHz), and higher-chip-rate cdma2000 links would typically occupy a bandwidth that is 1.25 MHz times the higher-chip-rate factor. A guard band of 625 kHz would typically be used on both sides of the operator's allocated band.

15.8.9 Key Characteristics of Reverse Link

The key characteristics of the reverse link are

- Channels are primarily code multiplexed.
- Separate channels are used for different QoS and physical layer characteristics.
- Transmission is continuous to avoid electromagnetic interference (EMI).
- Channels are orthogonalized by Walsh functions and I/Q split so that performance is equivalent to BPSK.
- Hybrid combination of QPSK and BPSK.
- Coherent reverse link with continuous pilot.
- Forward power control information is time multiplexed with the pilot.
- By restricting alternate phase changes of the complex scrambling, power peaking is reduced and side lobes are narrowed.
- Independent fundamental and supplemental channels with different transmit power and FER target.
- Forward error correction:

  - Convolutional codes ($k = 9$) are used for voice and data.
  - Parallel turbo codes ($k = 4$) are used for high data rates on supplemental channels.

- Fast-reverse power control: 800 times per second.
- Frame lengths:

  - 20-ms frames are used for signaling and user information.
  - 5-ms frames are used for control information.

15.9 Data Services in cdma2000

Two types of data services are being considered for cdma2000—packet and high-speed circuit data services. The packet service and the MAC layer are designed to support a large number of mobile stations using packet data services. Many packet data services exhibit highly bursty traffic patterns with relatively long periods of inactivity. Due to limited air-interface capacity, limited base station equipment, and constraints on mobile station power consumption, dedicated channels for packet service users are allocated on demand and released immediately after the end of the activity period.

Releasing the dedicated channels and reestablishing them introduces latency and signaling overhead due to the renegotiation process that has to take place between the BS and MS before user data exchange. The overhead of reestablishing the dedicated channels includes the cost of RLP synchronization and the signaling overhead associated with service negotiation to reconnect the packet service. The MAC avoids this latency and overhead by allowing the BS to save a set of state information after the initialization phase is completed.

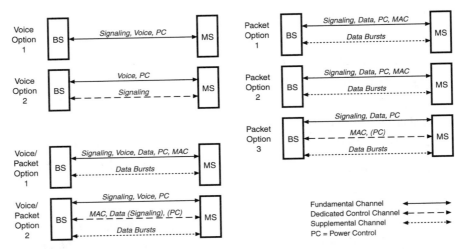

Figure 15-34 Logical-to-Physical Channels Mapping for Voice and Packet Services

To further reduce the overhead associated with assignment of dedicated channels, the packet service allows for exchange of short bursts of user data when no dedicated channels are present. This mode of operation may be suitable for mobile-IP registration, notification services (e.g., e-mail notification), and location tracking services where the volume of data to be exchanged is typically small.

Circuit services can be viewed as a special case of the packet services in the sense that dedicated traffic and control channels are typically assigned to the MS for extended periods of time during the circuit service sessions. This will lead to a less efficient use of the air-interface capacity. However, some delay-sensitive services such as video applications require a dedicated channel for the duration of the call.

15.10 Mapping of Logical Channels to Physical Channels

The mapping from the logical channel to the physical channel is not one to one—multiple logical channels from multiple service options may be mapped to a single physical channel. This multiplexing operation is performed at the PLDCF MUX and QoS sublayer. Fig. 15-34 shows the channel structure for bearer service profiles that include only packet and voice services.

15.10.1   Forward Link

15.10.1.1   Forward Link Dedicated Channel

The mapping of forward logical channels into forward physical channels is given in Table 15-6. The table shows attributes of each physical channel such as variable rate or fixed rate, physical channel frame size, and sharability of the channel.

Voice Services. In the **V1** mode, the F-SCH and F-DCCH are not used. In this mode the upper layer signaling (f-dsch), voice frames (f-dtch), and power control information are multi-

Table 15-6 Forward Dedicated Channel Structure for Different Services

Forward Link Physical Channel	Forward Link Logical Channel	Description	Services						
			V1	V2	V3	P1	P2	VP1	VP2
Fundamental channel (F-FCH) (soft handoff)	f-dsch	upper layer signaling messages	x		x		x	x	x
	f-dtch	RLP frames			x		x	x	
		voice frames	x	x				x	x
	f-dmch	MAC messages			x			x	
	pc	power control	x	x	x		x	x	x
		frame size	20	20	5/20		20	5/20	20
		rates	V	V	V		V	V	V
Supplemental channel (F-SCH) (with or without soft handoff)	f-dtch	RLP frames, voice frames (for newly defined services)			x	x	x	x	x
		rates			H	V/H	H	H	V/H
Dedicated control channel (F-DCCH) (with or without soft handoff)	f-dsch	upper layer signaling messages		x		x			x
	f-dtch	RLP frames				x	x		x
	f-dmch	MAC messages				x	x		x
	pc	power control				x			
		sharable		Yes		No	Yes		Yes
		frame size		20		5/20	5		5/20
		rates		F		F	F		F

Key: F = fixed (9.6 kbps or 0); V = variable (9.6, 4.8, 2.7, 1.5 kbps for RS1 and 14.4, 7.2, 3.6, 1.8 kbps for RS2); H = high data rates (scheduled rates).

plexed in F-FCH (e.g., by using dim-and-burst or blank-and-burst mechanisms). Forward link continuity and outer-loop power control in this mode is maintained by the F-FCH.

The **V2** mode is used to provide higher voice-quality service by transmitting the upper layer signaling frames on an F-DCCH (e.g., no blank-and-burst or dim-and-burst signaling). Forward link continuity and outer-loop power control in this mode is maintained by the F-FCH.

Packet Services. The packet data service, **P1**, is offered on the forward direction by using F-FCH and F-SCH. The upper layer signaling message (f-dsch), MAC messages (f-dmch), and user data frames (f-dtch) are time multiplexed in the F-FCH. In this mode, the control of MAC is performed in a centralized manner since MAC messages are carried on the F-FCH which is typically in soft handoff to ensure reliability of delivery for upper layer signaling messages. The 5-ms frames are used to carry short MAC messages in this mode. The F-SCH carries high-rate RLP frames containing packet data, and transmission on the F-SCH is always scheduled in this mode. Lower-rate RLP frames may be carried on the F-FCH. The transmission rate of F-SCH is predefined using MAC messages. Forward link continuity and outer-loop power control in this mode are maintained by the F-FCH.

The **P2** mode of operation is an alternate basic packet data service and is similar to the **P1** mode in the sense that upper layer signaling messages (f-dsch), MAC messages (f-dmch), and user data frames (f-dtch) are time multiplexed in one channel. However, the physical channel that is used to carry these logical channels is the F-DCCH, which may or may not be in soft handoff. Thus, the control of the MAC can be performed in either a distributed or centralized manner. To support the mixing of MAC signaling with RLP frames or upper layer signaling information, the F-DCCH supports dual-frame-size operation (5 and 20 ms). The F-SCH carries high-rate, scheduled RLP frames containing packet data as well as lower-rate RLP frames. The rate for the lower-rate frames carried on F-SCH may be dynamically determined, but the transmission rate of the high-rate scheduled frames carried on the F-SCH is prespecified using MAC messages. Forward link continuity and outer-loop power control are maintained by the F-DCCH and, therefore, the F-DCCH becomes unsharable in this mode.

The **P3** mode is used for highly optimized packet data service with the potential support for distributed control of the MAC layer (i.e., the f-dmch is carried in a physical channel that can be operated with a reduced active set while upper layer signaling information is carried in a channel with a full active set). In this mode, the F-FCH is primarily used to carry high-reliability, low-delay upper layer messages. Power control bits are carried by F-FCH. The F-DCCH may be shared to make a more efficient use of Walsh code resources. The F-DCCH carries the MAC signaling (f-dmch) and may not be in soft handoff. Control of MAC can be performed in either a distributed or centralized manner. The F-SCH carries high-rate RLP frames containing packet data, and transmission on the F-SCH is always scheduled in this mode. The transmission rate of F-SCH is prespecified using MAC messages. Lower-rate RLP frames may be carried on the F-FCH. Forward link continuity and outer-loop power control in this mode are maintained by the F-FCH.

Concurrent Voice and Packet Services. The **VP1** mode offers simultaneous basic voice and packet data service by multiplexing upper layer signaling (f-dsch), MAC messages (f-dmch), voice frames (f-dtch), and potentially low-rate RLP frames (f-dtch) into the F-FCH. Control of MAC is performed in a centralized manner. To support the mixing of MAC signaling with RLP frames or upper layer signaling information, the F-FCH supports dual-frame-size operation (5 and 20 ms). The F-SCH carries high-rate RLP frames containing packet data, and transmission on the F-SCH is always scheduled in this mode. Forward link continuity and outer-loop power control are maintained by the F-FCH.

The **VP2** mode also provides simultaneous voice and packet data service. To provide a higher-quality voice service in conjunction with packet data service, the MAC messages (f-dmch) and potentially upper layer signaling (f-dsch) are carried on the F-DCCH. Control of MAC can be performed in either a distributed (if F-DCCH is not in soft handoff) or centralized (if DCCH is in soft handoff) manner. Power control bits are carried by the F-FCH. The F-DCCH can be shared to make a more efficient use of Walsh code resources. To support the mixing of MAC signaling with RLP frames or upper layer signaling information, the F-DCCH supports dual-frame-size operation (5 and 20 ms). The F-SCH carries high-rate, scheduled RLP frames containing packet data as well as lower-rate RLP frames. The low-rate RLP frames may be sent on F-SCH to avoid the potential contention between voice and low-rate RLP frames on the F-FCH. Forward link continuity and outer-loop power control are maintained by F-FCH.

Consult cdma2000 RTT [7] for circuit services and their combinations with voice and packet services.

15.10.1.2 Forward Link Common Channels
When neither F-DCCH nor F-FCH is allocated to the MS (e.g., in the suspended or dormant states), the upper layer signaling and MAC messages are carried to the MS using F-PCH or F-CCCH. Messages sent on these channels may be encrypted and must include mobile station ID or the packet service identifier because both F-PCH and F-CCCH are point-to-multipoint channels in the sense that there is no one-to-one mapping between the ID of these channels and mobile station ID. In addition to the control information (i.e., MAC or upper layer signaling), short data bursts may be carried by the F-PCH or F-CCCH. Table 15-7 shows the mapping of forward common channels to forward common physical channels.

15.10.2 Reverse Link
15.10.2.1 Reverse Link Dedicated Channels
The mapping of reverse logical channels into reverse physical channels is given in Table 15-8. The table also shows the attributes of each physical channel such as variable rate or fixed rate and physical channel frame size.

Table 15-7 Mapping of Forward Common Logical Channels to Forward Common Physical Channels

Forward Link Physical Channels	Forward Logical Channels	Description
Common control channel (F-CCCH) or paging channel (F-PCH)	f-csch	upper layer signaling messages
	f-ctch	RBP frames
	f-cmch	MAC messages
Common pilot	—	common pilot
Auxiliary pilot	—	auxiliary pilot
Sync channel (F-SYNC)	—	sync channel information

Table 15-8 Reverse Dedicated Channel Structure for Different Services

Reverse Link Physical Channel	Reverse Link Logical Channel	Description	Services						
			V1	V2	V3	P1	P2	VP1	VP2
Fundamental channel (R-FCH) (soft handoff)	r-dsch	upper layer signaling messages	x	x	x		x	x	x
	r-dtch	RLP frames			x		x	x	
		voice frames	x	x				x	x
	r-dmch	MAC messages			x			x	
		frame size	20	20	5/20		20	5/20	20
		rates	V	V	V		V	V	V
Supplemental channel (R-SCH) (with or without soft handoff)	r-dtch	RLP frames, voice frames (for newly defined services)			x	x	x	x	x
		rates			H	V/H	H	H	V/H
Dedicated control channel (R-DCCH) (with or without soft handoff)	r-dsch	upper layer signaling messages		x		x			x
	r-dtch	RLP frames				x	x		x
	r-dmch	MAC messages				x	x		x
		frame size		20		5/20	5		5/20
		rates		F		F	F		F

Key: F = fixed (9.6 kbps or 0); V = variable (9.6, 4.8, 2.7, 1.5 kbps for RS1 and 14.4, 7.2, 3.6, 1.8 kbps for RS2); H = high data rates (scheduled rates).

Voice Services. In the **V1** mode, the R-SCH and R-DCCH are not used. In this case the upper layer signaling (r-dsch), voice frames (r-dtch), and power control information are multiplexed in R-FCH (e.g., using dim-and-burst or blank-and-burst mechanisms).

The **V2** mode provides a higher voice-quality service typically by transmitting the upper layer signaling frames on an R-DCCH (e.g., no blank-and-burst or dim-and-burst signaling). However, if the mobile station cannot provide sufficient power to transmit on the R-DCCH, the upper layer signaling information can be transmitted on the R-FCH.

Packet Services. **P1** is offered on the reverse direction by using R-FCH and R-SCH. The upper layer signaling message (r-dsch), MAC messages (r-dmch), and user data frames (r-dtch) are time multiplexed in R-FCH. To support the mixing of MAC signaling with RLP frames or upper layer signaling information, the R-FCH supports dual-frame-size operation (5 and 20 ms).

The R-SCH carries high-rate RLP frames containing packet data, and transmission on the R-SCH is always scheduled. Lower-rate RLP frames may be carried on R-FCH.

In the **P2** mode, upper layer signaling messages (r-dsch), MAC messages (r-dmch), and user data frames (r-dtch) are time multiplexed in R-DCCH. To support the mixing of MAC signaling with RLP frames or upper layer signaling information, the R-DCCH supports dual-frame-size operation (5 and 20 ms). The R-SCH carries high-rate scheduled RLP frames containing data as well as lower-rate RLP frames.

R-FCH is primarily used to carry high-reliability, low-delay upper layer messages as well as power control information. The R-DCCH carries the MAC signaling (r-dmch). The R-SCH carries high-rate RLP frames containing packet data, and transmission on the R-SCH is always scheduled. Lower-rate RLP frames may be carried on the R-FCH.

Concurrent Voice and Packet Data. The **VP1** mode (which offers simultaneous basic voice and packet data service) multiplexes upper layer signaling (r-dsch), MAC messages (r-dmch), voice frames (r-dtch), and, potentially, low-rate RLP frames (r-dtch) into the R-FCH. To support the mixing of MAC signaling with RLP frames or upper layer signaling information, the R-FCH supports dual-frame-size operation (5 and 20 ms). The R-SCH carries high-rate RLP frames containing packet data, and transmission on the R-SCH is always scheduled in this mode.

In the **VP2** mode, the MAC messages (r-dmch) and potentially upper layer signaling information (r-dsch) are carried on R-DCCH (thereby reducing potential disruption due to dim-and-burst and blank-and-burst signaling). However, if the mobile cannot provide sufficient power to transmit on the R-DCCH, the upper layer signaling information can be transmitted on the R-FCH. To support the mixing of MAC signaling with RLP frames or upper layer signaling information, the R-DCCH supports dual-frame-size operation (5 and 20 ms). The R-SCH carries high-rate, scheduled RLP frames containing packet data as well as lower-rate RLP frames. The low-rate RLP frames may be sent on the R-SCH to avoid the potential contention between voice and low-rate RLP frames on the R-FCH.

15.10.2.2 Reverse Link Common Channels

When neither R-DCCH nor R-FCH is allocated to the mobile (e.g., in the suspended or dormant state), the upper layer signaling and MAC messages are conveyed to the base station using R-ACH or R-CCCH. Messages sent on these channels may be encrypted and must include the mobile ID or the packet service identifier since there is no one-to-one mapping between the identity of the R-ACH and R-CCCH channel and the mobile ID. In addition to the control information (i.e., MAC or upper layer signaling), short data bursts may be carried by the R-ACH or R-CCCH. Table 15-9 shows the mapping of the reverse common logical channels to reverse common physical channels.

15.11 Evolution of cdmaOne (IS-95) to cdma2000

Data is the word as cdmaOne operators eye a host of new network capabilities enabling them to offer new value-added services that can exploit present and future generations of technology. With the Internet and corporate intranet becoming more essential to daily business activities, the

Table 15-9 Mapping of Reverse Common Logical Channels to Reverse Common Physical Channels

Reverse Link Physical Channels	Reverse Logical Channels	Description
Common control channel (R-CCCH) or access channel (R-ACH)	r-csch	upper layer signaling messages
	r-ctch[a]	RBP frames
	r-cmch	MAC messages
Pilot	—	pilot
	pc	power control bits

[a] r-ctch is carried on the same physical channel (R-CCCH), but can be physically separated by using a different long code.

rush is on to create the wireless office that can easily tie mobile workers to the enterprise. Further, there is great potential for push technologies that deliver news and other information directly to a wireless device—this could create entirely new revenue streams for operators.

Although cdmaOne networks were not the first to offer data access, these networks are uniquely designed to accommodate data. To start with, they handle data and voice transmissions in much the same way. cdmaOne's inherent variable-rate transmission capability allows data rate determination to accommodate the amount of information being sent, so system resources are used only as needed. Because cdmaOne systems employ a packetized backbone for voice, packet data capabilities are already inherent in the equipment. The cdmaOne packet data transmission technology uses a TCP/IP-compliant Cellular Digital Packet Data (CDPD) protocol stack to enable seamless connectivity with enterprise networks and to expedite third-party application development.

Adding data to the cdmaOne network will allow an operator to continue using its existing radios, backhaul facilities, infrastructure, and handsets while merely implementing a software upgrade with an interworking function. Upgrading to IS-95B allows for code or channel aggregation to provide data rates of 64–115 kbps, as well as offering improvements in soft handoffs and interfrequency hard handoffs. Equipment manufacturers have already announced IS-707 packet data, circuit-switched data, and digital fax capabilities on its cdmaOne infrastructure equipment.

Mobile IP, the proposed Internet standard for mobility, is an enhancement to basic packet data services. Mobile IP lets users maintain a continuous data connection and retain a single IP address while traveling between base station controllers (BSCs) or roaming on other CDMA networks.

One of the key objectives of ITU IMT-2000 is the creation of standards that will encourage a worldwide frequency band to promote a high degree of design commonality and to support high-speed data services. IMT-2000 will utilize small pocket terminals, an expanded range of operation environments, and the deployment of an open architecture that allows the graceful introduction of newly created technology. Furthermore, 3G systems promise to deliver wireless voice services with wireline quality levels, along with the speed and capacity needed to support multimedia and high-speed data applications. Location-based services, on-board navigation, emergency assistance, and other advanced services will also be supported.

The evolution of 3G systems will open the door of the wireless local loop (WLL) to PSTN and public data network access, while providing more convenient control of applications and network resources. It will also provide global roaming, service portability, zone-based ID and billing, and global directory access. The 3G technology is even expected to support seamless satellite interworking.

One of the technical requirements for cdma2000 includes cdmaOne backward compatibility for voice services, vocoders, and signaling structure, as well as for privacy, authentication, and encryption capabilities.

Phase one of the cdma2000 effort, also known as 1xRTT, employs 1.25 MHz of bandwidth and delivers a peak data rate of 144 kbps for stationary or mobile applications. Phase two of cdma2000, called 3xRTT, will use 5 MHz of bandwidth and is expected to deliver a peak data rate of 144 kbps for mobile and vehicular applications and up to 2 Mbps for fixed applications. Industry insiders predict that the 3xRTT phase will eventually yield up to 1 Mbps for each traffic or Walsh channel. By aggregating or bundling two channels, users can achieve the 2-Mbps peak data rate targeted for IMT-2000.

The primary difference between phase one and phase two of cdma2000 is bandwidth and the resulting throughput speed, or peak data rate capability. Phase two will introduce advanced multimedia capabilities and lay the foundation for popular 3G voice services and vocoders, such as voice over IP. Since the 1xRTT and 3xRTT standards essentially share the same baseband radio elements, operators can take a major step toward full 3G capabilities by implementing 1xRTT. cdma2000 phase two will include detailed descriptions of signal protocols, data management, and expected upscale requirements for moving from 5-MHz radios to 10- and 15-MHz radios in future interactions.

By migrating from the current IS-95 CDMA air-interface technology to 1xRTT of the cdma2000 standard, operators can reap a twofold increase in radio capacity and the ability to handle up to 144 kbps of packet data. Phase one capabilities of cdma2000 include a new physical layer for 1x and 3x 1.25-MHz channel sizes; support for direct-spread and multicarrier forward link 3x options; and definitions for the 1x and 3x numerology. Operators will also enjoy voice service enhancements that will produce two times the voice capacity.

In the area of extended battery life, phase one will employ a quick paging channel and gated transmission of 1/8 rate to produce gains of two times the battery life currently available. Hard-handoff enhancements between 2G and 3G systems and power control enhancements will also be key factors in the improvement of voice service.

Data services will also be improved with the advent of cdma2000 phase one. Phase one will feature a MAC framework and a packet data radio link protocol (RLP) definition to support packet data rates of at least 144 kbps.

Implementation of cdma2000 phase two will bring a host of new capabilities and service enhancements. Phase two will support all channel sizes (6×, 9×, and 12×) and associated numerology and a framework for advanced cdma2000 3G voice services and vocoders—including voice over IP. With phase two, true multimedia services will be available to bring additional revenue opportunities for operators. Multimedia services will be made possible through enhanced packet data MAC, full support for packet data services up to 2 Mbps, RLP support for all data rates up to 2 Mbps, and the advanced multimedia call model.

In the area of signaling and services, phase two cdma2000 will bring native 3G cdma2000 signaling structure to the link access control (LAC) and upper layer signaling structure. This structure will provide support for enhanced privacy, authentication, and encryption functionality. An operator's existing architecture and network equipment can greatly affect the ease of this migration. A network built on an open, advanced architecture with a clear upward migration pathway can attain 1xRTT capabilities with a simple modular upward movement of the H-band operation of the radio. Networks with a less flexible architecture may be required to take the more costly steps of replacing the entire base transceiver system (BTS). To achieve the expected 144-kbps peak data rate performance, operators can make software upgrades to networks and base stations to support 1xRTT data protocol.

Packet Data Service Node (PDSN) will be required to support data connectivity to the Internet/intranet. Many equipment vendors already offer solutions that incorporate PDSN elements, thus opening a smooth upward pathway to 3G technologies.

The recent agreement between Qualcomm and Ericsson proposes three optional CDMA modes and the eventual development of a global standard that is compatible with both ANSI IS-41 and GSM MAP. This approach envisions the use of multimode handsets and various market-driven solutions as the surest pathway to a unified CDMA 3G standard in the next generation of wireless communications. As subscribers demand greater wireless power and convenience, the migration to 3G technology will benefit operators by supporting higher capabilities, lowering network costs, and increasing overall profitability. Fig. 15-35 shows the cdmaOne evolution timeline.

cdmaOne operators will be able to upgrade to 3G system without acquiring additional spectrum, a key component to minimum time to market without additional, significant investment. The design of cdma2000 will allow for deployment of the 3G enhancements while maintaining existing 2G support for cdmaOne in the spectrum being used by an operator today.

Both cdma2000 phase one and phase two can be intermingled with cdmaOne to maximize the effective use of spectrum according to the needs of an individual operator's customer base. For example, an operator with a strong demand for high-speed data service may choose to deploy a combination of cdma2000 phase one and cdmaOne that uses more channels for cdma2000 (see Fig. 15-36). In another market, users may not be as quick to adopt high-speed data services, and more channels will remain dedicated to cdmaOne services. As cdma2000 phase two capabilities become available, an operator has even more choices of ways in which to use the spectrum to support the new services (see Figs. 15-37 and 15-38).

15.12 Major Technical Differences between cdma2000 and W-CDMA

Table 15-10 lists the major technical differences between the two wideband CDMA 3G proposals.

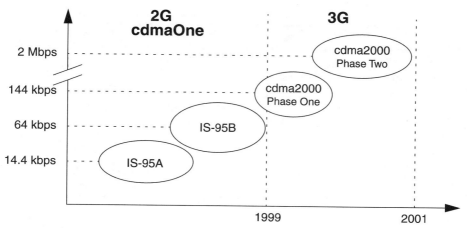

Figure 15-35 cdmaOne Evolution Timeline

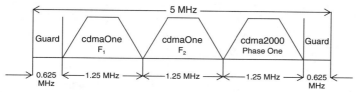

Figure 15-36 Intermixing of cdmaOne and cdma2000 Phase One

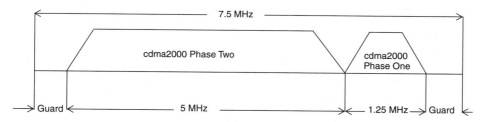

Figure 15-37 Intermixing of cdma2000 Phase One and cdma2000 Phase Two

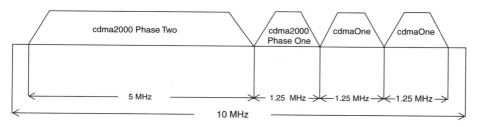

Figure 15-38 Intermixing of cdmaOne, cdma2000 Phase One, and cdma2000 Phase Two

Table 15-10   Major Technical Differences between cdma2000 and W-CDMA

	cdma2000	W-CDMA
Core network	ANSI-41	GSM MAP
Chip rate	3.6864 Mcps	4.096 Mcps (DOCOMO) 3.84 Mcps (UMTS)
Synchronized BS	yes	no/yes (optional)
Frame length	20 ms	10 ms
Multicarrier spreading option	yes	no
Voice coder	EVRC	new
Overhead	low (because of shared pilot code channel)	high (because of nonshared pilot code channel)

15.13 Summary

This chapter first presented cdma2000 layering structure and logical and physical channels and then concentrated on the cdma2000 physical layer, providing details of forward/reverse dedicated and common physical channel structures. We described forward and reverse link features and pointed out the improvements of cdma2000 over cdmaOne.

We briefly discussed data services in cdma2000 and considered the mapping of logical channels. Next we presented the evolution plans for cdmaOne to cdma2000. We concluded the chapter by providing the major technical differences between cdma2000 and W-CDMA.

15.14 References

1. "CDMA for Next Generation Mobile Communications Systems," *IEEE Communications Magazine* 36(9), September 1998.

2. Dahlman, E., Gumudson, B., Nilsson, M., and Skold, J., "UMTS/IMT-2000 Based Wideband CDMA," *IEEE Communications Magazine* 36(9), September 1998.

3. Garg, V. K., Halpern, S., and Smolik, K. F., "Third Generation (3G) Mobile Communication Systems," *1999 IEEE International Conference on Personal Wireless Communications,* Jaipur, India, February 1999.

4. Knisley, D., Quinn, L., and Ramesh, N., "cdma2000: A Third Generation Radio Transmission Technology," *Bell Labs Technical Journal* 3(3), July–September 1998.

5. Rao, Y. S., and Kripalani, A., "cdma2000 Mobile Radio Access for IMT-2000," *1999 IEEE International Conference on Personal Wireless Communications,* Jaipur, India, February 1999.

6. Shanker, B., McClelland, S., "Mobilising the Third-Generation [Cellular Radio]," *Journal of Telecommunications* (International Edition), August 1997.

7. TIA TR 45.5, "The cdma2000 ITU-RTT Candidate Submission," TR 45-ISD/98.06.02.03, May 15, 1998.

Traffic Tables

This appendix provides traffic tables for a variety of blocking probabilities and channels. The blocked-calls-cleared (Erlang B) call model is used. In Erlang B, when traffic arrives in the system, we assume that it either is served, with probability from the table, or is lost to the system. A customer attempting to place a call will therefore either see a call completion or will be blocked and will abandon the call. This assumption is acceptable for low blocking probabilities. In some cases, the call will be placed again after a short period of time. If too many calls reappear in the system after a short delay, the Erlang B model will no longer hold.

Table A-1   Offered Load for Given Blocking Probability

Number of Servers	Blocking Probability							
	0.005	0.01	0.015	0.02	0.03	0.05	0.07	0.1
1	—	0.01011	0.01524	0.02041	0.03093	0.05264	0.07527	0.1111
2	—	0.1527	0.1904	0.2235	0.2816	0.3814	0.4705	0.5955
3	—	0.4556	0.5352	0.6022	0.7152	0.8994	1.057	1.271
4	—	0.8693	0.9919	1.092	1.259	1.525	1.748	2.045
5	—	1.361	1.524	1.657	1.875	2.219	2.504	2.881
6	—	1.909	2.112	2.276	2.543	2.961	3.305	3.759
7	—	2.501	2.741	2.936	3.25	3.738	4.139	4.666
8	—	3.127	3.405	3.627	3.987	4.543	4.999	5.597
9	—	3.783	4.095	4.345	4.748	5.371	5.88	6.547
10	—	4.462	4.808	5.084	5.53	6.216	6.777	7.511

Table A-1   Offered Load for Given Blocking Probability (Continued)

Number of Servers	Blocking Probability							
	0.005	0.01	0.015	0.02	0.03	0.05	0.07	0.1
11	—	5.159	5.54	5.842	6.328	7.076	7.688	8.487
12	—	5.877	6.288	6.615	7.142	7.95	8.61	9.474
13	—	6.607	7.05	7.401	7.967	8.835	9.543	10.47
14	—	7.352	7.825	8.201	8.804	9.73	10.49	11.47
15	—	8.109	8.61	9.01	9.651	10.63	11.43	12.48
16	—	8.876	9.406	9.829	10.51	11.54	12.39	13.5
17	—	9.653	10.21	10.66	11.37	12.46	13.35	14.52
18	—	10.44	11.03	11.49	12.24	13.39	14.32	15.55
19	—	11.23	11.85	12.33	13.12	14.31	15.29	16.58
20	—	12.03	12.67	13.18	14	15.25	16.27	17.61
21	11.86	12.84	13.51	14.04	14.89	16.19	17.25	18.65
22	12.63	13.65	14.35	14.9	15.78	17.13	18.24	19.69
23	13.42	14.47	15.19	15.76	16.68	18.08	19.23	20.74
24	14.2	15.3	16.04	16.63	17.58	19.03	20.22	21.78
25	15	16.13	16.9	17.5	18.48	19.99	21.22	22.83
26	15.8	16.96	17.75	18.38	19.39	20.94	22.21	23.89
27	16.6	17.8	18.62	19.27	20.31	21.9	23.21	24.94
28	17.41	18.64	19.48	20.15	21.22	22.87	24.22	26
29	18.22	19.49	20.35	21.04	22.14	23.83	25.22	27.05
30	19.04	20.34	21.23	21.93	23.06	24.8	26.23	28.11
31	19.85	21.19	22.1	22.83	23.99	25.77	27.24	29.17
32	20.68	22.05	22.98	23.73	24.92	26.75	28.25	30.24
33	21.51	22.91	23.87	24.63	25.85	27.72	29.26	31.3
34	22.34	23.77	24.75	25.53	26.78	28.7	30.28	32.37
35	23.17	24.64	25.64	26.44	27.71	29.68	31.29	33.43
36	24.01	25.51	26.53	27.34	28.65	30.66	32.31	34.5
37	24.85	26.38	27.42	28.26	29.59	31.64	33.33	35.57
38	25.69	27.25	28.32	29.17	30.53	32.62	34.35	36.64
39	26.54	28.13	29.22	30.08	31.47	33.61	35.37	37.72
40	27.38	29.01	30.12	31	32.41	34.6	36.4	38.79

Table A-1   Offered Load for Given Blocking Probability (Continued)

Number of Servers	Blocking Probability							
	0.005	0.01	0.015	0.02	0.03	0.05	0.07	0.1
41	28.23	29.89	31.02	31.92	33.36	35.59	37.42	39.86
42	29.09	30.77	31.92	32.84	34.31	36.57	38.45	40.94
43	29.94	31.66	32.83	33.76	35.25	37.57	39.47	42.01
44	30.8	32.55	33.74	34.68	36.2	38.56	40.5	43.09
45	31.66	33.43	34.65	35.61	37.16	39.55	41.53	44.17
46	32.52	34.32	35.56	36.54	38.11	40.55	42.56	45.24
47	33.38	35.22	36.47	37.46	39.06	41.54	43.59	46.32
48	34.25	36.11	37.38	38.39	40.02	42.54	44.62	47.4
49	35.12	37.01	38.3	39.32	40.98	43.54	45.65	48.48
50	35.99	37.9	39.21	40.25	41.93	44.53	46.69	49.56
51	36.86	38.8	40.13	41.19	42.89	45.53	47.72	50.64
52	37.73	39.7	41.05	42.12	43.85	46.53	48.76	51.73
53	38.6	40.61	41.97	43.06	44.81	47.54	49.79	52.81
54	39.47	41.51	42.9	44	45.78	48.54	50.83	53.89
55	40.36	42.41	43.82	44.94	46.74	49.54	51.86	54.98
56	41.23	43.32	44.74	45.88	47.71	50.54	52.9	56.06
57	42.11	44.23	45.67	46.82	48.67	51.55	53.94	57.15
58	43	45.13	46.6	47.76	49.64	52.55	54.98	58.23
59	43.87	46.04	47.52	48.7	50.6	53.56	56.02	59.32
60	44.75	46.95	48.45	49.65	51.57	54.57	57.06	60.4
61	45.65	47.86	49.38	50.59	52.54	55.57	58.1	61.49
62	46.53	48.78	50.31	51.54	53.51	56.58	59.14	62.58
63	47.42	49.69	51.24	52.48	54.48	57.59	60.18	63.66
64	48.3	50.6	52.18	53.43	55.45	58.6	61.22	64.75
65	49.2	51.52	53.11	54.38	56.42	59.61	62.27	65.84
66	50.09	52.43	54.05	55.33	57.4	60.62	63.31	66.93
67	50.98	53.36	54.98	56.28	58.37	61.63	64.35	68.02
68	51.87	54.27	55.92	57.23	59.34	62.64	65.4	69.11
69	52.76	55.2	56.85	58.18	60.32	63.66	66.44	70.2
70	53.66	56.12	57.79	59.13	61.29	64.67	67.49	71.29

Table A-1   Offered Load for Given Blocking Probability (Continued)

Number of Servers	Blocking Probability							
	0.005	0.01	0.015	0.02	0.03	0.05	0.07	0.1
71	54.55	57.04	58.73	60.08	62.27	65.68	68.53	72.38
72	55.46	57.96	59.67	61.04	63.25	66.69	69.58	73.47
73	56.35	58.88	60.61	61.99	64.22	67.71	70.63	74.56
74	57.26	59.81	61.55	62.95	65.2	68.72	71.67	75.65
75	58.15	60.73	62.5	63.9	66.18	69.74	72.72	76.74
76	59.06	61.66	63.44	64.86	67.16	70.75	73.77	77.83
77	59.96	62.58	64.38	65.82	68.14	71.77	74.81	78.93
78	60.87	63.51	65.32	66.77	69.12	72.79	75.86	80.02
79	61.77	64.44	66.27	67.73	70.1	73.8	76.91	81.11
80	62.67	65.36	67.21	68.69	71.08	74.82	77.96	82.2
81	63.58	66.3	68.16	69.65	72.06	75.84	79.01	83.3
82	64.47	67.23	69.11	70.61	73.04	76.86	80.06	84.39
83	65.38	68.16	70.05	71.57	74.03	77.88	81.11	85.49
84	66.3	69.09	71	72.53	75.01	78.89	82.16	86.58
85	67.2	70.02	71.95	73.49	75.99	79.91	83.21	87.67
86	68.11	70.95	72.9	74.46	76.98	80.93	84.26	88.77
87	69.03	71.89	73.85	75.42	77.96	81.95	85.31	89.86
88	69.93	72.81	74.8	76.38	78.95	82.97	86.36	90.96
89	70.85	73.75	75.74	77.34	79.93	83.99	87.41	92.05
90	71.76	74.69	76.7	78.31	80.92	85.02	88.46	93.15
91	72.67	75.62	77.65	79.27	81.9	86.04	89.52	94.24
92	73.58	76.56	78.6	80.24	82.89	87.06	90.57	95.34
93	74.5	77.5	79.56	81.2	83.88	88.08	91.62	96.43
94	75.41	78.43	80.51	82.17	84.86	89.1	92.67	97.53
95	76.33	79.37	81.46	83.14	85.85	90.13	93.73	98.63
96	77.25	80.3	82.42	84.1	86.84	91.15	94.78	99.72
97	78.16	81.25	83.37	85.07	87.83	92.17	95.83	100.8
98	79.08	82.19	84.33	86.03	88.82	93.19	96.89	101.9
99	80	83.12	85.28	87.01	89.81	94.22	97.94	103
100	80.92	84.07	86.24	87.98	90.8	95.24	99	104.1

Abbreviations

A

AC	Authentication Center
ACCLOC	Access Overload Class
ACH	Access Channel
ACK	Acknowledgment
ACM	Address Complete Message
ACSE	Association Control Service Element
ADPCM	Adaptive Differential Pulse Code Modulation
AMPS	Advanced Mobile Phone Service
ANSI	American National Standard Institute
ARPA	Advanced Research Project Agency
ARQ	Automatic Repeat Request
AT	Prefix for dialing using a modem
AUC	Authentication Center
AWGN	Additive White Gaussian Noise

B

BB	Baseband filter
BCAF	Bearer Control Agent Function
BCCH	Broadcast Control Channel

BCF	Bearer Control Function
BCFr	Bearer Control Function for radio bearer
BER	Bit Error Rate
BHCA	Busy Hour Call Attempts
bps	bits per second
BPSK	Binary Phase Shift Keying
BS	Base Station
BSAP	Base Station Application Part
BSC	Base Station Controller
BSMAP	Base Station Management Application Part
BSS	Base Station System
BTS	Base Transceiver System

C

CASE	Common Application Service Element
CAPICH	Common Auxiliary Pilot Channel
CBSEED	Codebook Seed
CC	Connection Confirm
CC	Country Code
CCAF	Call Control Agent Function
CCCH	Common Control Channel
CCF	Call Control Function
CCR	Commitment Concurrency and Recovery
CCAF'	Call Control Agent Function (enhanced)
CCF'	Call Control Function (enhanced)
CDG	CDMA Development Group
CDMA	Code Division Multiple Access
CDPD	Cellular Digital Packet Data
CELP	Code-Excited Linear Predictor
CH	Channel
CI	Cell Identity
CGI	Cell Global Identification
CLIP	Connectionless Interworking Protocol
CM	Connection Management
COUNT	Call history parameter
COST	Committee On Standards and Technology
CR	Connection Request

CRC	Cyclic Redundancy Check
CSMA/CA	Carrier-Sense Multiple Access with Collision Avoidance

D

DAPICH	Dedicated Auxiliary Pilot Channel
dB	Decibels
dBm	Decibels with respect to 1 milliwatt
DCCH	Dedicated Control Channel
DLCI	Data Link Connection Identifier
DMH	Data Message Handler
DPC	Destination Point Code
DQPSK	Differential Quadrature Phase Shift Keying
DS	Direct Sequence
DSP	Digital Signal Processing
DSSS	Direct Sequence Spread Spectrum
DTAP	Direct Transfer Application Part
DTMF	Dual-Tone Multifrequency

E

EIR	Equipment Identity Register
ETSI	European Telecommunications Standards Institute
ESN	Electronic Serial Number
EVRC	Enhanced Variable Rate Codec

F

FAF	Floor Attenuation Factor
FCH	Fundamental Channel
FCC	Federal Communication Commission
FD	Full Duplex
FDD	Frequency Division Duplex
FDMA	Frequency Division Multiple Access
FE	Functional Entity
FEC	Forward Error Correction
FER	Frame Error Rate
FFPC	Fast-Forward Power Control

FHSS	**Frequency Hopping Spread Spectrum**
FLPC	**Forward Link Power Control**
FPICH	**Forward Pilot Channel**
FPLMTS	**Future Public Land Mobile Telephone System**

G

G	**Codebook Gain**
GBN	**Go-Back-N**
GFSK	**Gaussian Frequency Shift Keying**
GHz	**gigaHertz**
GSM	**Global System of Mobile Communications**
GoS	**Grade of Service**
GPRS	**General Packet Radio Service**

H

HCM	**Handoff Completion Message**
HD	**Half Duplex**
HDM	**Handoff Direction Message**
HLR	**Home Location Register**
Hz	**Hertz**

I

IAM	**Initial Address Message**
ICMP	**Internet Control Message Protocol**
IF	**Intermediate Frequency**
IMT-2000	**International Mobile Telecommunications in year 2000**
IMSI	**International Mobile Subscriber Identifier**
IN	**Intelligent Network**
INAP	**Intelligent Network Application Part**
IP	**Internet Protocol**
IPCP	**Internet Protocol Control Protocol**
IPR	**Intellectual Property Rights**
ISDN	**Integrated Services Digital Network**
ISM	**Industrial Scientific Medical**
ISO	**International Organization for Standardization**

ISUP	ISDN User Part (of Signaling System 7)
ITU	International Telecommunications Union
IWF	Interworking Function

J

| J | Joules |

K

K	Kelvin
kbps	kilobits per second
kHz	kilohertz
km	kilometers
km/h	kilometers per hour

L

LAC	Link Access Control
LAC	Location Area Code
LBT	Listen-Before-Talk
LCP	Link Control Protocol
LLC	Logical Link Control
LPC	Linear Predictive Coding
LPF	Low Pass Filter
LOS	Line-of-Sight
LS	Least Squares
LSP	Linear Spectral Pairs

M

MAC	Medium Access Control
MAN	Metropolitan Area Network
MAP	Mobile Application Part
MBS	Mobile Broadband Systems
Mbps	Megabits per second
Mcps	Megachips per second
MC	Message Center

MCC	Mobile Country Code
MCTD	Multicarrier Transmit Diversity
MDN	Mobile Directory Number
MDR	Medium Data Rate
MHz	megahertz
MIN	Mobile Identification Number
MM	Mobility Management
MMSE	Minimum Mean Square Error
MNC	Mobile Network Code
MOS	Mean Opinion Score
MPL	Mean Path Loss
mph	miles per hour
MRRC	Mobile Radio Resource Control
MRTR	Mobile Radio Transmission and Reception
MS	Mobile Station
ms	milliseconds
MSC	Mobile Switching Center
MSIN	Mobile Station Identification Number
MSS	Mobile Satellite Services
MSS	Mobile Switching Subsystem
MT	Mobile Termination
MTP	Message Transfer Part
MUD	Multiuser Detection
MUX	Multiplexer
mW	milliwatts

N

NID	Network Identification
NE	Network Element
NLUM	Neighbor List Update Message
NMSI	National Mobile Subscriber Identity

O

OA&M	Operation Administration and Maintenance
OMC	Operation Maintenance Center
OQPSK	Offset Quadrature Phase Shift Keying

OS	Operations System
OSI	Open System Interconnection
OSS	Operations Support System
OTAF	Over-the-Air Function
OTASP	Over-the-Air Service Provisioning
OTD	Orthogonal Transmit Diversity

P

PACA	Priority Access Channel Assignment
PBX	Private Branch Exchange
PCG	Power Control Group
PCH	Paging Channel
PCM	Pulse Code Modulation
PCMCIA	Personal Computer Memory Card International Association
PCS	Personal Communication Services
PDA	Personal Digital Assistance
PICH	Pilot Channel
PIN	Personal Identification Number
PMRM	Pilot Strength Measurement Message
PN	pseudonoise
PPDN	Public Packet Data Network
PPP	Point-to-Point Protocol
PSK	Phase Shift Keying
psd	power spectral density
PSMM	Power Measurement Report Message
PSTN	Public Switched Telephone Network

Q

QCELP	Qualcomm Code-Excited Linear Prediction
QoS	Quality of Service
QPSK	Quadrature Phase Shift Keying

R

RACE	Research in Advanced Communications Equipment
RACF	Radio Access Control Function

RAM	Random Access Memory
RAND	Random Number
RBP	Radio Burst Protocol
RCF	Radio Control Function
RCLP	Relaxed Code-excited Linear Prediction
RF	Radio Frequency
RFTR	Radio Frequency Transmission and Reception
RLP	Radio Link Protocol
rms	Root Mean Square
ROLPC	Reverse Outer Loop Power Control
ROSE	Remote Operation Service Element
RRC	Radio Resource Control
RS	Radio System
RS1	Rate Set 1
RS 2	Rate Set 2
RT	Random Time
RTF	Radio Terminal Function
RTT	Radio Transmission Technology

S

SACF	Service Access Control Function
SASE	Specific Application Service Element
SBS	Switched Beam System
SCCP	Signaling Connection Control Part
SCF	Service Control Function
SCH	Supplementary Channel
SCI	Synchronized Capsule Indicator
SCP	Switching Control Point
SDF	Service Data Function
SID	System Identification
SIR	Signal-to-Interference Ratio
SLS	Signaling Link Selection
SMF	Service Management Function
SMRS	Specialized Mobile Radio Services
SMS	Short Message Service
SNR	Signal-to-Noise Ratio
SNDCF	Sub-Network Dependent Convergence Function

SNDCP	Sub-Network Dependent Convergence Protocol
SOM	Start of Message
SP	Signaling Point
SRBP	Signaling Radio Burst Protocol
SRF	Specialized Resource Function
SRLP	Signaling Radio Link Protocol
SRP	Selective Repeat Request
SS	Spread Spectrum
SSD	Shared Secret Data
SSF	Service Switching Function
SSP	Switching System Platform
STP	Signal Transfer Point
SYNC	Sync Channel

T

TACAF	Terminal Access Control Agent Function
TACF	Terminal Access Control Function
TCAP	Transaction Capabilities Application Part
TCH	Traffic Channel
TCP	Transport Control Protocol
TDD	Time Division Duplex
TDM	Time Division Multiplex
TDMA	Time Division Multiple Access
TE	Terminal Equipment
THSS	Time-Hopped Spread Spectrum
TIA	Telecommunication Industry Association
TIMF	Terminal Identification Management Function
TMA	Telesystems Micro-cellular Architecture
TMSI	Temporary Mobile Subscriber Identity
TRAC	Telecommunication Research and Action Center

U

UDP	User Data Protocol
UIMF	User Identification Management Function
UMTS	Universal Mobile Telephone Service

V

| VLR | Visitor Location Register |
| VLSI | Very Large Scale Integration |

W

WARC	World Administration Radio Conference
W-CDMA	Wideband CDMA
WLAN	Wireless Local Area Network
WLL	Wireless Local Loop
WORM	Window control Operation-based Reception Memory
WPBX	Wireless Private Branch Exchange

X

| XC | Transcoder |

Additional References

Adachi, F., Sawahashi, M., and Suda, H., "Wideband DS-CDMA for Next Generation Mobile Communications Systems," *IEEE Communication Magazine* 36(9), September. 1998.

Applebaum, S. P., "Adaptive Arrays," *IEEE Transactions on Antenna and Propagation*, AP-24, September 1976, pp. 585–98.

Bang S. C., "Performance Analysis of a Wideband CDMA System for FPLMTS," *Proceedings of IEEE VTC*, May 1997.

Benedetto, S., and Montorsi, G., "Design of Parallel Concatenated Convolutional Codes," *IEEE Transactions on Communications* 44(5), May 1996.

Benedetto, S., Garello, R., and Montorsi, G., "A Search for Good Convolutional Codes to be Used in the Construction of Turbo Codes," *IEEE Transactions on Communications* 46(9), September 1998.

Braun, W. R., and Dersch, U., "A Physical Mobile Radio Channel Model, " *IEEE Transactions on VT*, VT-40(2), May 1991, pp. 472–82.

Chan, M., C., and Woo, T. Y. C., "Next Generation Wireless Data Services: Architecture and Experience," *IEEE Personal Communications* 6(1), February 1999, pp. 20–33.

Chang, C. R., Wan, J. Z., and Yee, M. F., "PN Offset Planning Strategies for Non-Uniform CDMA Networks," *Proceeding of IEEE VTC*, Arizona, May 1997, pp. 1543–47.

Chopra, M., Rohani, K., and Reed, J., "Analysis of CDMA Range Extension Due to Soft Handoff," *Proceedings of IEEE VTC*, Chicago, July 1995, pp. 917–21.

Compton, R. T., *Adaptive Antennas (Concepts and Performance),* Prentice Hall, Englewood Cliffs, NJ, 1988.

Compton, R. T., "An Adaptive Antenna in a Spread Communication Systems," *Proceedings of IEEE 66*, March 1978, pp. 289–98.

Feher, K., *Wireless Digital Communications,* Prentice Hall, Upper Saddle River, NJ, 1995.

Fleming, G., et al., "A Flexible Network Architecture for UMTS," *IEEE Personal Communications* 5(2), April 1998, pp. 8–15.

Gabriel, W. F., "Adaptive Arrays (an Introduction)", *Proc. IEEE* 64, February 1976, pp. 239–72.

Gilhousen, K. S., Jacobs, I. M., Padovani, R., Viterbi, A. J., Weaver, L. A., and Wheatley, C. E., "On the Capacity of a Cellular CDMA System," *IEEE Transaction of Vehicular Technology* 40, May 1991, pp. 472–80.

Goodman, D. J., "Trends in Cellular and Cordless Communications," *IEEE Communication Magazine* 29(6), June 1991, pp. 31–40.

Hall C. J., and Foose, W. A., "Practical Planning for CDMA Networks (A Design Process Overview)," *Proceedings of the 1996 Southcon Conference*, Orlando, FL, June 1996, pp. 66–71.

Hong, D., and Rappaport, S. S., "Traffic Model and Performance Analysis for Cellular Radio Telephone Systems with Prioritized and Non-Prioritized Handoff Procedures," *IEEE Trans. VT*, VT-35(3), August 1986, pp. 77–92.

Hu, L.-R., and Rappaport, S. S., "Personal Communications Systems Using Hierarchical Cellular Overlays," *Proc. IEEE ICUPC*, 1994, pp. 397–401.

"Harmonized Global 3G (G3G)," Technical Framework for ITU IMT-2000 CDMA proposal, OHG Technical Framework, June, 1998.

"IMT-2000: Standards Efforts of the ITU," Special Issue, *IEEE Personal Communications* 4(4), August 1977.

Manji, S., and Zhuang, W., "Capacity Analysis of an Integrated Voice/Data DS-CDMA Network," *Proceedings of IEEE International Conference of Communications,* 1997, pp. 979–83.

Markoulidakis, J. G., et al., "Mobility Modeling in Third-Generation Mobile Telecommunications Systems," *IEEE Personal Communications,* August 1997, pp. 41–56.

McClelland, S., "Europe's Wireless Future," *Microwave Journal* 42(9), September 1999, pp. 78–108.

Mende, W., "On the Hand-Over Rate in Future Cellular Systems," *IEEE VTC*, 1998, pp. 358–61.

Naguib, A., and Paulraj, A., "Performance of CDMA Cellular Networks with Base Station Antenna Arrays," *Proc. International Zurich Seminar on Digital Communications*, Zurich, Switzerland, March 1994, pp. 87–100.

Naguib, A., and Paulraj, A., "Performance of Wireless CDMA with M-ary Orthogonal Modulation and Cell Site Antenna Arrays," *IEEE Journal on Selected Areas in Communications* 14(9), December 1996, pp. 1770–83.

Naguib, A., Paulraj, A., and T. Kailath, "Capacity Improvement with Base Station Antenna Arrays in Cellular CDMA," *IEEE, Trans. on VT*, VT-43(3), August 1994, pp. 691–4.

Nanda, S., "Teletraffic Models for Urban and Suburban Microcells: Cell Sizes and Handoff Rates," *IEEE Trans. Vehicular Tech.* 42(4), November 1993, pp 673–82.

Nikula, E., et al., "FRAMES Multiple Access for UMTS and IMT-2000," *IEEE Personal Communications* 5(2), April 1998, pp. 16–24.

Nobelen, R. V., et al., "An Adaptive Radio Link Protocol with Enhanced Data Rates for GSM Evolution," *IEEE Personal Communications* 6(1), February 1999, pp. 54–63.

Ojanpera, Tero, and Prasad, R., "An Overview of Third-Generation Wireless Personal Communications: A European Perspective," *IEEE Personal Communications* 5 (6), December 1998, pp. 59–65.

Ojanpera, Tero, et al., "Comparison of Multiple Access Schemes for UMTS," *Proceedings of IEEE VTC*, May 1997.

Ojanpera, Tero, et al., "Analysis of CDMA and TDMA for Third Generation Mobile Radio Systems," *Proceedings of IEEE VTC*, May 1997.

Padgett, J. E., Gunther, C. G., and Hattori, T., "Overview of Wireless Personal Communications," *IEEE Communications Magazine* 33(1), January 1995, pp. 28–41.

Pollini, G. P., "Trends in Handover Design," *IEEE Communications Magazine*, March 1996, pp. 82–90.

Prasad, N. R., "GSM Evolution Towards Third Generation UMTS/IMT-2000," *1999 IEEE International Conference on Personal Wireless Communications,* Jaipur, India, February 1999.

Proakis, J. G., *Digital Communications,* McGraw-Hill, New York, 1989.

Skold, J., et al., "Cellular Evolution into Wideband Services," *Proceedings of IEEE VTC,* May 1997.

Soroushnejad, M., and Geraniotis E., "Multi-Access Strategies for an Integrated Voice/Data CDMA Packet Radio Network," *IEEE Transaction of Communications* 43, February 1995, pp. 934–45.

Suard, B., Naguib, A., and Paulraj, A., *Performance Analysis of CDMA Mobile Communication Systems Using Antenna Arrays,* Proceeding IC-ASSP'93, Volume VI, Minneapolis, MN, April 1994, pp. 153–56.

Tripathi, N. D., Reed, J. H., and Vanlandingham, H. F., "Handoff in Cellular Systems," *IEEE Personal Communications* 5(6), December 1998, pp. 26–37.

Viterbi, A. J., and Padovani, R., "Implications of Mobile Cellular CDMA," *IEEE Communication Magazine* 30(12), December 1992, pp. 38–41.

Viterbi, A. J., et al., "Soft Handoff Extends CDMA Cell Coverage and Increases Reverse Link Capacity," *IEEE JSAC* 12(8), October 1994, pp. 1281–87.

Winters, J. H., "Signal Acquisition and Tracking with Adaptive Arrays in Digital Mobile Radio System IS-54 with Flat Fading, *IEEE Trans. VT* 2(4), July 1993, pp. 377–84.

Wong, D., "Soft Handoff in CD," *IEEE Personal Communications,*" December 1997, pp. 6–17.

Wu, J. T.-H., and Geraniotis, E., "Power Control in Multi-Media CDMA Networks," *Proceeding IEEE Vehicular Technology Conference,* 1995, pp. 789–93.

Zou, J., and Bhargava, V. K., "On Soft Handoff, Erlang Capacity and Service Quality of a CDMA Cellular System: Reverse Link Analysis," *IEEE International Symp. Personal, Indoor, Mobile Radio Communications*, Toronto, Canada, September 1995, pp. 603–7.

Index

About the Author

Dr. Vijay Garg is an adjunct professor in the Electrical Engineering and Computer Science department at the University of Illinois at Chicago, where he teaches courses in wireless communications and wireless networking. Dr. Garg is also a Distinguished Member of Technical Staff at the Bell Labs of Lucent Technologies in Naperville, Illinois. Dr. Garg received his Ph.D. from the Illinois Institute of Technology, Chicago, in 1973 and his MS from the University of California at Berkeley, California, in 1966. Dr. Garg has co-authored several technical books, including four in Telecommunications; one in Rigid Body Dynamics; and one in Railway Vehicle Dynamics, which has been translated into Russian and Chinese. He is a Fellow of ASCE and ASME and a Senior Member of IEEE. Dr. Garg is a Registered Professional Engineer in the states of Maine and Illinois. He is an Academic Member of the Russian Academy of Transport. He has served on various technical committees of IEEE and ASME, and he is a feature editor for a PCS series in IEEE Communication Magazine.